Springer Texts in Statistics

Advisors:
Stephen Fienberg   Ingram Olkin

Springer Texts in Statistics

(continued after index)

Peter Whittle

Probability via Expectation

Third Edition

With 22 Illustrations

Springer-Verlag

New York Berlin Heidelberg London Paris
Tokyo Hong Kong Barcelona Budapest

Peter Whittle
Department of Pure Mathematics
  and Mathematical Statistics
University of Cambridge
16 Mill Lane
Cambridge CB2 1SB
England

Editorial Board

Mathematics Subject Classification: 60-01

First edition, Penguin 1970.   Second edition, Wiley, 1976   Russian translation, Nakua, 1982.

Whittle, Peter.
    Probability via expectation / by Peter Whittle. — 3rd ed.
      p.   cm. — (Springer texts in statistics)
    Rev. ed. of Probability. 1976.
    Includes bibliographical references and index.
    ISBN 0-387-97764-3 (U.S.). — ISBN 3-540-97764-3 (Berlin)
    1. Probabilities. I. Whittle, Peter. Probability. II. Title.
III. Series.
QA273.W59   1992
519.2—dc 20   91-40782

Printed on acid-free paper.

Production coordinated by Brian Howe and managed by Terry Kornak; manufacturing supervised by Jacqui Ashri.
Typeset by Asco Trade Typesetting Ltd., Hong Kong.
Printed and bound by R. R. Donnelley & Sons, Harrisonburg, VA.
Printed in the United States of America.

9 8 7 6 5 4 3 2 1

ISBN 0-387-97764-3 Springer-Verlag New York Berlin Heidelberg (softcover)
ISBN 3-540-97764-3 Springer-Verlag Berlin Heidelberg New York (softcover)
ISBN 0-387-97758-9 Springer-Verlag New York Berlin Heidelberg (hardcover)
ISBN 3-540-97758-9 Springer-Verlag Berlin Heidelberg New York (hardcover)

To my parents

Preface to the Third Edition

This book is a complete revision of the earlier work *Probability* which appeared in 1970. While revised so radically and incorporating so much new material as to amount to a new text, it preserves both the aim and the approach of the original.

That aim was stated as the provision of a 'first text in probability, demanding a reasonable but not extensive knowledge of mathematics, and taking the reader to what one might describe as a good intermediate level'. In doing so it attempted to break away from stereotyped applications, and consider applications of a more novel and significant character.

The particular novelty of the approach was that expectation was taken as the prime concept, and the concept of expectation axiomatized rather than that of a probability measure. In the preface to the original text of 1970 (reproduced below, together with that to the Russian edition of 1982) I listed what I saw as the advantages of the approach in as unlaboured a fashion as I could. I also took the view that the text rather than the author should persuade, and left the text to speak for itself. It has, indeed, stimulated a steady interest, to the point that Springer-Verlag has now commissioned this complete reworking.

In re-examining the approach after this lapse of time I find it more persuasive than ever. Indeed, I believe that the natural flow of the argument is now more evident to me, and that this revised version is much more successful in tracing that flow from initial premises to surprisingly advanced conclusions. At the risk I fear most—of labouring the argument—I would briefly list the advantages of the expectation approach as follows.

(1) It permits a more economic and natural treatment at the elementary level.
(2) It opens an immediate door to applications, because the quantity of interest in many applications is just an expectation.

(3) Precisely for this last reason, one can discuss applications of genuine interest with very little preliminary development of theory. On the other hand, one also finds that a natural unrolling of ideas leads to the development of theory almost of itself.

(4) The approach is an intuitive one, in that people have a well-developed intuition for the concept of an average. Of course, what is found 'intuitive' depends on one's experience, but people with a background in the physical sciences have certainly taken readily to the approach. Historically, the early analysts of games of chance found the question 'What is a fair price for entering a game?' quite as natural as 'What is the probability of winning it?' We make some historical observations in Section 3.4.

(5) The treatment is the natural one at an advanced level. However, as noted in the preface to *Probability*, here we do not need to make a case—the accepted concepts and techniques of weak convergence and of generalized processes are characterized wholly in terms of expectation.

(6) Much conventional presentation of probability theory is distorted by a preoccupation with measure-theoretic concepts which is in a sense premature and irrelevant. These concepts (or some equivalent of them) cannot be avoided indefinitely. However, in the expectation approach, they find their place at the natural stage.

(7) On the other hand, a concept which is notably and remarkably absent from conventional treatments is that of convexity. (Remarkable, because convexity is a probabilistic concept, and, in optimization theory, the necessary invocations of convexity and of probabilistic ideas are intimately related.) In the expectation approach convexity indeed emerges as an inevitable central concept.

(8) Finally, in the expectation approach, classical probability and the probability of quantum theory are seen to differ only in a modification of the axioms—a modification rich in consequences, but succinctly expressible.

The reader can be reassured that the book covers at least the material that would be found in any modern text of this level, and will leave him at least as well equipped in conventional senses as these. The difference is one of order and emphasis, although this cannot be dismissed, since it gives the book its point. The enhanced role of convexity has already been mentioned. The concept of least square approximation, fundamental in so many nonprobabilistic contexts, is found to pervade the treatment. In the discussion of stochastic processes one is led to give much greater importance than usual to the backward equation, which reveals both the generator of the process and another all-pervading concept, that of a martingale.

The conventions on the numbering of equations, etc. are not quite uniform, but are the most economical. Sections and equations are numbered consecutively through the chapter, so that a reference to 'Section 2' means Section 2 of the current chapter, whereas a reference to 'Section 4.2' is to Section 2 of Chapter 4. Correspondingly for equations. Figures are also numbered consecutively through a chapter, but always carry a chapter label; e.g. 'Fig. 12.3'. Theorems are numbered consecutively through a section, and always carry full chapter/section/number label; e.g. 'Theorem 5.3.2' for Theorem 2 of Section 5.3. Exercises are numbered consecutively through a section, and are given a chapter/section reference (e.g. Exercise 10.9.2) only when referred to from another section.

I am grateful to David Stirzaker for some very apt historical observations and references, also to Roland Tegeder for helpful discussion of the final two sections. The work was supported in various phases by the Esso Petroleum Company Ltd. and by the United Kingdom Science and Engineering Research Council. I am most grateful to these two bodies.

P. Whittle

Preface to *Probability* (1970)
(Section references amended)

This book is intended as a first text in theory and application of probability, demanding a reasonable, but not extensive, knowledge of mathematics. It takes the reader to what one might describe as a good intermediate level.

With so many excellent texts available, the provision of another needs justification. One minor motive for my writing this book was the feeling that the very success of certain applications of probability in the past has brought about a rather stereotyped treatment of applications in most general texts. I have therefore made an effort to present a wider variety of important applications, for instance, optimization problems, quantum mechanics, information theory and statistical mechanics.

However, the principal novelty of the present treatment is that the theory is based on an axiomatization of the concept of expectation, rather than that of a probability measure. Such an approach is now preferred in advanced theories of integration and probability; it is interesting that the recent texts of Krickeberg (1965), Neveu (1964) and Feller (1966) all devote some attention to it, although without taking it up whole-heartedly. However, I believe that such an approach has great advantages even in an introductory treatment. There is no great point in arguing the matter; only the text itself can provide real justification. However, I can briefly indicate the reasons for my belief.

(i) To begin with, people probably have a better intuition for what is meant by an 'average value' than for what is meant by a 'probability'.

(ii) Certain important topics, such as optimization and approximation problems, can then be introduced and treated very quickly, just because they are phrased in terms of expectations.

(iii) Most elementary treatments are bedevilled by the apparent need to ring the changes of a particular proof or discussion for all the special cases of

continuous or discrete distribution, scalar or vector variables, etc. In the
expectation approach these are indeed seen as special cases which can be
treated with uniformity and economy.

(iv) The operational approach—analysis of the type of assertion that is really
relevant in a particular application—leads one surprisingly often to a
formulation in expectations.

(v) There are advantages at the advanced level, but here we do not need to
make a case.

The mathematical demands made upon the reader scarcely go beyond
simple analysis and a few basic properties of matrices. Some properties of
convex sets or functions are required occasionally—these are explained—
and the spectral resolution of a matrix is used in Chapters 11 and 16. Because
of the approach taken, no measure theory is demanded of the reader, and any
set theory needed is explained. Probability generating functions and the like
are used freely, from an early stage. I feel this to be right for the subject and
a manner of thinking that initially requires practice rather than any extensive
mathematics. Fourier arguments are confined almost entirely to Section 15.6.

This project was begun at the University of Manchester, for which I have
a lasting respect and affection. It was completed during tenure of my current
post, endowed by the Esso Petroleum Company Ltd.

Preface to the Russian Edition of *Probability* (1982)

When this text was published in 1970 I was aware of its unorthodoxy, and uncertain of its reception. Nevertheless, I was resolved to let it speak for itself, and not to advocate further the case there presented. This was partly because of an intrinsic unwillingness to propagandize, and partly because of a conviction that an approach which I (in company with Huygens and other early authors) found so natural would ultimately need no advocate. It has then been a great pleasure to me that others have also shared this latter view and have written in complimentary terms to say so. However, the decision of the 'Nauka' Publishing House to prepare a Russian edition implies the compliment I value most, in view of the quite special role Russian authors have played in the development of the theory of probability.

I have taken the opportunity to correct some minor errors kindly pointed out to me by readers, but the work is otherwise unrevised. My sincere thanks are due to Professor N. Gamkrelidze for bringing to the unrewarding task of translation, not only high professional competence, but even enthusiasm.

Contents

CHAPTER 1

Uncertainty, Intuition and Expectation

1. Ideas and Examples

Probability is an everyday notion. This is shown by the number of common words related to the idea: chance, random, hazard, fortune, likelihood, odds, uncertainty, expect, believe. Nevertheless, as with many concepts for which we have a strong but rather vague intuition, the idea of probability has taken some time to formalize. It was in the study of games of chance (such as card games and the tossing of dice) that the early attempts at formalization were made, with the understandable motive of determining one's chances of winning. In these cases the basis of formalization was fairly clear, because one can always work from the fact that there are a number of elementary situations (such as all the possible deals in a card game) which can be regarded as 'equally likely'. However, this approach fails in the less tangible problems of physical, economic and human life which we should like to consider, and of which we shall give a few examples below.

Nevertheless, Man (and Nature before him) had evolved guards against uncertainty long before there was any formal basis for calculation. When a farmer decides to reap, he is weighing up the uncertainties. When an insurance company sets a premium it is assessing what fixed rate would be a fair equivalent (overheads being taken into account) of the rate of the varying flow of claims. When a banker makes a loan his charges will take account of his assessment of the risk. These examples are quite venerable; bankers and insurance companies have existed in some form for a very long time.

As is fortunately common, the absence of a firm basis did not prevent the subject of probability from flourishing, certainly from the seventeenth century onwards. However, in 1933 a satisfactory general basis was achieved by A.N. Kolmogorov in the form of an axiomatic theory which, while not necessarily the last word on the subject, set the pattern of any future theory.

The application of probability theory to games of chance is an obvious one. However, there are applications in science and technology which are just as clean cut. As examples, we can quote the genetic mechanism of Mendelian inheritance (Sections 5.5 and 9.2), the operation of a telephone exchange (Section 10.4) or the decay of radioactive molecules (Section 10.4). In all these cases one can make valuable progress with a simple model, although it is only fair to add that a deeper study will demand something more complicated.

In general, the physical sciences provide a rich source of interesting and well-defined probability problems: see, for example, some of the models of statistical mechanics (Sections 6.3 and 10.9) and Brownian motion (Section 10.11). The problems generated by the technological sciences can be just as interesting and no less fundamental: see, for example, the discussion of information theory (Sections 16.1–16.3) and of routing in a telephone network (Section 10.8). In encountering the 'natural variability' of biological problems one runs into rather more diffuse situations, but this variability makes a probabilistic approach all the more imperative, and one can construct probabilistic models of, say, population growth (Sections 6.4 and 10.4) and epidemics (Section 10.7) which have proved useful. One encounters natural variability in the human form if one tries to construct social or economic models but again such models, inevitably probabilistic, prove useful. See, for example, the discussion of the Pareto distribution in Section 10.5.

One of the most recent and fascinating applications of probability theory is to the field of control or, more generally, to that of sequential decision-making (Section 16.4). One might, for example, be wishing to hold an aircraft on course, despite the fact that random forces of one sort or another tend to divert it, or one might wish to keep a factory in a state of over-all efficient production despite the fact that so many future variables (such as demand for the product) must be uncertain. In either case, one must make a sequence of decisions (regarding course adjustment or factory management) in such a way as to ensure efficient running over a period, or even optimal running, in some well-defined sense. Moreover, these decisions must be taken in the face of an uncertain future.

It should also be said that probability theory has its own flavour and intrinsic structure, quite apart from applications, as may be apparent from Chapters 2, 3, 12, 13 and 14 in particular. Just as for mathematics in general, people argue about the extent to which the theory is self-generating, or dependent upon applications to suggest the right direction and concepts. Perhaps either extreme view is incorrect; the search for an inner pattern and the search for a physical pattern are both powerful research tools, neither of them to be neglected.

2. The Empirical Basis

Certain experiments are nonreproducible in that, when repeated under standard conditions, they produce variable results. The classic example is that of coin-tossing: the toss being the experiment, resulting in the observation of a head or a tail. To take something less artificial, one might be observing the response of a rat to a certain drug, observation on another rat constituting repetition of the experiment. However uniform in constitution the experimental animals may be, one will certainly observe a variable response. The same variability would be found in, for example, lifetimes of electric lamps, crop yields, the collisions of physical particles or the number of telephone calls made over a given line on a given day of the week. This variability cannot always be dismissed as 'experimental error', which could presumably be explained and reduced, but may be something more fundamental. For instance, the next ejection of an electron from a hot metal filament is a definite event, whose time is not predictable on any physical theory yet developed.

Probability theory can be regarded as an attempt to provide a quantitative basis for the discussion of such situations, or at least for some of them. One might despair of constructing a theory for phenomena whose essential quality is that of imprecision, but there is an empirical observation which gives the needed starting point.

Suppose one tosses a coin repeatedly, keeping a record of the number of heads $r(n)$ in the first n tosses ($n = 1, 2, 3, \ldots$). Consider now the proportion of heads after n tosses:

$$p(n) = \frac{r(n)}{n}. \tag{1}$$

It is an empirical fact that $p(n)$ varies with n much as in Fig. 1.1, which is derived from a genuine coin-tossing experiment. The values of $p(n)$ show

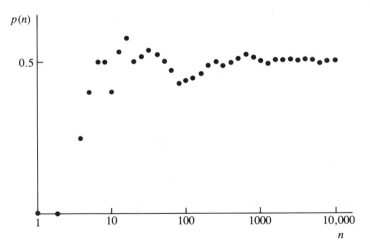

Figure 1.1. A graph of the proportions of heads thrown, $p(n)$, in a sequence of n throws, from an actual coin-tossing experiment. Note the logarithmic scale for n. The figures are taken from Kerrich (1946), by courtesy of Professor Kerrich and his publishers.

fluctuations which become progressively weaker as n increases, until ultimately $p(n)$ shows signs of tending to some kind of limit value, interpretable as the 'long-run proportion of heads'. Obviously this cannot be a limit in the usual mathematical sense, because one cannot guarantee that the fluctuations in $p(n)$ will have fallen below a prescribed level for all values of n from a certain point onwards. However, some kind of 'limit' there seems to be, and it is this fact that offers the hope of a useful theory: that beyond the short-term irregularity there is a long-term regularity.

The same regularity manifests itself if we examine, for example, lifetimes of electric lamp bulbs. Let the observed lifetimes be denoted $X_1, X_2, X_3, \ldots$ and suppose we keep a running record of the arithmetic average of the first n lifetimes:

$$\overline{X}_n = \frac{1}{n} \sum_1^n X_j. \tag{2}$$

Then, again, it is an empirical fact that, provided we keep test conditions constant, the graph of $\overline{X}_n$ against n will show a similar convergence. That is, fluctuations slowly die down with increasing n, and $\overline{X}_n$ appears to tend to a 'limit' value, interpretable as the 'long-run average' of lifetime for this particular make of bulb.

One sees the same phenomenon in human contexts, where it is a matter of observing frankly variable material rather than of a controlled experiment. The claims that an insurance company will receive in a definite category of risk (e.g. domestic fire in the winter months in a given town) will, as a record in time, be variable and unpredictable in detail. Nevertheless, the company knows that it can adopt 'expected liability per week' from this source as a working concept, because long-term averages of weekly claims do in fact seem to stabilize.

One might make the same point in a slightly different way. Suppose one conducted two separate sampling surveys, measuring in each the cholesterol levels of n people, say. Then, if pains have been taken to sample representatively, the average cholesterol levels from the two samples will be found to agree very much more closely than would the cholesterol levels of two randomly chosen individuals. Furthermore, the larger n, the better the agreement. There is a feeling, then, that the sample average measures 'something real' and approaches this 'something real' as the sample size increases. This is, of course, the justification of opinion polls and market surveys.

It is on this feature of empirical convergence that one founds probability theory; by postulating the existence of an idealized 'long-run proportion' (a *probability*) or 'long-run average' (an *expectation*).

Actually, the case of a proportion is a special case of that of an average. Suppose that in the coin-tossing experiment one defined an 'indicator variable' X_j which took the value 1 or 0 according as the jth toss resulted in a head or a tail. Then the average of X-values yielded by expression (2) would reduce just to the proportion given by expression (1). Conversely, one can build up an average from proportions; see Exercise 1.

Correspondingly, in constructing an axiomatic theory one has the choice of two methods: to idealize the concept of a proportion or that of an average. In the first case, one starts with the idea of a probability and later builds up that of an expectation. In the second, one takes expectation as the basic concept, of which a probability is to be regarded as a special case. In this text we shall take the second course, which, although less usual, offers substantial advantages.

Readers might feel that our 'nonreproducible experiments' could often be made reproducible if only they were sufficiently refined. For example, in the coin-tossing experiment, the path of the coin is surely mechanically determined, and, if the conditions of tossing were standardized sufficiently, then a standard (and theoretically predictable) result should be obtained. For the rat experiment, increased standardization should again produce increased reproducibility. More than this; a sufficiently good understanding of the biological response of rats to the drug should enable one to *predict* the response of a given animal to a given dose, and so make the experiment reproducible in the sense that the observed variation could be largely explained.

Whether it is possible in principle (it certainly is not in practice) to remove all variability from an experiment in this way is the philosophic issue of determinism which, if decidable at all, is certainly not decidable in naive terms. In probability theory, we shall simply start from the premise that there is a certain amount of variability which we cannot explain and must accept.

The practical point of coin- or die-tossing is that the coin or die acts as a 'variability amplifier': the dynamics are such that a small variability in initial position is transformed into a large variability in final position. This point, long recognized, has been given much greater substance by the recent development of the theory of chaos. This theory demonstrates simple deterministic models whose solution paths are not merely very sensitive to initial conditions, but continue to show variation of a superficially irregular character throughout time.

EXERCISES AND COMMENTS

1. Suppose that the observations X_j can take only discrete set of values: $X = x_k$ ($k = 1, 2, \ldots K$). Note that the average of formula (2) can then be written

$$\overline{X}_n = \sum_k x_k p_k(n),$$

where $p_k(n)$ is the proportion of times the value x_k has been observed in the first n readings.

3. Averages over a Finite Population

In this section we shall make our closest acquaintance with official statistics.

Suppose that a country of N people has a 100 per cent census and that, as a result of the information gathered, individuals can be assigned to one of K

mutually exclusive categories or cells, which we shall label $\omega_1, \omega_2, \ldots, \omega_K$. Thus, the specification characterizing a person in a particular category might conceivably be: 'born in the United Kingdom in 1955, male, married, three children, motor mechanic, with an income in the range £14,000–£14,999. If the specification were a very full one there would be relatively few people in each category—perhaps at most one, if the person's full name were specified, for example. On the other hand, a rather unspecific description would mean fewer categories, each with more members. The essential point is that we assume a level of specification has been fixed, by circumstance or design, and people are assigned to different categories if and only if they can be distinguished on the basis of that specification. So, the separation into categories ω_k represents the completest breakdown possible on the basis of census response.

The value of ω attached to an individual is what will more generally be termed the *realization*, the description of the individual at the level adopted. The motivation for the term is that, if one picks an individual 'at random', i.e. off the street or blindly from a directory, then examination of the value of ω tells one exactly in which category the individual fell in the particular case. The possible realization values $\omega_1, \omega_2, \ldots, \omega_K$ can be regarded as points in an abstract space Ω. This is termed the *sample space*, because the sampling of an individual can be regarded as the sampling of a point ω from Ω. Any property of an individual which is determined by the realization is termed a *random variable*—these are the variables which are meaningful at the level of description adopted.

So, in the example above, 'marital status' and 'size of family' are random variables, but 'type of dwelling' is not. 'Year of birth' is a random variable, but 'age in years' at a prescribed date is not—there is an indeterminacy of one year. Likewise, 'income tax band' is very nearly a random variable, but not quite, because to be able to determine this we would need rather more information, such as ages of children and reasons for special allowances.

Since a random variable is something whose value is determined by the realization ω we can write it as a function $X(\omega)$, where ω takes the values $\omega_1, \omega_2, \ldots, \omega_K$. That is, *a random variable is a function on the sample space*. For example, 'marital status' is the function which takes the values 'single', 'married', 'widowed', 'divorced', etc. as ω takes the different state values ω_k.

Now, consider a numerically-valued random variable, such as 'size of family'. In summarizing the results of the census one will often quote the population average of such variables, the average being the conventional arithmetic mean, with each individual equally weighted:

$$A(X) = \frac{1}{N} \sum_k n_k X(\omega_k) = \sum_k p_k X(\omega_k). \tag{3}$$

Here we have denoted the number of people in the kth category by n_k, and the proportion n_k/N in that category by p_k. The notation $A(X)$ emphasizes the fact that the average is a figure whose value depends on the particular random variable X we are considering. In fact, $A(X)$ is a *functional* of the function $X(\omega)$, a quantity whose value is determined from the values of $X(\omega)$ by the rule (3).

Although only numerical variables can be averaged, there is often a way of attaching a numerical value to a nonnumerical variable. Consider, for example, the random variable 'marital status'. We could define the random variable

$$X(\omega) = \begin{cases} 1 & \text{if the category } \omega \text{ is one of married people,} \\ 0 & \text{otherwise,} \end{cases} \tag{4}$$

and $A(X)$ would then be the proportion of married people in the country.

The function defined in (4) is an *indicator function*, a function taking the value 1 in a certain ω-set (the 'married' set) and 0 elsewhere. This is the point made in Section 2: the proportion of the population in a given set of ω-values is the average of the indicator function of that set.

We shall take the concept of an average as basic, so the properties of the functional $A(X)$ are important. The reader will easily confirm the following list of properties from the definition (3):

(i) If $X \geq 0$ then $A(X) \geq 0$.
(ii) If X_1 and X_2 are numerical-valued random variables and c_1 and c_2 are constants then

$$A(c_1 X_1 + c_2 X_2) = c_1 A(X_1) + c_2 A(X_2).$$

(iii) $A(1) = 1$.

In words, the averaging operator A is a positive linear operator, fulfilling the normalization condition (iii).

Instead of defining the averaging operator explicitly by formula (3) and then deducing properties (i)–(iii) from it we could have gone the other way. That is, we could have regarded properties (i)–(iii) as those which we would intuitively expect of an averaging operator, and taken them as the axioms for a theory of such operators. Actually, in the present case, the approach taken scarcely makes much difference, because it follows from (i)–(iii) that the operator must have the form (3) (see Exercise 3).

The realization ω is often characterized as a variable which describes the possible *elementary outcomes* of an experiment. In the present case, the experiment would be the collection of census data on an individual, which would, by definition, exactly specify the realization. In general, one could imagine an 'ideal experiment' which would reveal what the realization was in a particular case, so that the notions of realization and of experimental outcome would be coincident. However, we shall have to consider actual experiments which reveal very much less, so that, while the realization would determine the outcome of an actual experiment, the reverse does not necessarily hold; see Section 5. It seems best, then, to separate the idea of realization from that of an experimental outcome.

EXERCISES AND COMMENTS

1. The following questionnaire is posed to a subject:
   (a) Do you suffer from bronchitis?
   (b) If so, do you smoke?
   How many possible outcomes are there to the experiment? Can one always decide from the answers to these questions whether the subject is a smoker?

2. The set of values $X(\omega_k)$ of a random variable on Ω form the points of a new sample space Ω_X. Show that ω is a random variable on Ω_X (i.e. no information has been lost by the transformation) if and only if the K values $X(\omega_k)$ are distinct.

3. Suppose it is known that the sample space consists just of the K points $\omega_1, \omega_2, \ldots, \omega_K$, and that properties (i)–(iii) hold for the average $A(X)$ of any numerical-valued variable X on this sample space. Show, by choosing $X(\omega)$ as the indicator function of appropriate sets, that $A(X)$ must have the form (3), with the p_k some set of numbers satisfying

$$p_k \geq 0 \quad (k = 1, 2, \ldots, K), \qquad \sum_k p_k = 1.$$

   and p_k identifiable as the proportion of the population having realization value ω_k.

4. Show that, if c is a constant, then

$$A(X - c)^2 = A(X)^2 - 2cA(X) + c^2.$$

   One could regard this quantity as measuring the mean square deviation of X from the constant value c in the population; it is nonnegative for all c. The value of c minimizing this mean square deviation is just $A(X)$, the population average itself. The minimal value

$$A[(X - A(X))^2] = A(X^2) - A(X)^2$$

   measures the variability of the random variable in the population; it is termed the *population variance* of X. Note an implication: that

$$A(X^2) \geq A(X)^2. \tag{5}$$

5. Show that equality holds in (5) if and only if $X(\omega_k)$ has the same value for every k such that $p_k > 0$, i.e. for every category which actually occurs in the population.

4. Repeated Sampling: Expectation

Suppose the 'experiment', or the taking of a sample, reveals the value of the random variable X in the particular case. We can consider more imaginative examples than that of a census: the experiment might consist of the tossing of a die, of a shot at a target, of a count of the day's sales of newspapers, of a test for water pollution or of a count of plant species present in a unit area. We shall again assume for simplicity that the possible realizations ω behind the experiment, and at the level of description adopted, can take only the K values $\omega_1, \omega_2, \ldots, \omega_K$. More general situations will be considered in Section 5.

The experiment can be repeated (on different days, water samples or ground samples in the case of the last three examples, respectively). We might perform it n times, observing outcomes $X(\omega^{(1)})$, $X(\omega^{(2)})$, ..., $X(\omega^{(n)})$, where each $\omega^{(j)}$ is some point ω of Ω, the space of possible realizations. These n observations are rather like the N observations we made in taking a census of a country of N people in the last section.

There is an important difference, however. In the case of the census, we had made a complete enumeration of the population, and could go no further (if interest is restricted to the population of that one country). At least in the cases of the first three examples above we could go on repeating the experiment indefinitely, however, and each observation would be essentially new, in that its outcome could not be perfectly predicted from those of earlier experiments.

In a terminology and manner of thinking which have fallen out of fashion, but which are nevertheless useful, we imagine that we are sampling from a 'hypothetical infinite population', as compared with the physical finite population of Section 3. However, since the population is infinite and complete enumeration is impossible, we cannot write down the 'population average', as we did in formula (3). Nevertheless, the empirical fact that sample averages seem to 'converge' with increasing sample size leads us to postulate that the 'population average' or 'long-term average' of the random variable X does exist in principle. This idealized value we shall term the *expected value* or *expectation* of X, denoted $E(X)$, and we shall demand that the functional $E(X)$ should have just the properties (i)–(iii) which we required of a population average $A(X)$ in Section 3. Thus we have the basis of an axiomatic treatment, which is the modern approach to the subject.

We have appealed to the empirical fact that a sample average 'converges', i.e. that a partial average can approximate the total average, if only the sample is large enough. This is useful, not only as an indication that the concept of a population average is reasonable even when this average cannot be exactly evaluated, but also as a practical procedure. For example, although complete enumeration is certainly possible if a census of a country is required, one will often carry out only a partial census (the census of a sample), on the grounds of economy. If only the sample is large enough and precautions are taken to make it representative, then one can expect the results of the partial census to differ little from those of the complete census. This assertion has not merely the status of an empirical observation; we shall see in Section 2.7 that it is a *consequence* within the axiomatic framework we shall develop.

EXERCISES AND COMMENTS

1. The sample space Ω is that appropriate to the case of a single sample. If we consider an n-fold sample, i.e. the repetition of the experiment on n distinct cases, then we must consider a compound realization value which is an n-tuple of elements from Ω, and the sample space required will be the product of n copies of Ω.

2. Consider a committee of n people which is to vote on an issue; members are required to vote for or against the issue, and abstentions are not permitted. The realization for an individual committee member might be defined as his voting intention; the realization for the committee then consists of the voting intentions of all n identified individuals. The 'experiment' is the taking of a vote. How many possible experimental outcomes are there if the voting is (i) open? (ii) by secret ballot?

5. More on Sample Spaces and Variables

One certainly must go beyond finite sample spaces. Suppose, for example, that the realization is the result of a football match. One level of description is simply to report 'win, lose or draw' for the home team, so that Ω would contain just three points. A more refined level of description would be to give the score, so that ω would be a pair of nonnegative integers (s_1, s_2) and Ω the set of such integer pairs.

This sample space is already more general than the one considered in Section 3, in that it contains an infinite number of points. Of course, for practical purposes it is finite, since no team is ever going to make an infinite score. Nevertheless, there is no very obvious upper bound to the possible score, so it is best to retain the idea in principle that s_1 and s_2 can take any nonnegative integral value.

For another example, suppose the realization is wind-direction in degrees at a given moment. If we denote this by θ, then θ is a number taking any value in $[0, 360)$. The natural sample space is thus a finite interval (or, even more naturally, the circumference of a circle). Again Ω is infinite, in that it contains an infinite number of points. Of course, one can argue that, in practice, wind-direction can only be measured to a certain accuracy (to the nearest degree, say), so the number of distinguishable experimental outcomes is in fact finite. However, this is just the assertion that an actual experiment might be cruder than the ideal experiment which revealed the value of θ exactly, a point to which we return in Section 5.

A more refined description would be to consider both wind-direction θ and wind-speed v, so that Ω would be the set of values $\{\omega = (\theta, v); 0 \leq \theta < 360, v \geq 0\}$, a two-dimensional sample space.

The idea of a random variable of course transfers to these more general sample spaces. So, in the case of the football match, the 'winning margin' $|s_1 - s_2|$ is a random variable on the second sample space but not on the first. In the case of the wind measurement, the 'component of wind-velocity in direction α', $v \cos\{\pi(\alpha - \theta)/180\}$, is a random variable on the second space, but not on the first.

EXERCISES AND COMMENTS

1. The tossing of a coin requires a sample space of two points. As noted in Section 4, if one discusses a sequence of n tosses one requires a sample space of 2^n points. Is

the variable 'number of tosses before the first head is thrown' a random variable on this latter space?

2. Note that if we based the definition of Ω on the outcome of an experiment rather than on the abstract notion of realization, then the description of Ω would become heavily dependent on the precise method of experiment. For example, suppose we measure wind-velocity (i) by measuring components of velocity along two given orthogonal axes to the nearest mile per hour, or (ii) by measuring direction to the nearest degree and speed to the nearest mile per hour. Then the two sample spaces are different. In general, we encounter trouble then with variable transformations, and 'point densities' which vary over Ω. Yet it may be that in a fundamental study it would be the method and accuracy of observation which determined the structure of Ω.

3. Suppose one considered temperature θ as a function $\theta(t)$ of time over a period of 24 hours. Note that the sample space for this experiment must have a continuous infinity of dimensions, at least if one postulates no continuity properties for $\theta(t)$.

6. Ideal and Actual Experiments: Observables

For historical reasons we have termed Ω the sample space. This indeed remains the accepted term, a term indicating that one samples an ω value as one samples reality; i.e. carries out an experiment. However, only the ideal experiment will reveal the value of ω completely; most actual experiments reveal less.

So, if we consider the committee example of Exercise 1.4.2, the experiment which consists of a vote by secret ballot will not reveal the voting intentions of individual committee members. If we return to the measurement of wind-direction considered in the last section, then it is reasonable to accept that there *is* an actual direction θ, which could be taken as defining the realization, but that we can measure it only to limited accuracy.

If we consider the occurrence of a fault in a computer, then it is reasonable to take 'nature of the fault' as the realization ω. However, the tests (i.e. experiments) available may give only a partial diagnosis, and so only partially determine ω.

If we consider the response R of a rat to dosage D of a drug, then it may be that there is an exact relationship

$$R = f(C, D), \tag{6}$$

where C is the condition of the rat, specified sufficiently finely, and one could reasonably adopt the triple (C, D, R) as realization. In an ideal experiment one could measure all three variables C, D and R, and so verify relation (6). However, it may be that one can measure only D and R, and so obtain a much more diffuse relationship between these two quantities.

So, while the sample space, and the ideal experiment associated with it, set the level of one's description, the actual experiment sets the level of one's observation. Just as any quantity determined by the realization is a random variable, so any quantity determined by the outcome of the actual experiment is an *observable*.

One might ask: Why not simply define a new sample space listing the possible outcomes of the actual experiment? There are at least two answers. One is that we may wish to consider variables which are not observable (such as the condition of the rat) and form hypotheses in terms of these (such as (6)). Another is that we may wish to consider *alternative* actual experiments in a given situation. For example, consider, fault diagnosis for the computer. One has a given computer with a given fault, so that ω, the nature of the fault, is determined (although not yet known) and fixed. If one now carries out a series of diagnostic tests, then one is carrying out an experiment which is a real one, but which has several stages, so that one's state of knowledge increases as one moves through these stages. One has thus a shifting experimental outcome, and must retain the notion of realization if the experiments are to be related to a fixed hypothesis.

The notion that some variables are observable and some are not turns out to be a fundamental structural feature (see Section 2.8 and Chapters 5 and 11), as does the notion that the number of observables may increase as time progresses (see Chapters 9, 11 and 14).

EXERCISES AND COMMENTS

1. Suppose that one tries to observe a scalar variable y, but actually observes $y' = y + \varepsilon$, where ε is an observational error, of which one knows only that $|\varepsilon| < d$. The variable 'y to within $\pm d$' is thus observable by definition. Show, however, that 'y to the nearest multiple of δ' is not observable for any δ.

2. An electrical circuit consists of m elements in series, and exactly one element is known to be faulty. A test between any two points in the circuit shows whether the fault lies between those two points or not. Let r be the minimal number of tests needed to locate the fault, i.e. to make the variable 'location of the fault' observable. What is r if $m = 8$? Show that for large m the minimal value of r is $\log_2 m$ plus terms of smaller order. Note that the r tests are not to be regarded as the r-fold repetition of an experiment, but as stages in a single compound experiment.

3. One presumes that any actual experiment will always have an outcome, and that the outcome is determined by the realization ω. The outcome is thus a random variable $X(\omega)$, and points of the space Ω_X of possible outcomes of the experiment correspond to the sets of ω in Ω within which $X(\omega)$ is constant. 'Error' in the observation of ω can then only take the form that ω is located to within membership of one of these sets.

CHAPTER 2

Expectation

1. Random Variables

We shall now take up in earnest the axiomatic treatment sketched in Section 1.3. We assume a sample space Ω, setting a level of description of the realization ω of the system under study. In addition, we postulate that to each numerical-valued observable $X(\omega)$ can be attached a number $E(X)$, the *expected value* or *expectation* of X. The description of the variation of ω over Ω implied by the specification of these expectations will be termed a *probability process* or *random process*. The introduction of a probabilistic element justifies the term *random variable*, which we shall consistently abbreviate to r.v.

We are in effect changing our viewpoint, from considering what has happened in an experiment already performed, to considering what might happen in an experiment yet to be performed. The expectation $E(X)$ is the idealized average of X, the average being taken over all the outcomes that might result if the experiment were actually performed. The idea of such an average is given empirical support by the fact that sample averages seem to 'converge' to a limit with increasing sample size. This is, of course, only an empirical fact, which motivates the concept of an idealized average, but does not justify it. However, the achievement of a self-consistent and seemingly realistic theory is justification enough in itself. As an example of 'realism' we shall find that convergence of sample averages to expectation values is a feature which is reproduced in quite general models, under plausible assumptions concerning the joint outcomes of repeated experiments (see Sections 2.7, 2.9, 7.3 and 14.2).

We shall postulate in Section 2 that the expectation operator has certain axiomatic properties, essentially the properties (i)–(iii) asserted for the averaging operator A in Section 1.3. These axioms, although few and simple, will take us quite a long way.

In particular, they will enable us to develop the consequences of the basic assumptions of a physical model, and this is the usual aim. For example, if one makes some probabilistic assumptions concerning the fission of uranium nuclei by incident neutrons, what can one then infer concerning the behaviour of an atomic reactor core? If one makes some probabilistic assumptions concerning the development of the economy from day to day, what can one then infer concerning the state of the economy three months hence.?

From an abstract point of view the situation is as follows. By physical or other arguments one is given the expectation values of a family $\mathscr{F}$ of r.v.s Y. From this information one would wish to determine as closely as possible the expectation values of other r.v.s of interest, X, by appeal to the axioms.

Can an expectation be defined for any r.v.? We shall take the view that it can, insofar as that, once we have established that the given expectations $E(Y)$ for Y in $\mathscr{F}$ are internally consistent with the axioms, we shall thereafter accept any statement concerning an expectation $E(X)$ which can be derived via the axioms from the given expectations, for any r.v. Y. Such an approach can be faulted only if such derived statements turn out to contain an inconsistency, and this we shall not find. It does not imply that $E(X)$ can be simultaneously and consistently prescribed for all X on Ω; cases are in fact known for which this is impossible.

Of course, if X is, relative to the Y of $\mathscr{F}$, rather bizarre as a function of ω, then the bounds on $E(X)$ derivable from the given expectations will presumably be rather wide. The deduction of bounds (and, in particular, best possible bounds) on expectation values $E(X)$ from knowledge of $E(Y)$ for Y in $\mathscr{F}$ is the *extension problem*, considered in Chapters 12 and 15.

Note that the values $\pm\infty$ for an expectation have not been excluded as in any sense improper.

2. Axioms for the Expectation Operator

We shall refer indiscriminately to E as the expectation operator, or to $E(X)$ as the expectation functional; see Exercise 8. The point is that there is a rule for determining a number $E(X)$ from a function $X(\omega)$.

The exact form of the operator, i.e. the actual rule for attaching a value to $E(X)$ for a r.v. X, must be determined by special arguments in individual cases, and these arguments will usually determine $E(X)$ only for certain X. Much of the rest of the book will be concerned with such particular cases; physical processes of one sort or another. However, in this section we shall concern ourselves with the general rules which E should obey if it is to correspond to one's intuitive idea of an expectation operator. These rules will take the form of axioms, relating the expectations of r.v.s.

We restrict ourselves for the moment to scalar-valued r.v.s, and assume that $E(X)$ is then also a scalar, satisfying the following axioms.

Axiom 1. *If $X \geq 0$ then $E(X) \geq 0$.*

Axiom 2. *If c is a constant then $E(cX) = cE(X)$.*

Axiom 3. $E(X_1 + X_2) = E(X_1) + E(X_1)$.

Axiom 4. $E(1) = 1$.

Axiom 5. *If a sequence of r.v.s $\{X_n(\omega)\}$ increases monotonically to a limit $X(\omega)$ then*

$$E(X) = \lim E(X_n).$$

The first four axioms state that E is a positive linear operator with the normalization $E(1) = 1$, just as was the averaging operator of equation (1.3). Axiom 5 is a continuity demand, stating that for a monotone sequence of r.v.s the operations E and lim commute. Although this condition is also satisfied by the averaging operator (1.3), as an axiom it appears somewhat less natural than the others, especially since a weak form of it can be derived from the other axioms (see Exercise 5). In fact, one can go a long way without it, and there are interesting physical situations for which the axiom does not hold (see Exercise 2.4.3). However, some condition of this type becomes necessary when one considers limits of infinite sequences, as we shall have occasion to do later (see Sections 15.1 and 15.4).

The axioms have certain immediate consequences that the reader can verify; for example, that

$$E\left(\sum_{1}^{n} c_j X_j\right) = \sum_{1}^{n} c_j E(X_j), \tag{1}$$

if the c_j are constants and n is finite. Also, if $X_1 \leq Y \leq X_2$, then $E(Y) \leq E(Y) \leq E(X_2)$.

The equations in the axioms are all to be understood in the sense that, if the right-hand member is well defined, then the left-hand member is also well defined, and the two are equal. There are occasional failures. For example, suppose that $E(X_1) = +\infty$ and $E(X_2) = -\infty$. Then Axiom 3 would give $E(X_1 + X_2)$ the indeterminate value $+\infty - \infty$.

We can avoid such indeterminacies by restricting the class of r.v.s considered. For example, suppose we separate X into positive and negative parts

$$X = X_+ - X_-,$$

where

$$X_+ = \begin{cases} X & (X \geq 0), \\ 0 & (\text{otherwise}), \end{cases} \tag{2}$$

and require that both parts have finite expectation. Since $|X| = X_+ + X_-$, this is equivalent to requiring that

$$E(|X|) < \infty. \tag{3}$$

If we restrict ourselves to r.v.s for which (3) holds (r.v.s with *finite absolute expectation*), then $\sum_1^n E(X_j)$ will always be well defined, at least for finite n. This is a convenient restriction, on the whole, and we shall henceforth adopt it, unless the contrary is stated.

We postulated scalar-valued r.v.s for definiteness, but extension to the vector or matrix case is immediate; see Exercise 7.

EXERCISES AND COMMENTS

1. Show that if Axioms 2 and 3 are assumed, then Axioms 1 and 4 are jointly equivalent to the single statement:

   $$a \le X \le b \quad \text{implies that} \quad a \le E(X) \le b, \quad \text{for constant } a, b.$$

   Although such a single axiom would be more economic, it is helpful to separate the properties of positivity and normalization.

2. Show that equation (1) holds for n infinite if the $c_j X_j$ are all of the same (constant) sign.

3. Show that $|E(X)| \le E(|X|)$.

4. Show that $E(|X_1 + X_2|) \le E(|X_1|) + E(|X_2|)$.

5. Show, without appeal to Axiom 5, that if $|X_n - X| \le Y_n$ and $E(Y_n) \to 0$ with increasing n, then $E(X_n) \to E(X)$.

6. Note that $E(X)$ is defined only for numerical-valued X. So, if X were 'fingerprint', then $E(X)$ would be undefined. Of course, one could speak of the expectation of a numerical-valued function of X (e.g. 'number of whorls' of a specified type).

7. 'Numerical-valued' certainly includes 'scalar-valued'. However, if X is a vector whose components are r.v.s X_j, then one can consistently define $E(X)$ as the vector of expectations, i.e. the vector with components $E(X_j)$. Similarly for matrices of r.v.s, or arrays of r.v.s in general. One is simply requiring validity of the axioms for each element of the array individually.

8. One minor point of notation which might as well be clarified. What we have written as $E(X)$ is sometimes also written simply as EX. One might say that the two conventions correspond to the view of E as a functional or as an operator, respectively. We shall adhere to the first convention, for definiteness. Thus, $E(X)^2$ is to be understood as $[E(X)]^2$, not as $E(X^2)$.

3. Events: Probability

Intuitively, an event A is something which, in a given case, either occurs or does not occur. In our framework it corresponds then to a set of realizations

ω, i.e. to a subset of Ω, which we shall also denote by A. If $\omega \in A$ then 'the event occurs' in the particular case ω; if $\omega \notin A$ then it does not.

Thus, for the football example of Section 1.4 we might consider the event 'the home team won'. In the first space suggested this would correspond to a single point; in the second it would correspond to the set of points satisfying $s_1 > s_2$.

For the wind-measurement example we might consider the event 'the wind-speed exceeds 50 m.p.h.'. This would be defined only on the second sample space, when it would correspond to the set of points $v > 50$.

The *probability of* A, denoted $P(A)$, will be defined as

$$P(A) = E[I(A)], \tag{4}$$

where $I(A, \omega)$ is the *indicator function* of the set A

$$I(A, \omega) = \begin{cases} 1 & (\omega \in A), \\ 0 & (\omega \notin A). \end{cases} \tag{5}$$

$P(A)$ is to be regarded as the expected proportion of cases in which the event A actually occurs. The motivation for the definition comes from the finite population census of Section 1.3, where we saw that the proportion of the population falling in a set A was just the average of the indicator variable for that set.

We shall not investigate the concepts of events or probability to any extent before Chapter 3, but it is helpful to have them formulated.

The probability measure $P(A)$ is a function with a *set* A as argument. However, it is sometimes very convenient notationally to take just a description of the event as the argument, regarding this as equivalent to the corresponding set. Thus, we write the probability that X is greater than Y as $P(X > Y)$ rather than the more correct but cumbersome $P(\{\omega: X(\omega) > Y(\omega)\})$. The same goes for more verbal descriptions: we would write $P(\text{rain})$ rather than $P(\text{the set of } \omega \text{ for which it rains})$. Nevertheless, the true argument of $P(\cdot)$ is always a subset of Ω.

This rather loose convention will also be transferred to indicator functions. We shall often suppress the ω-argument and write $I(A, \omega)$ simply as $I(A)$, the random variable which is 1 or 0 according as the event A does or does not occur. We shall again often substitute for A a verbal or informal description of the event, so that $I(X > Y)$ is the random variable which is 1 or 0 according as X exceeds Y or not.

4. Some Examples of an Expectation

Before proceeding further we should show that expectation operators satisfying the axioms of Section 2 really do exist. It is sufficient to find some examples, and for these we can choose the types of process arising in applications.

One says that the process is *discrete* (or, the probability distribution is discrete) if ω can take only a countable set of values; say $\{\omega_1, \omega_2, \ldots, \omega_K\}$.

Theorem 2.4.1. *The process is discrete with ω confined to the set $\{\omega_1, \omega_2, \ldots, \omega_K\}$ if and only if the expectation functional takes the form*

$$E(X) = \sum_k p_k X(\omega_k), \qquad (6)$$

where the numbers p_k satisfy the constraints $p_k \geq 0$, $\sum_k p_k = 1$. One can then make the identification $p_k = P(\omega_k)$.

Relation (6) can thus be written more pleasingly as

$$E(X) = \sum_{\omega \in \Omega} P(\omega) X(\omega).$$

PROOF. To prove necessity, note first that we can write

$$X(\omega) = \sum_k I(\omega_k) X(\omega_k), \qquad (7)$$

where the sum is meaningful, since only one summand is ever nonzero. (Recall that $I(\omega_k)$ is the indicator function of the point set ω_k.) Taking expectations in (7) we deduce expression (6) with $p_k = E[I(\omega_k)] = P(\omega_k) \geq 0$. Setting $X = 1$ we deduce that $\sum p_k = 1$.

To prove sufficiency, assume that (6) holds. One has then, for any event (set) A,

$$P(A) = E(I(A)) = \sum_{k \in a} p_k, \qquad (8)$$

where a is the set of k for which $\omega_k \in A$. Taking A as the point set ω_k we deduce that indeed $p_k = P(\omega_k)$. Taking A as a set containing none of the values ω_k ($k = 1, 2, \ldots, K$), we find that $P(A) = 0$. That is, these are the only values of realization which can occur. $\qquad\square$

Relation (8) could be rewritten more attractively as

$$P(A) = \sum_{\omega \in A} P(\omega) \qquad (9)$$

for A a subset of Ω. In this form it expresses what is often termed the *additive law of probability*: that the probability of an event is the sum of probabilities of realizations consistent with that event.

As the standard example of a discrete process, suppose that one throws a die and observes the number occurring; this number can itself be taken as defining the realization ω. Let Ω be taken as the real line; that is, let us allow the possibility that ω can take any real value. In fact, however, the only possible values are $\omega = 1, 2, 3, 4, 5$ and 6. If we assume that the die is fair, then, by symmetry, all the p_k should be equal, and so equal to $\frac{1}{6}$. Hence, in this case,

$$E(X) = \frac{1}{6} \sum_{k=1}^{6} X(k).$$

The fact that in this formula $X(\cdot)$ has no argument other than the values 1, 2, ..., 6 indicates that these are the only values possible; the fact that all the p_k are equal expresses the symmetry of the die.

For a second type of process, let us again suppose that Ω is the real line, so that ω is a real scalar. Suppose that

$$E(X) = \int_{-\infty}^{\infty} X(\omega)f(\omega) \, d\omega \tag{10}$$

for all $X(\omega)$ for which the integral is defined and absolutely convergent. Then E wil obey the axioms (at least for r.v.s of this class; see the note at the end of the section) if f obeys the conditions

$$f(\omega) \geq 0, \tag{11}$$

$$\int_{-\infty}^{\infty} f(\omega) \, d\omega = 1. \tag{12}$$

The relation analogous to (9) is

$$P(A) = E[I(A)] = \int_{A} f(\omega) \, d\omega, \tag{13}$$

so that f can be regarded as a *probability density* on Ω. In this case, one speaks of a *continuous probability distribution* on Ω. The idea can be extended to more general sample spaces than the real line, provided one has an appropriate definition of the integral (10).

As an example, consider the spinning of a roulette wheel. If ω is the angle in radians that the pointer makes with some reference radius on the wheel, then ω can only take values on the range $[0, 2\pi)$. If the wheel is a fair one, all these values will be equally likely, so, by symmetry, the expectation formula must be

$$E(X) = \frac{1}{2\pi} \int_{0}^{2\pi} X(\omega) \, d\omega.$$

That is, we have a continuous probability distribution with density

$$f(\omega) = \begin{cases} 1/(2\pi) & (0 \leq \omega < 2\pi), \\ 0 & (\text{otherwise}). \end{cases}$$

This example can help to clarify a point: the difference between impossible events, and events of zero probability. Impossible events (e.g. the throwing of a seven with a die) have zero probability; the converse is not necessarily true. For example, consider the event that the roulette wheel comes to rest within an angle δ of a prescribed direction θ; the probability of this is

$P(\theta - \delta < \omega < \theta + \delta) = \delta/\pi$ for $\delta \leq \pi$. As δ tends to zero this probability also tends to zero. In other words, the event $\omega = \theta$, that the rest-angle ω has a prescribed value, has zero probability. Yet the event is plainly not impossible. The event has zero probability, not because it is impossible, but because it is just one of an infinite number of equally probable realizations.

If a particular r.v. $X(\omega)$ is such that

$$E[H(X)] = \int H(x)f(x)\,dx$$

for any function H for which the integral is meaningful, then we have a distribution which is continuous on the sample space Ω_X constituted by the x-axis. In this case the r.v. X is said to be continuously distributed with *probability density function* (or *frequency function*) $f(x)$. This is very much the same situation as before; we have, for example

$$P(X \in A) = \int_A f(x)\,dx, \tag{14}$$

except that the continuous distribution is now a property of the r.v. X rather than of the realization ω. It is conventional to use an uppercase letter X for the r.v. and a corresponding lowercase letter x for particular values that the r.v. may adopt, and on the whole this is a helpful distinction. Thus, we write $f(x)$ rather than $f(X)$.

Note. Representation (10) is restricted to some class of r.v.s because the integral is presumably to be interpreted in some classic sense, such as the Riemann sense, and Xf must then be Riemann integrable. However, use of the axioms will enable one to construct bounds for expectations of r.v.s which are not Riemann integrable, even possibly to the point of determining $E(X)$ completely. So $E(X)$ is not necessarily representable as a Riemann integral for all expectations which can be derived from representation (10).

EXERCISES AND COMMENTS

1. Suppose that a person joining a queue has to wait a time τ before he is served, and that

$$E[H(\tau)] = pH(0) + \int_0^\infty H(t)f(t)\,dt$$

for all functions H for which this expression is defined. Find the conditions on p and f for this formula to represent an expectation on Ω_τ, and interpret the formula.

2. An electron oscillating in a force field has energy ε which can take values $\varepsilon_k = \alpha(k + \frac{1}{2})$ with probabilities proportional to $\exp(-\beta\varepsilon_k)$ $(k = 0, 1, 2, \ldots)$, where α and β are constants. Determine $E(\varepsilon)$ and $E(\varepsilon^2)$.

3. Suppose that the expectation operator is defined by

$$E(X) = \lim_{D\to\infty} \frac{1}{2D}\int_{-D}^D X(\omega)\,d\omega.$$

Show that this satisfies the first four axioms of Section 2, but not the fifth. (Consider the sequence of r.v.s

$$X_n(\omega) = \begin{cases} 1 & (|\omega| \le n), \\ 0 & (\text{otherwise}), \end{cases}$$

for $n = 0, 1, 2, \ldots$.) This process would correspond to a uniform distribution over the whole infinite axis, and so might, for example, be used to represent the position of a star equally likely to lie anywhere within an infinite universe.

4. The *distribution function* $F(x)$ of a scalar r.v. X is defined as $P(X \le x)$. Show that, if F is differentiable, then X has a probability density

$$f(x) = \frac{\partial F(x)}{\partial x} = \frac{\partial}{\partial x} P(X \le x).$$

This is the relation inverse to (14).

5. Let θ_1 and θ_2 be the rest-angles observed in two consecutive spins of a roulette wheel, and suppose that expectations on the two-dimensional sample space space thus generated are given by

$$E[X(\theta_1, \theta_2)] = \frac{1}{4\pi^2} \int_0^{2\pi} \int_0^{2\pi} X(\theta_1, \theta_2) \, d\theta_1 \, d\theta_2.$$

Show that $P(\theta_1 \in A_1, \theta_2 \in A_2) = P(\theta_1 \in A_1)P(\theta_2 \in A_2)$. If $X = \theta_1 + \theta_2 - 2\pi$, then show, either from Exercise 4 or by calculating $E[H(X)]$ for arbitrary H, that X is continuously distribution over the interval $[-2\pi, +2\pi]$ with density $(2\pi - |x|)/(4\pi^2)$.

5. Moments

If X is a scalar r.v., then one set of expectations of interest is provided by the *moments* $\mu_j = E(X^j)$ for $j = 0, 1, 2, \ldots$.

The first moment $\mu_1 = E(X)$ is the *mean* of X, and is by definition the expected value of X and the obvious characterization of its 'typical' value (although there are others; see Exercise 3).

From the first two moments one derives the measure of spread in the distribution of X known as the *variance* of X, denoted by var(X) and variously evaluated as

$$\text{Var}(X) = \min_c E[(X - c)^2] = E[(X - E(X))^2]$$

$$= E(X^2) - E(X)^2 = \mu_2 - \mu_1^2. \tag{15}$$

(see Exercise 1.3.4). A value of zero for var(X) would seem to imply that X is identically constant, and equal to $E(X)$. This very nearly the case, but we give a more careful statement of the situation in Section 9.

The *standard deviation* $\sigma = \sqrt{\text{var}(X)}$ is often taken as a measure of dispersion in X; it is a natural scale parameter in that it is measured in the same scale as X itself.

EXERCISES AND COMMENTS

1. Note an implication of the middle pair of relations in (15): that $\mu_2 \geq \mu_1^2$. Equality implies that $\text{var}(X) = 0$.

2. Consider a linear transformation $Y = a + bX$. Show that $E(Y) = a + bE(X)$ and $\text{var}(Y) = b^2 \text{var}(X)$. The variable

$$Y = \frac{X - E(X)}{\sqrt{\text{var}(X)}}$$

is thus a linear transformation of X which has zero mean and unit variance; it is referred to as the *standardized* version of X.

3. Different measures of the 'location' of a distribution (i.e. of the value of X which is in some sense central, or typical) are the *mean* $\mu_1 = E(X)$, the *mode* (the most probable value of X) and the *median* (the value m for which $P(X \leq m)$ is as near to $\frac{1}{2}$ as the discreteness of the distribution will allow). If the distribution has a reasonably symmetric and *unimodal* (i.e. single-humped) form then these measures will tend to agree. If the distribution is strongly *skew* (i.e. asymmetric) then they will not.

    For example, income has a strongly skew distribution, and there are reasons (see Exercise 10.5.4) for supposing that it follows the *Pareto distribution*, with density

$$f(x) = \begin{cases} 0 & (x < h), \\ \alpha h^\alpha x^{-\alpha-1} & (x \geq h). \end{cases}$$

Here h is a minimum income and α is the *Pareto parameter*, found in practice to have values between 1.6 and 2.4. The mode of the distribution is at h, the mean at $[\alpha/(\alpha - 1)]h$ and the median at $2^{1/\alpha}h$. The first two are unrealistically low and high, respectively; the median is the figure generally quoted as 'typical' for income distributions.

6. Applications: Optimization Problems

The theory we have developed, slight though it is as yet, is enough to help us to useful conclusions in a variety of problems. Of particular interest are those problems concerned with *optimization*, for which one is trying to achieve maximal expected return in some enterprise.

A simple such problem is that of the newsagent who stocks N copies of a daily paper, and wishes to choose N so as to maximize his daily profit. Let a be the profit on a paper which is sold, b the loss on an unsold paper and c the loss if a customer wishes to buy a paper when stocks are exhausted. The quantities a and b can be determined immediately from the wholesale and retail prices of the paper. The quantity c is less easy to determine, because it measures 'loss of goodwill due to one lost sale' in monetary terms. However, an estimate of it must be made if the situation is to be analysed.

If the newsagent stocks N papers and has X customers on a given day, then the components of his profit (or negative loss) are as follows:

Item	Profit	
	$X \leq N$	$X > N$
Sales	aX	aN
Unsold papers	$-b(N - X)$	0
Unsatisfied demand	0	$-c(X - N)$.

Thus his net profit is

$$g_N(X) = \begin{cases} (a + b)X - bN & (X \leq N), \\ (a + c)N - cX & (X > N). \end{cases}$$

If X were known then the newsagent would obviously maximize profit by choosing $N = X$. However, the demand X will certainly be variable, and can realistically be regarded as a r.v., so that one has to work instead with an *expected* profit

$$G_N = E[g_N(X)]$$

and choose N so as to maximize this. If expected profit is virtually identical with long-term average profit (the intuitive foundation of our theory; see also Sections 4.5, 7.3 and 14.2) then to maximize G_N is a reasonable procedure.

The increment in expected profit when an extra paper is stocked is

$$G_{N+1} - G_N = E[g_{N+1}(X) - g_N(X)] = E[-b + (a + b + c)I(X > N)]$$
$$= -b + (a + b + c)P(X > N). \quad (16)$$

For small enough N this quantity is positive, but as N increases it ultimately turns negative, and the first value for which it does so is the optimal one. Roughly, one can say that the optimal N is the root of the equation $G_{N+1} - G_N \approx 0$, or

$$P(X > N) \approx \frac{b}{a + b + c}.$$

To complete the solution of the problem one needs to know $P(X > N)$ as a function of N. In practice, one would use records of past sales to obtain an estimate of this function. For example, $P(X > N)$ could be estimated directly by the actual proportion of times over a long period that potential sales would have exceeded N. More refined methods are possible if one can restrict the form of the function $P(X > N)$ on theoretical grounds.

The treatment of this problem has not followed quite the course we promised in Section 1, in that we have simply ploughed ahead with the maximization of expected profit, assuming that all expectations required for this purpose, such as $P(X > N)$, were known. For cases where one has to make the best of less information, see Exercise 12.5.3.

The example was simplified by the fact that a newspaper can be assumed to have commerical value only on the day of issue, so that there is no point in carrying stock over from one day to another; each day begins afresh. However, suppose that the newsagent also stocks cigarettes. For this commodity he will certainly carry stock over from day to day, and the decisions made on a given day will have implications for the future as well as for the present. This much more difficult and interesting situation amounts to a *sequential decision problem*; we return to the topic in Section 16.4.

EXERCISES AND COMMENTS

1. Suppose that one incurs a loss as if early for an appointment by time s, and a loss bs if late by time s ($s \geq 0$). The time taken to reach the place of appointment is a continuously distributed r.v. T. Suppose that one allows a time t to reach the appointment. Show that the value of t minimizing expected loss is determined by

$$aP(T < t) = bP(T > t) \qquad \text{or} \qquad P(T < t) = \frac{b}{(a + b)}.$$

   Note that we can also write the first characterization as $P(\text{early})/P(\text{late}) = b/a$.

2. A steel billet is trimmed to a length x and then rolled. After rolling, its length becomes $y = \beta x + \varepsilon$, where ε is a r.v. expressing the variability of rolling. It is then trimmed again to a final length z. If y is greater than z then there is a loss proportional to the excess: $a(y - z)$. If y is less than z then the billet must be remelted, and there is a flat loss b. The original trim-length x must now be chosen so as to minimize expected loss. Show that if ε has probability density $f(\varepsilon)$ then the equation determining the optimal value of x is

$$bf(z - \beta x) = a \int_{z-\beta x}^{\infty} f(\varepsilon) \, d\varepsilon.$$

7. Equiprobable Outcomes: Sample Surveys

There are cases where one can place constraints upon the expectation functional by appealing to some underlying symmetry in the situation. The most extreme such case is that in which one has a finite sample space $\Omega = \{\omega_1, \omega_2, \ldots, \omega_K\}$, and assert that the expectation should be invariant under any permutation of the elements of Ω. The expectation functional then necessarily has the symmetric form

$$E(X) = K^{-1} \sum_{k=1}^{K} X(\omega_k) \tag{17}$$

(see Exercise 3). That is, all realizations have the same probability, K^{-1}.

Calculation of expectations amounts then just to the calculation of arithmetic averages (17), and calculation of probabilities is a matter of enumeration:

$$P(A) = K^{-1} \#(A),$$

the enumeration of the number $\#(A)$ of realizations ω_k which fall in A.

Gaming problems throw up such situations very readily, and calculation of $\#(A)$ can be a challenging combinatorial problem (see, e.g. Section 4.7).

An example of some practical significance concerns the drawing of a sample of size n from a population of N identified individuals. Consider first the rather degenerate case $n = 1$, the sampling of a single individual. Let the identities of the N individuals be denoted $\omega_1, \omega_2, \ldots, \omega_N$, so that $\omega = \omega_k$ means that the kth individual has been chosen. Suppose that we measure some numerical characteristic $X(\omega)$ of this individual (e.g. weight, income or cholesterol level). If all draws are equally likely then

$$E(X) = N^{-1} \sum_{k=1}^{N} X(\omega_k) = A(X), \tag{18}$$

so the expected reading $E(X)$ is just the population average $A(X)$ of that variable.

If we now choose a sample of n then the outcome of the experiment will be an ordered list $\zeta = (\zeta_1, \zeta_2, \ldots, \zeta_n)$, where ζ_j is the identity of the jth individual sampled, taking possible values $\omega_1, \omega_2, \ldots, \omega_N$. It is these n-tuples of ω-values which now provide the points of the compound sample space $\Omega^{(n)}$ adequate to describe the results of the experiment. We imagine that, once we have sampled the individuals, we can measure any of their attributes which are of interest (e.g. income, political affiliation) and so determine derived r.v.s $X(\omega)$. In this sense, the 'experiment' then consists of the drawing of the sample, i.e. of the revealing of the identities of the individuals sampled, rather than of the subsequent measuring of attributes.

One can envisage two sampling rules: *without replacement* (when all individuals in the sample must be distinct) and *with replacement* (when there is no such constraint; individuals may be sampled repeatedly). In the first case $\Omega^{(n)}$ contains $N(N-1)\ldots(N-n+1)$ points; in the second it contains N^n. The notation $N^{(n)}$ for the factorial power $N(N-1)\ldots(N-n+1)$ is a convenient one, which we shall use consistently. If the sample has been drawn in a blind fashion then there is no reason to expect any possible outcome to be more likely than any other, so, whether sampling is with or without replacement, one would expect all elements of the appropriate $\Omega^{(n)}$ to be equiprobable.

Suppose now that we are interested in some numerical characteristic X of individuals; say, income. This has value $X(\omega_k)$ for individual k, but we will find it more convenient to denote the value simply by x_k, so that

$$A(X) = N^{-1} \sum_k x_k, \qquad A(X^2) = N^{-1} \sum_k x_k^2,$$

etc. We shall also find it useful to define the *population variance*

$$V(X) = A(X^2) - A(X)^2$$

of X (see Exercise 1.3.4 and Section 5).

Let $\xi_j = X(\zeta_j)$ be the value of X for the jth member of the sample, and define the sample average

$$\bar{\xi} = n^{-1} \sum_{j=1}^{n} \xi_j.$$

Then the ξ_j are r.v.s on $\Omega^{(n)}$, and so is $\bar{\xi}$. We could regard $\bar{\xi}$ as an estimate of $A(X)$, the population mean income, and we could ask if it is a successful one. It could be regarded as successful if $E[(\bar{\xi} - A(X))^2]$ were small. Let us then consider calculation of moments, under the assumption of sampling without replacement, for definiteness.

Theorem 2.7.1. *If sampling is without replacement then the sample mean has expectation*

$$E(\bar{\xi}) = A(x) \tag{19}$$

and variance

$$\text{var}(\bar{\xi}) = E[(\bar{\xi} - A(X))^2] = \frac{1}{n} \frac{N-n}{N-1} V(X). \tag{20}$$

A statistician would say that (19) implied that $\bar{\xi}$ was an *unbiased* estimate of $A(X)$. Relation (20) states that $\bar{\xi}$ in fact converges to $A(X)$ with increasing n, in that the mean square difference $E[(\bar{\xi} - A(X))^2]$ tends to zero. It of course exactly zero if $n = N$, because then the whole population has been sampled. However, if N were 100,000, say, and n were 1000, then expression (20) would be small, of order $1/1000$, despite the fact that the fraction of the population sampled is only $n/N = 1/100$. This is the verification in our model of the empirical fact remarked upon in Section 1.2: that the mean of n observations seems to stabilize in value as n increases. We find this in our model because of the assumption that all possible samples were equiprobable for given n.

PROOF. The order in which individuals can occur in the sample actually has no significance, and the ξ_j all have the same distribution. In fact, each takes the values $x_1, x_2, \ldots, x_N$ with equal probability, so that

$$E(\xi_j) = A(X)$$

as already noted in (18). Assertion (19) thus follows. Correspondingly, $E(\xi_j^2) = A(X^2)$, so that

$$E[(\xi_j - A(X))^2] = V(X). \tag{21}$$

If we consider $E(\xi_j \xi_k)$ for $j \neq k$ then $\xi_j \xi_k$ can take the $N(N-1)$ values $x_h x_i$

$(h \neq i)$ with equal probability, so that

$$E(\xi_j\xi_k) = \frac{1}{N(N-1)} \sum_{h \neq i}\sum x_h x_i = \frac{1}{N(N-1)} [(\sum x_k)^2 - \sum x_k^2]$$

and

$$E[(\xi_j - A(X))(\xi_k - A(X)] = E(\xi_j\xi_k) - A(X)^2 = -\frac{1}{N-1} V(X). \quad (22)$$

We then deduce from (21) and (22) that

$$E[(\bar{\xi} - A(X))^2] = \frac{1}{n^2} E\left[\sum_j \sum_k (\xi_j - A(X))(\xi_k - A(X))\right]$$

$$= \frac{1}{n^2}\left[nV(X) - \frac{n(n-1)}{(N-1)} V(X)\right],$$

which indeed reduces to expression (20)   □

EXERCISES AND COMMENTS

1. Suppose that sampling is without replacement. Show that assertion (19) still holds and that (20) is replaced by

$$\text{Var}(\bar{\xi}) = V(X)/n. \quad (23)$$

Thus $E[(\bar{\xi} - A(X))^2]$ still tends to zero as n^{-1} with increasing sample size. On the other hand, it is not exactly equal to zero for $n = N$, because only a fraction $N!/N^N$ of the samples of size N will sample every member of the population exactly once.

2. Suppose that two samples of size n are taken, with sample means for X equal to $\bar{\xi}$ and $\bar{\xi}'$, respectively. The same rule is used for each sample (i.e. with or without replacement) but individuals are allowed to appear in both samples. Suppose that all possible sample compositions under these constraints are equally likely. It is then evident that $\bar{\xi}$ and $\bar{\xi}'$ both have expectations equal to $A(X)$ and the same variance. Show that $E(\bar{\xi}\bar{\xi}') = A(X)^2$, and hence that $E[(\bar{\xi} - \bar{\xi}')^2]$ has twice the value of var$(\bar{\xi})$ given by (20) or (23), depending on whether sampling is with or without replacement. Thus $E[(\bar{\xi} - \bar{\xi}')^2]$ also tends to zero as n^{-1} with increasing n, reflecting increasing consistency of the two sample averages.

3. We know from Theorem 2.4.1 that, in the case of a discrete sample space, the expectation functional necessarily has the form $E(X) = \sum_\omega X(\omega)P(\omega)$. Suppose that the expectation is known to be invariant under a permutation σ of the elements of Ω. That is, $X(\omega)$ and $X(\sigma\omega)$ have the same expectation (for any X), where $\sigma\omega$ is the value to which ω goes under the permutation σ. Show then that $P(\omega) = P(\sigma\omega)$ for all ω of Ω.

8. Applications: Least Square Estimation of Random Variables

Let us consider an optimization problem of rather a different character to that of Section 6. Commonly one is in the position of being able to observe the values of a number of r.v.s $Y_1, Y_2, \ldots, Y_m$ and wishes to infer from these the value of another r.v. X, which cannot itself be observed. For example, one may wish to predict the future position X of an aircraft on the basis of observations Y_j already made upon its path. This is a prediction problem, such as also occurs in economic contexts, numerical weather forecasting and medical prognosis. The element of prediction in time need not always be present, so it is perhaps better to speak of *estimation* or *approximation*. For instance, the r.v.s Y_j might represent surface measurements made by an oil prospector, and X a variable related to the presence of oil at some depth under the earth's surface. Even in the weather context one has the problem of 'smoothing' weather maps, i.e. of spatial interpolation of the readings at an irregular grid of weather stations.

In any case, one wishes to find a function of the observation

$$\hat{X} = \phi(Y_1, Y_2, \ldots, Y_m)$$

which estimates X as well as possible. That is, the function ϕ is to be chosen so that the estimate $\hat{X}$ is as close as possible to the true value X, in some sense.

Suppose, for the moment, that all variables are scalar-valued. An approach very common in both theoretical and practical work is to restrict onself to *linear estimates*

$$\hat{X} = \sum_{j=1}^{m} a_j Y_j \tag{24}$$

and to choose the coefficients a_j so as to minimize the *mean square error*

$$E[(X - \hat{X})^2] = E(X^2) - 2 \sum_j a_j E(X Y_j) + \sum_j \sum_k a_j a_k E(Y_j Y_k). \tag{25}$$

The r.v. $\hat{X}$ is then known as the *linear least square estimate* (or *approximant*, or *predictor*) of X in terms of the observations Y_j. We shall use the term 'linear least square' so frequently that we shall abbreviate it to LLS.

We can write relations (24) and (25) in matrix form as

$$\hat{X} = a^{\mathsf{T}} Y \tag{24'}$$

and

$$D(a) = E[(X - \hat{X})^2] = U_{XX} - 2a^{\mathsf{T}} U_{YX} + a^{\mathsf{T}} U_{YY} a, \tag{25'}$$

where we have defined a scalar U_{XX}, column vectors a and U_{YX} and a matrix U_{YY}. The matrix U_{YY} has jkth element $E(Y_j Y_k)$, so it is called a *product moment matrix*. In (25') we have written the mean square error also as a function $D(a)$ of the coefficient vector a.

Theorem 2.8.1 *Relation* (24′) *yields an LLS estimate of X in terms of Y if and only if the coefficient vector a satisfies the linear relations*

$$U_{YY}a = U_{YX}. \qquad (26)$$

PROOF. Relation (26) is certainly necessary, as the condition that expression (25′) be stationary in a. To verify that it determines a minimum of $D(a)$, note first that, for any vector c,

$$c^{\mathsf{T}}U_{YY}c = E[(c^{\mathsf{T}}Y)^2] \geq 0. \qquad (27)$$

Let us denote any solution of (26) by $\hat{a}$. Then, by setting $a = \hat{a} + (a - \hat{a})$ in (25′) and appealing to relation (26) we find that, for any a,

$$D(a) = D(\hat{a}) + (a - \hat{a})^{\mathsf{T}}U_{YY}(a - \hat{a}) \geq D(\hat{a}). \qquad \square$$

Relation (26) will have a unique solution only if the matrix U_{YY} is non-singular. However, the proof of the theorem indicates that *any* solution of (26) yields an optimal estimate. In fact, essentially the same estimate; see Exercise 2.9.9. We shall return to these matters in Section 9, and again in Chapter 11.

An estimate which incorporates a constant term

$$\hat{X} = a_0 + a^{\mathsf{T}}Y \qquad (28)$$

may seem to be more general than (24), but in fact is formally included in the earlier form. The righthand member of expression (28) can be written $\sum_0^m a_j Y_j$ where Y_0 is a r.v. identically equal to 1, so that (28) is of the same form as (24). Moreover, we can quickly reduce the second case to the first.

The extra minimization equation (with respect to a_0) is

$$E(X) = a_0 + a^{\mathsf{T}}E(Y)$$

so that we can eliminate a_0 from this relation and (28) to obtain the estimation relation

$$\hat{X} - E(X) = a^{\mathsf{T}}(Y - E(Y)). \qquad (29)$$

As far as the optimal determination of $a_1, a_2, \ldots, a_m$ is concerned, it is then simply a matter of repeating the whole previous analysis with X and Y replaced by $X - E(X)$ and $Y - E(Y)$, respectively. The effect on the determination of a is then that a product moment such as $E(X Y_j)$ is replaced by the *covariance*

$$\mathrm{cov}(X, Y_j) = E[(X - E(X))(Y_j - E(Y_j))] = E(X Y_j) - E(X)E(Y_j),$$

etc. The covariance is to the product moment as the variance $\mathrm{var}(X)$ is to the mean square $E(X^2)$. It provides a measure of the extent to which the deviations of the two r.v.s from their respective means tend to vary together. The quantities U_{XX}, U_{YX} and U_{YY} will now be replaced by the scalar $V_{XX} = \mathrm{var}(X)$, the vector $V_{YX} = (\mathrm{cov}(X, Y_j))$ and the matrix $V_{YY} = (\mathrm{cov}(Y_j, Y_k))$, respectively. Thus V_{xx} is just the variance of X, the matrix V_{YY} is the *covariance matrix* of the random vector Y and V_{YX} is the vector of covariances between X and the elements of Y.

We have now effectively proved

Theorem 2.8.2. *Relation* (29) *yields an LLS estimate of X in terms of Y, allowing for a constant term, if and only if the coefficient vector a satisfies the linear relations*

$$V_{YY}a = V_{YX}. \tag{30}$$

In most cases one assumes that the r.v.s have been *reduced to zero mean* in that their mean values have been subtracted, so that X is replaced by $X - E(X)$, etc. In this case, there is no need to include a constant term explicitly in the estimating relationship, and product moments are replaced by covariances.

Least square approximation is an important and recurrent topic. It is obviously important practically, since it enables one to construct prediction formulae and the like on the basis of a minimal probabilistic specification (first and second moments). These moments can be estimated from long-term averages of past data. It is also a crucial simplification that the relations determining the optimal coefficients are linear.

LLS theory also permeates mathematics generally, as L_2 theory, with its associated concepts of Euclidean distance, inner product, projection, etc. We give a more thoroughgoing account of L_2 theory in our context in Chapter 11, but this section has already set the scene. Note that there is virtually no invocation of the probability concept; one requires only certain expectations—the first- and second-order moments of the relevant r.v.s. Nevertheless, L_2 theory lies behind many of the central concepts of probability theory: conditional expectation (Section 5.3) the normal distribution (Sections 7.5 and 11.4) and martingale theory (Section 14.1).

EXERCISES AND COMMENTS

1. Note that the determining relations (26) and (30) can be written as $E[(\hat{X} - X)Y_j] = 0$ ($j = 1, 2, \ldots, m$) and $\text{cov}(\hat{X} - X, Y_j) = 0$ ($j = 1, 2, \ldots, m$), respectively.

2. Suppose that a constant term has indeed been included in the estimate, and that the covariance matrix V_{YY} is nonsingular. Show that the minimal value of the mean square error can be written in the various forms

$$E[(\hat{X} - X)^2] = V_{XX} - a^T V_{YX} = V_{XX} - V_{XY}V_{YY}^{-1}V_{YX} = \frac{\begin{vmatrix} V_{XX} V_{XY} \\ V_{YX} V_{YY} \end{vmatrix}}{|V_{YY}|}.$$

Here $V_{XY} = V_{YX}^T$ and a is a solution of (30).

3. We see from Exercise 2 that $V_{XY}V_{YY}^{-1}V_{XX}$ is the 'amount of variance of X explained by Y'. Consider the case of scalar Y. The proportion of the variance explained is

then

$$\frac{V_{XY}V_{YX}}{V_{XX}V_{YY}} = \frac{[\mathrm{cov}(X, Y)]^2}{\mathrm{var}(X)\,\mathrm{var}(Y)}.$$

Thus Y would explain nothing if $\mathrm{cov}(X, Y)$ were zero, when X and Y are said to be *uncorrelated*.

4. Note that expression (22) is a covariance. The reason why it is negative is because sampling is without replacement: the individual sampled on one draw is ineligible for the other. This leads to a negative correlation between different observations in the sample (admittedly, only of order N^{-1}).

5. Suppose that X is the height of a mountain and Y_j an observation on X from position j, which we can write as $Y_j = X + \varepsilon_j$, where ε_j is the error of that observation ($j = 1, 2, \ldots, m$). Suppose that it is known that X and the observation errors are all mutually uncorrelated, and that $E(X) = \mu$, $\mathrm{var}(X) = v$, $E(\varepsilon_j) = 0$. $\mathrm{var}(\varepsilon_j) = v_j$ ($j = 1, 2, \ldots, m$). (This information must be supposed to come from *prior information* on X and the errors, i.e. from previous experiments.)

   Then $\bar{Y}_n = n^{-1}\sum Y_j$ is a possible estimate of X, with mean square error $E[(\bar{Y}_n - X)^2] = n^{-2}\sum v_j$. Show that the LLS estimate is

$$\hat{X} = \frac{(\mu/v) + \sum(Y_j/v_j)}{(1/v) + \sum(1/v_j)}$$

with mean square error $((1/v) + \sum(1/v_j))^{-1}$. The case in which one has no prior information on X is that in which $v = +\infty$, in which case $\hat{X}$ reduces to the *minimum variance unbiased linear estimate*. (The notion of unbiasedness requires an appeal to the concept of the conditional expectation $E(\hat{X}|X)$ of $\hat{X}$ for given X; it implies that $E(\hat{X}|X) = X$—see Chapter 5.)

6. *Mean square convergence.* For the example of Exercise 5 we shall have $E[(\hat{X} - X)^2] \to 0$ as $n \to \infty$ if $((1/v) + \sum_{j=1}^n(1/v_j))^{-1} \to 0$. In this case, we say that $\hat{X}$ *converges in mean square* to X as $n \to \infty$, and shall write this as $\hat{X} \xrightarrow{\text{m.s.}} X$.

   For the most usual special case, suppose that $Y_1, Y_2, \ldots$ are uncorrelated r.v.s with common mean value μ and variance v. Then $E(\bar{Y}_n) = \mu$ and $E[(\bar{Y}_n - \mu)^2] = v/n$, so that the sample mean $\bar{Y}_n$ is an unbiased estimate of the expectation value μ, and moreover converges to this in mean square with increasing sample size n. This is a first example of theoretical confirmation of the empirically observed convergence; one that we shall greatly strengthen.

7. *Prediction.* Denote the midday temperature on day t by Y_t (t integral). Suppose that records support the assumptions $E(Y_t) = \mu$ and $\mathrm{cov}(Y_s, Y_t) = \alpha\beta^{|s-t|}$ where μ, α and β are constants ($|\beta| < 1$). Verify, by appeal to Theorem 2.8.2, that $\hat{Y}_{t+s} = \mu + \beta^s(X_t - \mu)$ is the LLS predictor of Y_{t+s} ($s \geq 0$) based on $Y_t, Y_{t-1}, \ldots$. This is an example of prediction s steps ahead in a time series.

8. *Signal extraction.* Suppose that a scalar-valued signal X_t is transmitted at time t and received as $Y_t = X_t + \varepsilon_t$, where the error ε_t is usually referred to as *noise* in this context (t integral). Suppose that all variables have expectation zero, that noise variables are uncorrelated with each other and with the transmitted signal, and that $\mathrm{cov}(X_s, X_t) = \alpha\beta^{|s-t|}$, $\mathrm{var}(\varepsilon_t) = \gamma$ for all integral s, t. Show, by appeal to Theorem

2.8.2, that the LLS estimate of the transmitted signal X_t for a prescribed t-value, based upon the received signal over the whole time axis, is

$$\hat{X}_t = \frac{\alpha\xi}{\beta\gamma(1-\xi^2)} \sum_s \xi^{|s|} Y_{t-s},$$

where ξ is the smaller root of the equation $\alpha + \gamma(1-\beta\xi)(1-\beta/\xi) = 0$.

9. Some Implications of the Axioms

As emphasized in Section 2, it is typical that in the enunciation of a problem one is given the expectation values for r.v.s Y of some class $\mathscr{F}$, this class being in some cases quite small, in others larger. The axioms then appear as *consistency conditions* among the given expectations. For r.v.s outside $\mathscr{F}$ they still appear as consistency conditions, which restrict the possible values of the unknown expectations, sometimes, indeed, to the point of determining them completely. For example, if $\mathscr{F} = \{Y_1, Y_2, \ldots\}$, then the expectation of the linear combination $X = \sum_j c_j Y_j$ is determined as $\sum_j c_j E(Y_j)$, at least if all the $c_j Y_j$ have finite absolute expectation and the number of terms in the sum is finite. The question of limits of such sums is more delicate, and will be treated in Chapter 13.

Again, if $Y_1 \le X \le Y_2$ then we have the implication $E(Y_1) \le E(X) \le E(Y_2)$. If, in particular, $E(Y_1) = E(Y_2)$ then the value of $E(X)$ is fully determined. This does not imply that the three random variables are identically equal, when the assertion would be trivial—rather that they differ only on a set of zero probability measure.

These two consistency conditions have consequences which are not immediately obvious, some of which we shall now follow up. For instance, suppose that X is a nonnegative scalar r.v., and consider the inequality

$$I(X \ge a) \le X/a \qquad (X \ge 0),$$

obvious from Fig. 2.1. Taking expectations on both sides of this inequality, we obtain then

$$P(X \ge a) \le E(X)/a. \tag{31}$$

This simple result, known as the *Markov inequality*, is extremely useful. It implies that one can set an upper bound on the probability that X exceeds a certain value if X is nonnegative and has a known expectation. The smaller $E(X)$, the smaller this bound, as one might expect.

If $E(X) \le a$ then the inequality is also sharp, in the sense that one can find a process for which $E(X)$ has the prescribed value and equality is attained in (31). The process is that in which X takes only the two values 0 and a, with probabilities $1 - E(X)/a$ and $E(X)/a$, respectively.

Consider now a scalar r.v. X which is not restricted in sign and for which

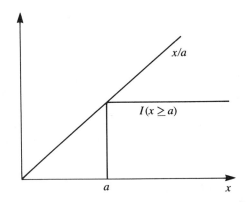

Figure 2.1. The graph illustrates the bounding relationship behind the Markov inequality.

the first two moments $E(X)$ and $E(X^2)$ are known. Then, evidently,

$$I(|X - b| \geq a) \leq \left[\frac{X - b}{a}\right]^2$$

(see Fig. 2.2). By taking expectations in this inequality we obtain the upper bound

$$P(|X - b| \geq a) \leq E[(X - b)^2]/a^2.$$

The bound is minimal when $b = E(X)$; setting b equal to this value we obtain *Chebyshev's inequality*

$$P(|X - E(X)| \geq a) \leq \text{var}(X)/a^2. \tag{32}$$

This is a most useful result. We can often prove that a variance can be made small, and Chebyshev's inequality then implies that the probability that X

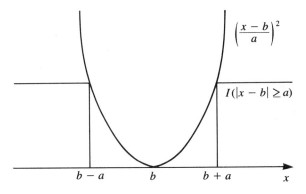

Figure 2.2. The graph illustrates the bounding relationship behind the Chebyshev inequality.

deviates by more than a given amount from its mean value is of a similar order of smallness. For example, we can conclude from the analysis of the survey sample model of Section 7 that the probability that the sample mean $\bar{\xi}$ deviates from the population mean $A(X)$ by more than any prescribed amount tends to zero at least as fast as n^{-1} with increasing sample size n. (In fact, the Chebyshev bound is a very conservative one in the this situation, and we can greatly strengthen the probability assertion; see Exercise 5.)

Suppose that $E(X) = \mu$ and that $\mathrm{var}(X) = 0$, so that $E[(X - \mu)^2] = 0$. We will then say that X *is equal to μ in mean square*, written

$$X \overset{\text{m.s.}}{=} \mu. \tag{33}$$

This relation has the implication $X = \mu$ in the sense that $E[H(X)] = H(\mu)$ for any $H(x)$ bounded each way by a quadratic function in x taking the value $H(\mu)$ at $x = \mu$. (More specifically, the statement that $E[(X - \mu)^2]$ is arbitrarily small implies that $|E[H(X)] - H(\mu)|$ is arbitrarily small for such a function, although this is not necessarily true for other functions; see Exercise 3.) In particular, as we see from (32), $P(|X - \mu| \le a) = 1$ for arbitrarily small a, or $X = \mu$ *with probability one*.

The Markov and Chebyshev inequalities, (31) and (32), are relatively direct consequences of the axioms. However, already in the last section we saw consequences which are less evident. Suppose that X is a random vector with elements X_j, the product moment matrix U of its elements can then be written

$$U = E(XX^{\mathrm{T}}). \tag{34}$$

Theorem 2.9.1. *A product moment matrix is symmetric and nonnegative definite. It is singular if and only if a relation*

$$c^{\mathrm{T}}X \overset{\text{m.s.}}{=} 0 \tag{35}$$

holds for some nonzero constant vector c. In particular, $|U| \ge 0$, with equality if and only if a relation of type (35) holds, and

$$|E(X_1 X_2)|^2 \le E(X_1^2)E(X_2^2) \tag{36}$$

(Cauchy's inequality) *with equality if and only if a nontrivial relation $c_1 X_1 + c_2 X_2 \overset{\text{m.s.}}{=} 0$ holds.*

PROOF. It follows from the definition (34) that U is symmetric. As noted in the previous section we have

$$c^{\mathrm{T}}Uc = E[(c^{\mathrm{T}}X)^2] \ge 0 \tag{37}$$

which implies positive definiteness. If equality holds in (37) then the relation (35) indeed holds, and U is also singular. If U is singular then a relation $Uc = 0$ holds for some nonzero c, whence equality holds in (37).

Finally, the determinant $|U|$ is the product of the eigenvalues of U, and so nonnegative. The statement of this fact for $m = 2$ is just Cauchy's inequality, which has a special importance, and is much used. $\qquad\square$

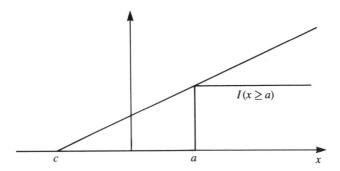

Figure 2.3. The weakening for the Markov inequality as the range of the variable increases.

See Exercise 8 for an alternative proof which does not appeal to the notion of an eigenvalue.

EXERCISES AND COMMENTS

1. Note that case in which the Markov inequality (31) was sharp was just that in which one could find a distribution with the assigned value of $E(X)$ which was concentrated on the values x for which equality held in $I(x \geq a) \leq x/a$. Show correspondingly that Chebyshev's inequality is sharp if $\text{var}(X) \leq a^2$.

2. Suppose that it is known that $E(X) = \mu$ and $X \geq c$. What is the best upper bound one can find for $P(X \geq a)$, by the methods adopted in the text? Note that this bound becomes useless as c becomes ever more negative, essentially because the straight line one uses to bound the graph of $I(x \geq a)$ from above in the interval $x \geq c$ approaches the horizontal as $c \to -\infty$. See Fig. 2.3.

3. Consider a r.v. X taking the values 0, $n^{1/4}$ and $-n^{1/4}$ with respective probabilities $1 - 1/n$, $1/(2n)$ and $1/(2n)$. Show from this example that one may have $E(X^4) = 1$ although $\text{var}(X)$ is indefinitely small. Show that indeed $E(X^4)$ may adopt any value from zero to plus infinity, consistently with indefinitely small $\text{var}(X)$.

4. Show that if $H(x)$ is an increasing nonnegative function of x then

$$P(X \geq a) \leq E[H(X)]/H(a).$$

5. Consider the sample survey problem of Section 7, and let us try and calculate an upper bound for $P(\bar{\xi} \geq A(X) + a)$ for prescribed positive a. Appealing to the slight generalization of the Markov inequality asserted in Exercise 4 we deduce that

$$P(\bar{\xi} \geq A(X) + a) \leq E(e^{\alpha n \bar{\xi}})e^{-\alpha n[A(X)+a]}$$

for any nonnegative α. Consider the case of sampling with replacement. Since the ξ_j take the values $x_1, x_2, \ldots, x_N$ equiprobably, and since $\xi_1, \xi_2, \ldots, \xi_n$ take any combination of these values equiprobably, then

$$E(e^{\alpha n \bar{\xi}}) = E\left[\exp\left(\alpha \sum_1^n \xi_j\right)\right] = A(e^{\alpha X})^n,$$

so that we have a bound

$$P(\bar{\xi} \geq A(X) + a) \leq [A(e^{\alpha X})e^{-\alpha(A(X)+a)}]^n = Q(\alpha)^n, \tag{38}$$

say. Thus, if we can find a nonnegative value of α such that $Q(\alpha) < 1$, then we have shown that $P(\bar{\xi} \geq A(X) + a)$ actually tends to zero *exponentially fast* in n.

Show that $Q(\alpha) = 1 - a\alpha + O(\alpha^2)$ for small α, so that $Q(\alpha)$ indeed decreases from unity as α increases from zero, and we can certainly find a value of α such that $Q(\alpha) < 1$. (Indeed, there is a unique value of α which minimizes $Q(\alpha)$; see Exercise 13.)

6. We are in danger of running ahead of ourselves, but the last example leads us on ineluctably. Note that

$$A(e^{\alpha}X) = 1 + \alpha A(X) + \tfrac{1}{2}\alpha^2 A(X^2) + O(\alpha^3)$$

for small α, so that

$$\log A(e^{\alpha X}) = \alpha A(X) + \tfrac{1}{2}\alpha^2 V(X) + O(\alpha^3),$$

where $V(X)$ is the population variance of X. Suppose we take $a = cn^{-1/2}$, for given nonnegative c. Appealing to (38) and the last relation we see that

$$P(\bar{\xi} \geq A(X) + cn^{-1/2}) \leq \exp[-n^{1/2}c\alpha + \tfrac{1}{2}n\alpha^2 V(X) + O(n\alpha^3)].$$

The value of α which minimizes the quadratic part of the exponent is $\alpha = c/[n^{1/2}V(X)]$; inserting this we derive the bound

$$P(\bar{\xi} \geq A(X) + cn^{-1/2}) \leq \exp\left[-\frac{c^2}{2V(X)} + O(n^{-1/2})\right].$$

This is getting us very close to a number of the central topics of probability theory: including the normal distribution (Section 7.5) and the central limit theorem (Section 7.4).

7. Suppose that X and Y are scalar variables. Show that

$$[\mathrm{cov}(X, Y)]^2 \leq \mathrm{var}(X)\,\mathrm{var}(Y). \tag{39}$$

Show also that equality can hold in (39) if and only if there is a relationship of the form $c_1 X + c_2 Y \overset{\mathrm{m.s.}}{=} c_0$.

8. Consider the following alternative proof of Theorem 2.9.1. Write $|U|$ as d_m, to emphasize its dependence on the size m of the vector X. Define also D_m as the mean square error in the LLS estimation of X_m in terms of $X_1, X_2, \ldots, X_{m-1}$. Now show that $d_m = D_m d_{m-1}$ and hence derive an inductive proof of the theorem.

9. Suppose that equation (30) has more than one solution for a. Show that the corresponding LLS estimates of X are all *mean square equivalent*, in that the mean square difference of any two of them is zero. In fact, they differ by a term $c^{\mathrm{T}}(Y - E(Y))$, where $V_{YY}c = 0$.

10. Recall the definition $\mu_j = E(X^j)$ of the jth moment of a scalar r.v. X. Show that

$$\begin{vmatrix} 1 & \mu_1 \\ \mu_1 & \mu_2 \end{vmatrix} \geq 0, \qquad \begin{vmatrix} 1 & \mu_1 & \mu_2 \\ \mu_1 & \mu_2 & \mu_3 \\ \mu_3 & \mu_3 & \mu_4 \end{vmatrix} \geq 0,$$

and generalize.

11. *Jensen's inequality.* A *convex* function $\phi(x)$ of a scalar argument x is one for which

$$\phi(px + qx') \leq p\phi(x) + q\phi(x')$$

for any pair of arguments x, x', where p and q are nonnegative and add to unity. From this it follows that the derivative $d\phi(x)/dx$ is nondecreasing (if it exists) and that for any fixed x_0 there exists a constant α such that

$$\phi(x) \geq \phi(x_0) + \alpha(x - x_0).$$

Show that

$$E[\phi(X)] \geq \phi(E(X)).$$

12. Since $|X|^r$ is a convex function of X for $r \geq 1$ we have

$$E(|X|^r) \geq [E(X)]^r$$

by Jensen's inequality. Deduce from this that $[E|X|^r]^{1/r}$ is an increasing function of r for $r \geq 0$.

13. Define the *moment-generating function* $M(\alpha) = E(e^{\alpha X})$ of a scalar r.v. X, and suppose that this exists for some range of real α. Then, in this range,

$$\left(\frac{d}{d\alpha}\right)^2 \log M(\alpha) = \frac{E(X^2 e^{\alpha X})}{E(e^{\alpha X})} - \left[\frac{E(X e^{\alpha X})}{E(e^{\alpha X})}\right]^2 \geq 0.$$

Nonnegativity follows because the central expression can be seen as the variance of X under a transformed expectation operator $\tilde{E}$ defined by

$$\tilde{E}[H(X)] = \frac{E(H(X)e^{\alpha X})}{E(e^{\alpha X})}.$$

It follows then that $\log M(\alpha)$ is convex, and that there is a unique value of α which minimizes $M(\alpha)e^{-h\alpha}$ for any prescribed h.

14. *Convergence of sample means in probability.* Suppose that scalar r.v.s Y_1, Y_2, Y_3, ... are uncorrelated and have common mean μ and common variance. Then we have already seen from Exercise 2.8.6 that the sample average $\bar{Y}_n = (1/n)\sum_1^n Y_j$ converges to μ in mean square. Show, by appeal to Chebyshev's inequality, that it then also converges to μ *in probability*, in that

$$P(|\bar{Y}_n - \mu| > \varepsilon) \to 0$$

for any prescribed positive ε.

CHAPTER 3

Probability

1. Events, Sets and Indicators

An event A is something which either occurs or does not occur in a particular case. It often has some kind of verbal description: e.g. 'I win the game' or 'the patient survives'. As we saw in Section 2.3, the formal characterization of the event is as a set of realizations ω, i.e. a subset of Ω, also denoted by A. This is just the set of realizations for which the event occurs. The probability of the event is defined as

$$P(A) = E[I(A)], \tag{1}$$

where $I(A, \omega)$ is the indicator function of the set A; the random variable which takes value 1 or 0 according as the event occurs or not. Thus $P(A)$ can be seen as the expected proportion of times that A occurs.

As we emphasized in Section 2.3, $P(A)$ is to be regarded as a function of the set A; it constitutes the *probability measure* to be attached to this set.

There is not a single property of probability measures that does not follow from the defining relation (1) and the expectation axioms. However, there is a whole formal framework of ideas implicit in this relation which we should make explicit. There are always three alternative descriptions to be carried in parallel: in the languages of events, of sets and of indicator functions.

For any event A there is the event 'not A', denoted $\bar{A}$. So, if A is the event 'rain' (more specifically, that a given rain-gauge records more than a specified amount over some specified time interval), then $\bar{A}$ is the event 'no rain' (based on the same definition). In set terms, $\bar{A}$ is the complement of A in Ω; the set of ω which do not lie in A (see Fig. 3.1). In indicator terms

$$I(\bar{A}) = 1 - I(A),$$

(with the ω-argument understood, as always).

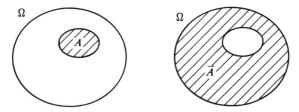

Figure 3.1. A Venn diagram illustrating the complement $\bar{A}$ of a set A.

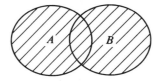

Figure 3.2. A Venn diagram illustrating the union $A \cup B$ of sets A and B.

Suppose A and B are two events; say 'rain' and 'wind' (with some agreed definition of 'wind'). Then the union of these events, written $A \cup B$, is the event that at least one of A or B occurs. That is, that it either rains or blows, or both. One might be interested in this compound event if one were planning a picnic, but would be deterred if it either rained or blew. In set terms, the union $A \cup B$ is the set of ω which lie in at least one of A or B (and so possibly both); see Fig. 3.2. The indicator function of this set is then

$$I(A \cup B) = \max[I(A), I(B)], \tag{2}$$

which is also sometimes written as

$$I(A \cup B) = I(A) \vee I(B). \tag{2'}$$

We can consider the union of several events $A_1, A_2, \ldots, A_n$; both the event and the corresponding set can be written $\bigcup_i A_i$. This set, the set of ω belonging to at least one of the designated sets, has indicator function

$$I\left(\bigcup_i A_i\right) = \max[I(A_1), I(A_2), \ldots, I(A_n)], \tag{3}$$

which again we write as

$$I\left(\bigcup_i A_i\right) = \bigvee_i I(A_i). \tag{3'}$$

One might be interested in the event that it rains *and* blows, because that is when rain can be driven under the roof-tiles. This is the *intersection* of A and B, written $A \cap B$, or sometimes just as AB. In set terms, the intersection $A \cap B$ is the set of ω belonging to both A and B; see Fig. 3.3. The indicator function of the set is

$$I(A \cap B) = I(A)I(B). \tag{4}$$

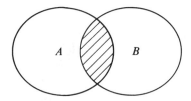

Figure 3.3. A Venn diagram illustrating the intersection $A \cap B$ of sets A and B

We achieve symmetry with the case of a union, and perhaps some significance, if we write this rather as

$$I(A \cap B) = \min[I(A), I(B)] = I(A) \wedge I(B). \tag{4'}$$

One can also consider the intersection of several events (or sets) $A_1, A_2, \ldots, A_n$, written $\bigcap_i A_i$. The set would be the set of ω belonging to all the designated sets; its indicator function can variously be written

$$I\left(\bigcap_i A_i\right) = \prod_i I(A_i) = \min[I(A_1), I(A_2), \ldots, I(A_n)] = \bigwedge_i I(A_i). \tag{5}$$

The set Ω is the set of all possible realizations. Its complement is the *empty set* $\emptyset$, the set containing no elements at all. If events A and B are *mutually exclusive*, in that there is no realization for which they both occur, then the set $A \cap B$ is empty. That is, $A \cap B = \emptyset$, and the sets A and B are said to be *disjoint*. Correspondingly, $I(A)I(B) = 0$.

For example, if the outcome of a game is classified as 'win', 'lose' or 'draw' and no other possibility is admitted (such as cancellation, abandonment or a referee of advanced progressive views) then 'win' and 'lose' are certainly mutually exclusive.

If A and B are disjoint then one often simply writes their union $A \cup B$ as $A + B$. Correspondingly, use of the summation notation $A + B$ or $\sum_i A_i$, will be taken to imply disjointness of the summands.

If occurrence of event A implies occurrence of event B (e.g. as 'suffering from influenza' implies 'being unwell') then the set A lies totally within the set B. One then writes $A \subset B$ (for 'A is included in B') and has the indicator function relation

$$I(A) \leq I(B).$$

In this case, one could equally well write $B \supset A$ ('B includes A').

A sequence of sets (or events) $\{A_i\}$ is *monotone* if either $A_{i+1} \supset A_i$ for all i (when the sequence is *increasing*) or $A_{i+1} \subset A_i$ for all i (when the sequence is *decreasing*). For example, if A_i is the event 'the dodo is extinct by year i' then $\{A_i\}$ is an increasing sequence. The limit event A_∞ in this case would be interpretable as the event 'ultimate extinction'. That is, that the dodo becomes extinct at some time. Alternatively, the sequence of events $\{\bar{A}_i\}$ is decreasing, and $\bar{A}_\infty$ is interpretable as the event 'indefinite survival' of the dodo.

EXERCISES AND COMMENTS

1. Sets $A_1, A_2, \ldots, A_n$ constitute a *partition* of Ω if every ω belongs to exactly one of them, so that $\sum_i A_i = \Omega$. The events are then *mutually exclusive* and *exhaustive*; i.e. only one can happen and one of them must happen. Note that the condition

$$\sum_i I(A_i) = 1$$

is necessary and sufficient for $\{A_i\}$ to constitute a partition.

2. Note that

$$I(A \cup B) = I(A) + I(B) - I(AB),$$

since subtraction of $I(AB)$ corrects for those realizations which are counted twice in $I(A) + I(B)$. The systematic way to derive this is to note that $A \cup B$ is the complement of the event $\bar{A} \cap \bar{B}$; that neither A nor B happen. Thus

$$1 - I(A \cup B) = I(\bar{A}\bar{B}) = I(\bar{A})I(\bar{B}) = (1 - I(A))(1 - I(B)).$$

3. More generally,

$$I\left(\bigcup_i A_i\right) = 1 - \prod_i [1 - I(A_i)] = \sum_i I(A_i) - \sum\sum_{i<j} I(A_i)I(A_j)$$
$$+ \sum\sum\sum_{i<j<k} I(A_i)I(A_j)I(A_k) + \cdots.$$

When would it be legitimate to terminate this expansion after the single sum? after the double sum?

4. Note that $A \cup B = A + \bar{A}B$, equivalent to $I(A \cup B) = I(A) + (1 - I(A))I(B)$. More generally,

$$\bigcup_{i=1}^n A_i = A_1 + \bar{A}_1 A_2 + \bar{A}_1 \bar{A}_2 A_3 + \cdots.$$

Interpret. What is the corresponding relation in indicator functions?

5. Note the bound

$$I\left(\bigcup_i A_i\right) \le \sum_i I(A_i) \tag{6}$$

and its complementary version

$$I\left(\bigcap_i A_i\right) \ge 1 - \sum_i I(\bar{A}_i). \tag{7}$$

These are crude but useful.

6. The *symmetric difference* $A \triangle B$ is the set of ω lying in one of A or B but not both. Show that

$$I(A \triangle B) = [I(A) - I(B)]^2.$$

7. The algebra of events (or sets) is a Boolean algebra, as is also the algebra of switching circuits. Consider, for example, the circuit of Fig. 3.4, consisting of switches numbered 1, 2 and 3 between terminals a and b. Let A_i be the event that switch i is closed, and A the event that there is a closed circuit between a and b. Then

$$A = [A_1 \cup A_2] \cap A_3. \tag{8}$$

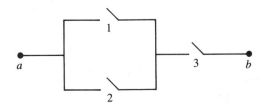

Figure 3.4. The switching circuit corresponding to the compound event defined in Exercise 3.1.7.

8. For the model of Exercise 7, $\bar{A}_i$ is the event that switch i is open and $\bar{A}$ the event that the circuit between a and b is open. Verify the relation

$$\bar{A} = [\bar{A}_1 \cap \bar{A}_2] \cup \bar{A}_3. \tag{9}$$

Relations (8) and (9) illustrate a general theorem in Boolean algebra: if an identity holds, then the identity also holds in which $\cup$ and $\cap$ are interchanged and sets are replaced by their complements.

9. What switching circuit would yield the relation (in the notation of Exercise 7)

$$A = [A_1 \cap A_2] \cup A_3?$$

2. Probability Measure

The reader will easily verify that the expectation axioms of Section 2.2 have the following implications for the probability measure defined in (1).

Theorem 3.2.1. *The probability measure $P(A)$ is a scalar-valued function of subsets A of Ω with the properties:*

(i) $P(A) \geq 0$.
(ii) *For disjoint A, B*

$$P(A + B) = P(A) + P(B). \tag{10}$$

(iii) $P(\Omega) = 1$.
(iv) *If $\{A_n\}$ is a monotonic sequence increasing to A then $P(A) = \lim_n P(A_n)$.*

These are consequences of the expectation axioms, and the only ones, in that all other possible assertions concerning probability measures are consequences of these (see Exercise 1).

If one were taking probability measure as the prime concept then one could take properties (i)–(iv) above as the axioms for such a measure, and this is the usual approach. For reasons already explained we have instead axiomatized the concept of expectation, so that relations (i)–(iv) are deductions rather than postulates. However, the two approaches are of course mutually consistent, and in the end one covers much the same material.

There are familiar further immediate consequences: we see from (ii) and (iii) that $P(A) + P(\overline{A}) = 1$ and so, by appeal to (i), that $P(A) \le 1$. Suppose Ω discrete with elements ω_k $(k = 1, 2, \ldots)$. It follows then from the 'additive law of probability' (10) that

$$P(A) = \sum_{k: \omega_k \in A} P(\omega_k), \tag{11}$$

so determining $P(A)$ for any A in terms of the probability distribution over realizations $P(\omega)$.

For more general sample spaces one would expect the analogue of (11) to be

$$P(A) = \int_A P(d\omega) \tag{12}$$

and this indeed must be one's mental picture. However, there may be trouble with the definition of either the integral or the measure for arbitrary A, whence the intrusion of measure theory into probability theory.

As for the expectation case, one normally assumes $P(A)$ known for some family $\mathscr{F}$ of sets A. One convenient case is that in which $\mathscr{F}$ constitutes a partition of Ω, but it is not quite clear what a partition means in the non-discrete case. A more general idea is that of a *field*: $\mathscr{F}$ is a field if closed under the operations of intersection and complementation; that is, if $\mathscr{F}$ contains also the intersections and complements of its elements. In a way this generalizes the idea of a partition; see Exercise 3. If $\mathscr{F}$ contains also the limit elements obtained by indefinite continued application of the intersection and complementation operations then it is termed a *σ-field*. One of the first products of Kolmogorov's establishment of an axiomatic theory of probability in 1933 was his *extension theorem*: that if $P(A)$ is prescribed on a field then it is determined on the σ-field generated from that field. We shall consider the 'expectation version' of these matters in Section 15.3.

EXERCISES AND COMMENTS

1. Note that, in deriving properties (i)–(iv) from Axioms 1–5 of Section 2.2, no appeal was made to Axiom 2: that $E(cX) = cE(X)$. This is an axiom specific to the expectation. We could recover the axiomatic properties of an expectation from properties (i)–(iv) only if we adopted a definition of expectation in terms of probabilities, just as (1) gives the definition in the reverse direction.

2. Note that a field is also closed under union.

3. The field generated from an arbitrary collection of sets is the smallest field containing those sets. So, the field generated from two sets A and B consists of the four sets AB, $\overline{A}B$, $A\overline{B}$ and $\overline{A}\overline{B}$ (which constitute a decomposition of Ω) and unions of these. Generalize.

4. Consider the σ-field of sets on the real line $(-\infty < x < +\infty)$ generated from the sets $x \le a$ for all a. Then this contains the intervals, and all countable intersections of intervals.

5. By taking expectations of inequalities (6) and (7) we deduce the useful *Boole's inequalities*

$$P\left(\bigcup_i A_i\right) \le \sum_i P(A_i),\tag{13}$$

$$P\left(\bigcap_i A_i\right) \ge 1 - \sum_i P(\bar{A}_i).\tag{14}$$

6. Correspondingly, from the assertions of Exercises 3.1.3 and 3.1.4 follow the identities

$$P\left(\bigcup_{i=1}^n A_i\right) = P(A_1) + P(\bar{A}_1 A_2) + P(\bar{A}_1 \bar{A}_2 A_3) + \cdots$$

and

$$P\left(\bigcup_i A_i\right) = \sum_i P(A_i) - \sum\sum_{i<j} P(A_i A_j) + \sum\sum\sum_{i<j<k} P(A_i A_j A_k) - \cdots.$$

Show that the expression obtained by breaking off the second expansion after the r-fold sum constitutes an upper or a lower bound to $P(\bigcup_i A_i)$ according as r is odd or even.

7. Suppose that X can assume only the values $0, 1, 2, \ldots$. By using indicator functions, or otherwise, show that

$$\sum_{n=0}^{\infty} P(X > n) = E(X).$$

8. By minimizing the left-hand member of the inequality

$$E\left\{\left[I\left(\bigcup_i A_i\right) - \sum_i c_i I(A_i)\right]^2\right\} \ge 0$$

with respect to the coefficients c_i show that

$$P\left(\bigcup_i A_i\right) \ge \Pi_1^{\mathsf{T}} \Pi_2^{-1} \Pi_1,$$

where Π_1 is the column vector with elements $P(A_i)$ and Π_2 is the matrix with elements $P(A_i A_j)$. This is useful in the study of maxima of r.v.s.: if A_i is the event $X_i \ge x$ then $\bigcup_i A_i$ is the event $\max_i X_i \ge x$.

3. Expectation as a Probability Integral

We have seen in Theorem 2.4.1 that, if Ω is discrete, then the expectation functional necessarily has the form

$$E(X) = \sum_{\omega \in \Omega} X(\omega) P(\omega).\tag{15}$$

Correspondingly, if one considers a discrete decomposition $\{A_k\}$ of a general sample space and $X(\omega)$ takes the constant value x_k in A_k (all k) then necessarily

$$E(X) = \sum_k x_k P(A_k).\tag{16}$$

One would expect, then, that in the general case the expectation would take the form

$$E(X) = \int X(\omega)P(d\omega), \tag{17}$$

this being the limit form of (16) as the decomposition becomes ever finer.

However, in writing representation (17) we have all the difficulties of defining the integral of a function $X(\omega)$ with respect to a measure P on a general type of space. By taking expectation as the prime concept, rather than probability measure, we have avoided this. Nevertheless, one is always aware that an expectation does have the nature (17). When we come to discuss, in Chapter 12, the consistency conditions which must hold between expectations we shall see that these conditions are equivalent to the requirement that a representation (17) holds for the specified r.v.s X for some P. On the other hand, discussion is simplified by the fact that if one is prescribing the expectation values of n r.v.s, then representations (17) exist for which the distribution P is confined to at most $n + 1$ elements of Ω.

However, we certainly know that a representation (17) holds for sufficiently regular $X(\omega)$. It is often convenient to unify all the cases of discrete and continuous distribution by writing this as

$$E(X) = \int X(\omega)f(\omega)\mu(d\omega), \tag{18}$$

where μ is a fixed measure which is natural to the sample space. It is fixed in that the distribution P is varied, not by varying μ, but by varying $f(\omega)$, which one characterizes as the *probability density of ω relative to μ*. So, for the discrete case (15), $P(\omega)$ is the density relative to *counting measure* $\#(A)$, the number of realizations in A. For the continuous distribution on the real line (2.10), $f(\omega)$ is the density relative to Lebesgue measure (i.e. to the obvious measure which specifies an interval's measure as its length).

4. Some History

In the early days, problems were posed as freely in terms of expectation as in terms of probability, and the two concepts were regarded as having much the same standing. For example, the celebrated 'problem of points' drew sustained attention from the days of Pacioli (1494). This concerns the situation of players who, for reasons at which we can only guess, must abandon a game before it is completed. What would then be a fair division of the stakes, given the positions of the players at the moment of abandonment? This we would now view as a question of calculating expectations conditional on these positions.

One of the celebrated earliest works on probability was Christian Huygens' *De Ratiociniis in Ludo Aleae* (1657). He starts from an axiom of the 'value' of a fair game, and thence derives three theorems on expectations.

For an amusing potted history and personal view of these matters I cannot do better than quote verbatim from a letter written to me by David Stirzaker. It is not as he would have written for publication, but that is its virtue. He begins by recommending the recent *History of Probability and Statistics and Their Applications before 1750* by Anders Hald (1990).

'Apart from the book by Ian Hacking, this is the first major history of probability to give expectation much of a look-in, and he makes clear that, at the beginning, the ideas of probability and statistics were inextricably intertwined, with equal status. (Expectation was called 'value' at that stage.) For example, Pascal's solution of the problem of points is essentially an exercise in conditional expectation, whereas Fermat's is purely probabilistic. Christian Huygens you know about, but Hald discusses the work of his brother *Ludwig* Huygens, who calculated life tables by conditional expectation in 1669. This was later used by de Witt (whom Keynes thought was the first to use expectation) in 1671. Of course, Leibniz had earlier discussed the *average*, but not I think in the context of problems that we would think of as probabilistic.

'And so on. The question is, Why by the late nineteenth century had expectation come to be of secondary importance? (Of use only for analytical purposes or by statisticians.) Why did Hilbert ask for axioms of probability, ahead of expectation? I don't know (and it would make a good thesis to get the answer!). For what it is worth, I have an unsubstantiated theory that it might have been something to do with the enormous amount of futile agonizing that went on over the St. Petersburg paradox. For many years after its appearance (1713, N. Bernoulli) most workers in the field had a shot at resolving it, but it is clear that no one was happy about infinite expectation, and they were even less happy with the suggestions for getting rid of it (utility, D. Bernoulli, 1738). Even the great polymath D'Alembert was still droning on about it and confusing the issue in 1780. (Of course D'Alembert was equally confused about probability, but that one was more readily dismissed.) Maybe this had something to do with Laplace starting with probabilistic axioms. And everyone followed Laplace.

'At any rate, for whatever reason, if you read the nineteenth century theorists (Venn, Ellis, de Morgan, Boole and later, Bernstein, Keynes, Jeffreys) it is clear that it is not really crossing anyone's mind that expectation is a kind of dual of probability. It is something you get from a distribution. And of course Kolmogorov (1933) nailed the lid down.'

The St. Petersburg paradox referred to is a method of staking money in a succession of fair games in a way that seems to offer a guaranteed return (which is disturbing in itself), this guaranteed return being the difference between an accumulated loss and a final win both of which have infinite expectation (which is even more disturbing). We discuss the model and its analysis in Section 4.5. A small modification in the direction of realism clarifies everything, resolves the paradox, and (one hopes) removes the unwarranted shadow Dr. Stirzaker believes may have been cast on the expectation concept.

The mention of a utility refers to something quite different. If X is the monetary return from a game or venture, then it was appreciated quite early that $E(X)$ was not necessarily the proper measure that one would wish to maximize. There is a law of decreasing returns—an extra pound, dollar or yen means less to a rich man than it does to a poor one. One would rather choose to maximize $E[U(X)]$ where U is an increasing concave function; what an economist would term a 'utility function'. This expectation was termed a 'moral expectation' by D. Bernoulli (1738). However, it is not a new expectation concept, but simply a more appropriate choice of the random variable which is to measure one's utility.

It was the loaded term 'value' used for an expectation which made the St. Petersburg paradox particularly painful—nothing has infinite value. Nowadays, we would distinguish clearly between the expectation operation and the utility function to which it is applied.

The formalization of probability could have gone other than it did had not someone as powerful as Kolmogorov set his impress. In further correspondence Dr. Stirzaker recalls the success at the turn of the century of the troika Chebyshev/Lyapunov/Nekrasov in preparing the ground for a good treatment of expectation by establishing the concept of a random variable. He also quotes from Hilbert (1900): 'As for the axioms of the probability calculus, it seems desirable that any logical study of them should go hand in hand with a rigorous and satisfactory development of the method of mean values in mathematical physics'.

One should not conclude without briefly mentioning two developments in this century. One is the recognition by pure mathematicians that the linear functionals constituted by integrals with respect to a measure are objects which are technically dual to that of a measure, and offer all the advantages as the prime concept that we are claiming. The other is the axiomatization by von Neumann (1932) of quantum probability, almost simultaneous with Kolmogorov's axiomatization of classical probability. The structure of quantum probability forced von Neumann to take the 'expectation route'; there is scarcely an evident alternative. The two theories are generally regarded as mysteriously different; we shall see in Sections 16.5 and 16.6 that the common axiomatization of expectation makes their origins seem quite close, and reveals clearly where the divergence begins.

5. Subjective Probability

One can lay down axioms, but one can then argue how these should be interpreted in 'real life' (or, equivalently, motivated). Our approach has been that the notion of a long-term average of a variable seems to be empirically meaningful, and that expectation is the idealization of this. We then also demand that the expectation functional $E(X)$ should have the same properties as does a finite population average $A(X)$.

This approach is equivalent to the 'frequentist' approach to probability: that the probability $P(A)$ of an event A is an idealization of the concept of a long-run proportion, or frequency, which again seems to be empirically meaningful.

There is another view, however: that of 'subjective probability'. In this $P(A)$ is regarded, not as an idealized version of the actual frequency of occurrence of A in a large number of cases, but as a measure of the 'degree of belief' on the part of an observer that A will occur. This degree of belief is regarded as quantifiable, by asking the observer how much he would be prepared to bet on the occurrence of A in a particular case.

The notion of an expectation is then particularly natural to a subjectivist. If the subjectivist has the opportunity to pay a price to enter a venture which will give him an uncertain return X, then he would argue that $E(X)$ is just the 'fair price' he should pay. However, for a subjectivist, a 'fair price' does not mean the price that would just balance average returns in a large number of repeated runs of the venture. He regards the venture as a one-off occurrence, and the expectation as a measure of his personal assessment of the financial worth of the venture. Of course, the two views may not be so very different: the human mind considers the repeated trials of past experience rather than of a hypothetical future.

The approach is one that has been developed particularly by de Finetti (1970). He used another language (e.g. 'prevision' rather than 'expectation'). However, by considering, on subjectivist grounds, what properties an expectation should have, he deduced just Axioms 1–4 of Section 2.2 and based a treatment on these parallel in many respects to ours.

It is probably true that most people find the frequentist motivation more compelling than the subjectivist one, if only because expectations and probabilities are then seen as properties of the system rather than of the observer. Nevertheless, one can only regard it as reassuring that differing motivations should lead to both the same choice of expectation as the prime concept and the same formal axiomatization.

CHAPTER 4

Some Basic Models

With the concepts of expectation and probability established, it is time to discuss a number of basic models. These are of interest both in themselves and because they indicate the way forward.

The spatial model of Section 1 provides the most natural physical framework for the introduction of several of the standard distributions. The concept of independence emerges immediately, and is discussed in a preliminary fashion in Section 3, before the formal treatment of Section 5.5. The expectation approach also invites immediate introduction of the natural and effective tool: the probability generating function.

1. A Model of Spatial Distribution

Suppose that N gas molecules are distributed over a region of volume V. We start from a model in which all configurations are assumed equally likely, this being in fact equivalent to the assumptions that the region is uniform in its properties and that the molecules do not interact.

Let us discretize the model initially, by supposing that the region is divided into M cells, each of volume V/M. Let us suppose that the molecules can be individually identified, so that there are then M^N possible allocations of the N molecules to the M cells. We shall suppose these all equally likely.

More specifically, let ξ_k be the position of the kth molecule, i.e. the number of the cell within which it lies. It thus takes values $j = 1, 2, \ldots, M$. Then a complete description of the configuration (at the level of location only to within a cell) is given by $\omega = (\xi_1, \xi_2, \ldots, \xi_N)$, and the expectation of a random variable $X(\omega)$ is given by

$$E(X) = M^{-N} \sum_{\xi_1=1}^{M} \sum_{\xi_2=1}^{M} \cdots \sum_{\xi_N=1}^{M} X(\xi_1, \xi_2, \ldots, \xi_N). \tag{1}$$

This supposition implies more than might be apparent.

Theorem 4.1.1. *The form* (1) *of the expectation operator implies that the r.v.s* ξ_k *are individually uniformly distributed over the set* $\{1, 2, \ldots, M\}$, *and that they are* statistically independent *in that*

$$E\left[\prod_k H_k(\xi_k) \right] = \prod_k E[H_k(\xi_k)] \tag{2}$$

for any functions H_k.

PROOF. The first assertion follows immediately if one chooses X to be a function of ξ_k alone in (1). The second implies the *definition* (2) of what is meant by statistical independence (which we shall simply refer to as 'independence' if the probabilistic context is clear). That (2) holds in the present case follows simply from the fact that

$$M^{-N} \sum_{\xi_1} \sum_{\xi_2} \cdots \sum_{\xi_N} \prod_k H_k(\xi_k) = \prod_k \left[M^{-1} \sum_{\xi} H_k(\xi) \right],$$

which is just a statement of (2) for the case (1). □

One can see from the characterization (2) what is meant by independence: roughly, that the r.v.s ξ_k do not interact. The concept is one of the most significant in probability. We shall consider it in a more general context in Section 3, and shall see in Section 5.5 that it is equivalent to the property which perhaps more obviously expresses independence: that the statistics of the positions of a prescribed subset of molecules is unaffected if the positions of the remaining molecules are prescribed.

One might consider a less detailed description in which the identity of molecules is not specified. That is, one concerns oneself only with the *numbers* $N_1, N_2, \ldots, N_M$ of molecules in the M cells, and not with their *identities*. We shall denote this vector of *occupation numbers* by $\mathbf{N}$, and denote its *distribution* $P(\mathbf{N} = \mathbf{n})$ simply by $P(\mathbf{n})$.

Theorem 4.1.2. *For the uniform spatial model* (1) *the distribution of the vector of occupation numbers is given by*

$$P(\mathbf{n}) = \frac{N!}{\prod_j n_j!} M^{-N} \tag{3}$$

if $\sum_j n_j = N$, *zero otherwise.*

We take for granted that the n_j can adopt only nonnegative integer values. Distribution (3) is a special case of the *multinomial distribution*. We shall see that it has a single maximum, the most probable value of $\mathbf{n}$ being specified by $n_j \approx N/M$ $(j = 1, 2, \ldots, M)$.

PROOF. The usual line of proof is to appeal to the combinatorial result that there are $N!/(\prod_j n_j!)$ configurations with occupation number vector $\mathbf{n}$, each of which has probability M^{-N}.

However, we can avoid appeal to this combinatorial result by taking another approach, which is both fast and powerful. Define the random variable ζ_k which labels the position of molecule k, in that it takes the value w_j if $\xi_k = j$ (i.e. if the kth molecule falls in the jth cell). It is then a function of ξ_k with expectation $E(\zeta_k) = M^{-1} \sum_j w_j$. Furthermore,

$$\prod_{k=1}^{N} \zeta_k = \prod_{j=1}^{M} w_j^{N_j}, \tag{4}$$

since exactly N_j of the factors ζ_k have value w_j. Taking expectations in (4) and appealing to the independence of the ζ_k, we have then

$$E\left(\prod_j w_j^{N_j}\right) = E\left(\prod_k \zeta_k\right) = \prod_k E(\zeta_k) = \left(\sum_j w_j/M\right)^N. \tag{5}$$

But since we can identify $E(\prod_j w_j^{N_j})$ with $\sum_{\mathbf{n}} P(\mathbf{n}) \prod_j w_j^{n_j}$, formula (3) for $P(\mathbf{n})$ follows from (5). □

This is a modest example of a situation which is almost the universal one. From physical or other considerations one deduces the probability of a given 'micro-description' (in fact, the probability $P(\omega)$ of any given realization ω). In the present case, the micro-description is the configuration of the N identified molecules over the M cells, the M^N possible configurations being assumed equiprobable. One then coarsens the description to a 'macro-description' and would like to integrate the statistics of the micro-description to those of the macro-description. In the present case the macro-description amounts to specification of the occupation numbers N_j only, and integration amounts to a summation over all permutations of molecules.

EXERCISES AND COMMENTS

1. Note the parallel of the model of this section with that of 'sampling with replacement' discussed in Section 2.7. One could regard the taking of a sample of n (with replacement) from a population of N as a distribution of n 'choices' over N population members, just as the N molecules are distributed over M cells. (Conflicts of notation are regrettable but inevitable: the n and N of the sampling model correspond to the N and M of the spatial model.)

2. The probability that cell 1 is unoccupied is $[(M - 1)/M]^N$. More generally, the probability that cells 1, 2, ..., s are unoccupied is $[(M - s)/M]^N$. So, if we de-

fine the random variable S, the number of the first cell which is occupied, then
$P(S > s) = [(M - s)/M]^N$, and S has the distribution

$$P(s) = P(S > s - 1) - P(S > s) = \left(\frac{M - s + 1}{M}\right)^N - \left(\frac{M - s}{M}\right)^N \quad (a = 1, 2, \ldots, M).$$

3. The calculation of Exercise 2 can be given a different cast. Consider a pack of M
cards, individually numbered 1, 2, ..., M. Suppose that one draws a sample of n
from this with replacement, all possible draws being equally likely. Show that if U
is the r.v. 'largest serial number occurring in the sample' then

$$P(U = u) = \left(\frac{u}{M}\right)^n - \left(\frac{u - 1}{M}\right)^n \quad (u = 1, 2, \ldots, M).$$

If N is large then one can approximate the sum in the expression for the expectation
by an integral, and deduce that

$$E(U) \approx \frac{nM}{n + 1}, \qquad var(U) \approx \frac{nM^2}{(n + 1)^2(n + 2)}.$$

Hence, if the number of cards M in the pack is unknown, one might estimate it from
$(n + 1)U/n$. This has applications to the observation of registration numbers, etc.;
see Exercise 5.2.4.

2. The Multinomial, Binomial, Poisson and Geometric Distributions

Continuing the example of the last section, let us coarsen the macro-description still further, by partitioning the set of M cells into m sets, the ith set
containing M_i cells, say. These sets then constitute also a partition of the
region into subregions A_i $(i = 1, 2, \ldots, m)$, where A_i contains a proportion

$$p_i = M_i/M \qquad (6)$$

of the volume of the region.

Suppose we now concern ourselves only with the numbers of molecules in
these subregions. For economy of notation we shall again denote these by N_i
$(i = 1, 2, \ldots, m)$.

Theorem 4.2.1. *The joint distribution of the numbers of molecules N_i in the
subregions A_i $(i = 1, 2, \ldots, m)$ has the multinomial form*

$$P(\mathbf{n}) = N! \prod_{i=1}^{m} \left(\frac{p_i^{n_i}}{n_i!}\right) \qquad \left(\sum_i n_i = N\right), \qquad (7)$$

where p_i is the proportion (6) of the whole volume falling in A_i.

PROOF. The result is in fact a specialization of that asserted in Theorem 4.1.2.
Redefine the r.v.s ζ_k so that $\zeta_k = z_i$ if molecule k falls into the set A_i. Relation

(4) is thus replaced by

$$\prod_{k=1}^{N} \zeta_k = \prod_{i=1}^{m} z_i^{N_i}.$$

Now, the redefinition of ζ_k simply has the effect of replacing w_j in relation (5) by z_i if cell j falls in A_i. Since it does so for M_i values of j, relation (4) simply becomes

$$E\left(\prod_i z_i^{N_i}\right) = E\left(\prod_k \zeta_k\right) = \left(\sum_i M_i z_i/M\right)^N = \left(\sum_i p_i z_i\right)^N,$$

whence (7) follows.   □

Formula (7) describes the *multinomial distribution*, one of the standard distributions, to which we shall return in Section 4. Suffice it to note for the moment that it is a unimodal distribution with a maximum near $n_j = Np_j$ ($j = 1, 2, \ldots, m$); see Exercise 1. The distribution is parameterized by the quantities $p_1, p_2, \ldots, p_m$ (which describe the probability distribution of a single molecule over the regions A_i) and N.

In our model the p_i are restricted to taking the rational values (6). However, we can well go to the *continuum limit*, in which the cells become arbitrarily small and M arbitrarily large. In our model the p_i would then take the values

$$p_i = V_i/V, \tag{8}$$

where V_i is the volume of A_i, and so could take any set of nonnegative values adding up to unity.

If we divide the region into just two sets then we have the special case of the *binomial distribution*, for which

$$E(z_1^{N_1} z_2^{N_2}) = (p_1 z_1 + p_2 z_2)^N. \tag{9}$$

However, since it is plain that $N_1 + N_2$ has the fixed value N then we may as well consider just the variable N_1. Let us denote it by R and set $p_1 = p$ and $p_2 = 1 - p = q$, so that R is the number of molecules in a region A whose volume is a fraction p of the full volume V. It follows then from (9) that

$$E(z^R) = (pz + q)^N \tag{10}$$

and that R has the distribution

$$P(r) = \binom{N}{r} p^r q^{N-r} \qquad (r = 0, 1, \ldots, N). \tag{11}$$

We graph the binomial distribution in Fig. 4.1. It is unimodal with its mode (its single maximum) near the value $r = Np$. The distribution is parameterized by the constants N and p.

Formula (11) describes the distribution of the number of molecules in a specimen subregion A. However, the distribution is still constrained by the fact that the total volume V and the total number of molecules N are finite.

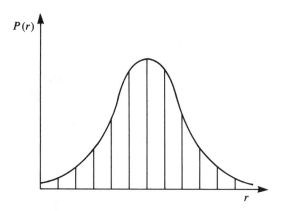

Figure 4.1. A graph of the binomial distribution.

One would be interested in the case when the specimen subregion could be seen as part of an infinite region over which the gas is distributed with a prescribed density of ρ molecules per unit volume. Consider then the *thermodynamic limit*, in which N and V are allowed to become indefinitely large, but in such a way that the density

$$\rho = N/V \tag{12}$$

is held fixed.

Theorem 4.2.2. *In the thermodynamic limit with molecular density ρ the number of molecules R in a given region A of unit volume follows the* Poisson *distribution*

$$P(r) = e^{-\rho}\rho^r/r! \qquad (r = 0, 1, 2, \dots). \tag{13}$$

PROOF. If the region A is assumed to have unit volume then $p = 1/V$ and $q = 1 - 1/V$, so that the expectation (10) can be written

$$E(z^R) = [1 + (z - 1)/V]^{\rho V}.$$

In the thermodynamic limit $V \to \infty$ this relation becomes

$$E(z^R) = e^{\rho(z-1)}, \tag{14}$$

which corresponds to distribution (13). $\qquad\qquad\qquad\qquad\qquad\square$

This proof is brief, but does raise points of rigour; see Exercise 3.

The Poisson distribution has just the single parameter ρ. It is decreasing in r if $\rho < 1$; otherwise, it has a single maximum near ρ.

All these distributions have arisen naturally from our spatial model. However, they are standard distributions and we shall see in the next section that, on plausible assumptions, they arise in all kinds of contexts.

EXERCISES AND COMMENTS

1. Consider expression (7) for the multinomial distribution. The arguments n_j are integer-valued, so the idea of a 'stationary point' can only be approximate. It is applicable, however, and we can determine such stationary points by seeking values $\mathbf{n}$ at which $P(\mathbf{n}) \approx P(\mathbf{n} + \Delta)$ where Δ is a small permissible perturbation. By considering the perturbation at which n_j increases by unity and n_k decreases by unity, show that at a stationary point $p_j/n_j \approx p_k/n_k$. That is, the only stationary point is at

$$n_j \approx N p_j \qquad (j = 1, 2, \ldots, m).$$

Show that this point is a maximum of the distribution.

2. Under the suppositions $p = V_1/V = 1/V$ and $N = \rho V$, the binomial distribution (11) has the form

$$P(r) = \binom{\rho V}{r}(1/V)^r(1 - 1/V)^{\rho V - r}. \qquad (15)$$

Verify directly that this converges to the Poisson form (13) as $V \to \infty$. (This calculation implies the assumption that ρV is integral and an appeal to the Stirling approximation to the factorial

$$N! \sim \sqrt{2\pi} e^{-N} N^{N+1/2}$$

for large N.)

3. Note that the discussion of a limit distribution (as we approach the thermodynamic limit) might need care. Let the expectation operators corresponding to the distributions (15) and (13) be denoted E_V and E_∞, respectively. The question is: Does the fact that one has established that $E_V[H(R)] \to E_\infty[H(R)]$ as $V \to \infty$ for some functions H imply the same convergence for all H of interest? In the proof of Theorem 4.2.2 we demonstrated the convergence for the functions $H(R) = z^R$; in Exercise 2 for $H(R) = I$ $(R = r)$. In the present case there is really no problem, but we return to the issue in Chapters 7 and 15.

4. Consider the distribution of the number of molecules in a fixed region of volume V_1 in the thermodynamic limit. Show that this is Poisson with parameter ρV_1.

5. One should distinguish the continuum limit and the thermodynamic limit. In fact, let us consider the thermodynamic limit in the discrete version by allowing M and N to go to infinity simultaneously in such a way that $M/N \to \gamma$, so that γ is the limiting molecular density in molecules per cell. Then the probability that the first s cells are unoccupied is

$$\left(\frac{M - s}{M}\right)^N = (1 - s/M)^{\gamma M + o(M)} \to e^{-\gamma s}.$$

If we define $\alpha = e^{-\gamma}$ (interpretable as the probability that any given cell is unoccupied) and define S as the number of the first occupied cell, then we have $P(S > s) = \alpha^s$. Thus S has distribution

$$P(s) = P(S > s) - P(S > s - 1) = (1 - \alpha)\alpha^{s-1} \qquad (s = 1, 2, \ldots). \qquad (16)$$

Formula (16) then gives the distribution of the distance one has to go along a sequence of cells to find the first one which is occupied, when the constraining effects of finiteness of the total system have been removed. This is one of the classic distributions, the *geometric distribution*, so-called because of the geometric decline of probability with increasing s.

3. Independence

It has been apparent from Section 1 that the essential factor which eased our treatment of the spatial model was that the molecular positions ξ_k were statistically independent; a consequence of the form (1) assumed for their distribution. We must now recognize the property of independence as one which occurs widely and is deeply significant; even a pattern of dependence is often best seen in terms of random variables which are independent.

Formally, random variables $X_1, X_2, \ldots, X_N$ are independent if and only if

$$E\left[\prod_{k=1}^{N} H_k(X_k) \right] = \prod_{k=1}^{N} E[H_k(X_k)] \tag{17}$$

for all scalar-valued functions H_k for which the right-hand member of (17) is defined. The full force of the concept will emerge in Chapter 5, when we consider conditioning of r.v.s, but characterization (17) does seem to imply that there is no interaction between the r.v.s X_k. Since it implies that

$$P(X_k \in B_k; k = 1, 2, \ldots, N) = \prod_{k=1}^{N} P(X_k \in B_k) \tag{18}$$

for largely arbitrary sets B_k it certainly implies that the joint probability distribution of the X_k factorizes correspondingly.

For all that it does not occur in the axioms, independence is one of the basic structural features of probability. By appealing to it we can see which are the features of the spatial model of Section 1 that transfer to a wider context. For example, we can transfer the model in which N molecules are distributed independently over a number of regions A_i to that in which one has a sequence of N independent *trials*, each of which can have *outcome i* with probability p_i ($i = 1, 2, \ldots, m$). The multinomial distribution (7) then gives the probability that n_i of the trials should have outcome i, jointly for all i. For example, n_i might be the number of voters out of N in an election who declare themselves for candidate i, the number of molecules in an assembly of N which are at energy level i, or the number of consumers out of N who opt for brand i of a certain product. Of course, in any particular case, the assumption of independence is one that must be examined, and which must have a good physical basis. For example, one would not expect pollen counts on consecutive days to be independent as r.v.s.

One often deals with random variables which are identically and independently distributed; they are often just said to be IID. The multinomial distribution is based upon a model in which the outcomes of different trials are IID.

The binomial distribution refers to the special case of only two outcomes; the stock outcomes are head or tail (in a coin toss), success or failure (in a trial). If p is the *success probability* then expression (11) gives the probability of r successes in a sequence of N independent trials.

The usual combinatorial derivation of the binomial distribution starts from a consequence of independence: that the probability that a sequence of N trials contains r successes and $N - r$ failures *in a given order* is $p^r q^{N-r}$. If one is interested only in the number R of successes and not in where they occur in the sequence then one sums this probability over the $\binom{N}{r}$ configurations which yield $R = r$, hence deriving expression (11) for $P(R = r)$.

Let S_1 be the number of the trial at which the first success occurs. This will be equal to s if one has a sequence of $s - 1$ failures followed by a success, an event which has probability $q^{s-1}p$. Thus S_1 follows the geometric distribution

$$P(S_1 = s) = pq^{s-1} \qquad (s = 1, 2, \dots) \tag{19}$$

already encountered in (16).

The Poisson law (13) is the limit case of the binomial when $N \to \infty$ and $p \to 0$ in such a way that $Np \to \rho$. We saw the physical motivation for this limit in the last section, but it has been used historically for less clear-cut situations in which there are a large number of 'trials', each with a small 'success' probability. Thus, it has been used to describe the distribution of the number of misprints per page of a book, or the number of Prussian cavalrymen dying each year from being kicked by a horse. Hence a name often previously attached to it: the 'law of rare events'.

Events $A_1, A_2, \dots$ are termed independent if their indicator functions are independent. This implies then that

$$P\left(\bigcap_k A_k\right) = \prod_k P(A_k), \tag{20}$$

and also that this relation remains true if any of the events are replaced by their complements. Relation (20) is sometimes termed the 'multiplicative law of probability' in analogue to the additive law

$$P\left(\sum_k A_k\right) = \sum_k P(A_k) \tag{21}$$

for disjoint A_k. The two relations differ very much in status, however. Relation (21) always holds; it is either itself an axiom or the consequence of axioms. Relation (20) holds only when the A_k are independent; it amounts to a *definition* of independence for events.

EXERCISES AND COMMENTS

1. One might imagine from the product form of the multinomial distribution (7) that the r.v.s N_i were independent. This is not true; the r.v.s are subject to the constraint

$\sum_i N_i = N$, and expression (7) has an implicit factor $\delta(\sum_i n_i - N)$. See Section 8, however.

2. Consider the following situation, in which independence is proved rather than assumed. Consider again the spatial model of Section 1, and show that the probability that cell s_1 contains r_1 molecules, cell s_2 (where $s_2 > s_1$) contains r_2 molecules and that all other of the first s_2 cells are empty is

$$P(r_1, r_2, s_1, s_2) = \frac{M^{-N} N!}{r_1! r_2!} \frac{(M - s_2)^{N - r_1 - r_2}}{(N - r_1 - r_2)!}.$$

Thus, if S_1 is the number of the first nonempty cell, and S_2 is the number of the second, then S_1 and S_2 have the joint distribution

$$P(s_1, s_2) = \sum_{r_1 > 1} \sum_{r_2 > 1} P(r_1, r_2, s_1, s_2).$$

This expression is unpromising. Show, however, that in the thermodynamic limit of Exercise 4.2.5 it converges to

$$P(s_1, s_2) = (1 - \alpha)\alpha^{s_1 - 1}(1 - \alpha)\alpha^{s_2 - s_1 - 1},$$

in the notation of that exercise. In other words, S_1 and $S_2 - S_1$ are independent in the thermodynamic limit, both following the same geometric law. What is perhaps more surprising than independence is the fact that the two r.v.s should follow the same distribution, since $S_2 - S_1$ is the distance between two successive occupied cells, whereas S_1 is the distance from an arbitrary starting point to the first occupied cell. One might have expected S_1 to be smaller than $S_2 - S_1$ in some sense. The geometric distribution is unique in that this is not the case; see Section 5.1.

3. Show that events $A_1, A_2, \ldots, A_N$ are independent if and only if the multiplicative relation (20) holds for any subset of these N events.

4. Suppose scalar variables are independent and have finite mean squares. Show that they are mutually uncorrelated. The reverse is not necessarily true: the classic example is the pair of variables $X = \cos \theta$, $Y = \sin \theta$, where the angle θ is a r.v. uniformly distributed on $[0, 2\pi)$. Then X and Y are uncorrelated but not independent. The relation between them is $X^2 + Y^2 = 1$, and so is of a highly nonlinear nature.

5. We require the full product of N terms in (17) if all N variables are to be independent. It is possible, for example, for r.v.s to be independent two at a time but not three at a time. This is true of the r.v.s X_1, X_2 and X_3 which take the possible triples of values $(1, 0, 0)$, $(0, 1, 0)$, $(0, 0, 1)$ and $(1, 1, 1)$ each with probability $\frac{1}{4}$.

6. *Personal records.* Suppose that a high-jumper achieves heights $X_0, X_1, X_2 \ldots$ on successive attempts, and that these variables are distributed independently with probability density $f(x)$, and so with probability distribution function $F(x) = \int_0^x f(u)\, du$. The first attempt X_0 thus represents his first record; he achieves his second record after a further n attempts if $X_j > X_0$ first for $j = n$. Show that this event has probability

$$P(n) = \int_0^\infty F(x)^{n-1}[1 - F(x)]f(x)\, dx = \frac{1}{n} - \frac{1}{n+1} = \frac{1}{n(n+1)} \qquad (n = 1, 2, 3, \ldots).$$

The distribution of n, the number of attempts needed for improvement of the initial performance, is thus so slowly convergent to zero with increasing n that it gives the random variable infinite expectation. Personal records are hard to improve! Serious improvement can come about only by a change in condition or technique; i.e. by a change in $f(x)$. Improvement in records from a *population* can come about because an increasing proportion of the population takes part.

One could also deduce the formula for $P(n)$ simply by noting that all permutations of $n + 1$ performances are equally likely, and then calculating the probability that the best and second-best performances were achieved on the last and first attempts, respectively. However, this calculation does depend on the assumption that all performances are different, which is true with probability one if X is distributed with a density.

4. Probability Generating Functions

It may have been independence that simplified the distributional results of Sections 1 and 2, but it was a particular technique which simplified their derivation. This was the approach in which the distribution of the random variable R, say, was determined by first calculating the expectation $E(z^R)$. This expectation, as a function of z, is known as the *probability generating function* of R, usually abbreviated to p.g.f. and denoted by $\Pi(z)$. The term is appropriate, since if R is an integer-valued variable with distribution $P(r)$ then

$$\Pi(z) = \sum_r P(r)z^r, \tag{22}$$

so that $P(r)$ is the coefficient of z^r in the expansion of $\Pi(z)$ in powers of z.

When we speak of a p.g.f. we shall always take it for granted that the corresponding r.v. is integer-valued. For other variables one considers rather the characteristic function or moment generating function; see Chapter 7. One may ask for what values of z the p.g.f. is defined, whether knowledge of the p.g.f. determines the distribution, and whether one may properly differentiate expression (22) with respect to z under the summation sign. These are all matters to which we shall return in Chapter 7. However, there is certainly no problem if $P(r)$ converges sufficiently quickly to zero with increasing r. Sometimes, for greater explicitness, we shall write expression (22) as $\Pi_R(z)$, to emphasize that it is the p.g.f. pertaining to the r.v. R.

The fundamental property of the p.g.f. is one that we have already exploited repeatedly.

Theorem 4.4.1. *Suppose that X and Y are independent integer-valued random variables. Then*

$$\Pi_{X+Y}(z) = \Pi_X(z)\Pi_Y(z). \tag{23}$$

That is, the p.g.f. of the sum of independent r.v.s is the product of the p.g.f.s of the summands. The proof follows immediately from

$$E(z^{X+Y}) = E(z^X z^Y) = E(z^X)E(z^Y). \tag{24}$$

This result has been implicit in all that we have done. For example, consider the variable R of the binomial distribution (11); the number of molecules falling into a subregion A. We could write this as

$$R = I_1 + I_2 + \cdots + I_N, \tag{25}$$

where I_k is the indicator function of the event that molecule k falls into A. Then I_k takes values 1 and 0 with respective probabilities p and q, so that

$$E(z^{I_k}) = pz + q.$$

It follows then from (25) and the fact that the I_k are independently and identically distributed that

$$E(z^R) = \prod_k E(z^{I_k}) = (pz + q)^N$$

as we already know from (9).

One can also define and use multivariate p.g.f.s

$$\Pi(z) = E\left(\prod_j z_j^{N_j}\right)$$

as we have in fact done for the case of the multinomial distribution. Theorem 4.4.1 still holds. That is, if $X = (X_1, X_2, \ldots)$ and $Y = (Y_1, Y_2, \ldots)$ are vectors of integers, the two vectors being independent as r.v.s, then

$$\Pi_{X+Y}(z) = E\left(\prod_j z_j^{X_j+Y_j}\right) = E\left(\left(\prod_j z_j^{X_j}\right)\left(\prod_j z_j^{Y_j}\right)\right)$$

$$= E\left(\prod_j z_j^{X_j}\right) E\left(\prod_j z_j^{Y_j}\right) = \Pi_X(z)\Pi_Y(z).$$

For the multinomial case we could write (25) in vector form, or write it more explicitly as

$$N_j = \sum_{k=1}^N I_{jk} \qquad (j = 1, 2, \ldots, m),$$

where I_{jk} is the indicator function of the event that molecule k falls into subregion A_j. We have then, as in the binomial case above,

$$E\left(\prod_j z_j^{I_{jk}}\right) = \sum_j p_j z_j,$$

$$E\left(\prod_j z_j^{N_j}\right) = \left(\sum_j p_j z_j\right)^N,$$

consistent with the calculation of Theorem 4.2.1.

Application of Theorem 4.4.1 can lead to quite striking results very economically. For example:

Corollary 4.4.2. *If X and Y are independent Poisson variables with respective parameters λ and μ, then $X + Y$ is a Poisson variable with parameter $\lambda + \mu$.*

One might have expected the result from our derivation of the Poisson distribution. It follows formally from the fact that $X + Y$ has p.g.f.

$$e^{\lambda(z-1)}e^{\mu(z-1)} = e^{(\lambda+\mu)(z-1)},$$

which is the p.g.f. of the distribution asserted.

There is yet another useful property of the p.g.f.: expressions for moments can be derived from it very easily.

Theorem 4.4.3. *If $\Pi(z)$ is the p.g.f. of a scalar r.v. X then, formally,*

$$[(\partial/\partial z)^{\nu}\Pi(z)]_{z=1} = E(X^{(\nu)}) \qquad (\nu = 0, 1, 2, \ldots). \qquad (26)$$

Equivalently, one has the formal expansion

$$\Pi(1 + \theta) = \sum_{\nu=0}^{\infty} \frac{\theta^{\nu}}{\nu!} E(X^{(\nu)}). \qquad (27)$$

We recall the definition $X^{(\nu)} = X(X - 1)(X - 2)\ldots(X - \nu + 1)$ of the factorial power.

PROOF. Assertion (26) follows from differentiation of $E(z^X)$ with respect to z under the expectation sign. Assertion (27) follows from an expansion of $E[(1 + \theta)^X]$ in powers of θ under the expectation sign. The formal calculation is always justified if the resultant expectation is absolutely convergent. □

So, if X follows a binomial distribution, $E(z^X) = (pz + q)^N$, then

$$E(X^{(\nu)}) = N^{(\nu)}p^{\nu} \qquad (\nu = 0, 1, 2, \ldots).$$

In particular,

$$E(X) = Np, \qquad E[X(X - 1)] = N(N - 1)p^2,$$

whence

$$\text{var}(X) = Npq.$$

In the same way we find that if X follows a Poisson distribution with parameter ρ, then

$$E(X^{(\nu)}) = \rho^{\nu}, \qquad E(X) = \text{var}(X) = \rho.$$

The multinomial version of Theorem 4.4.3 follows by immediate analogue. Here, if we set $z = 1$, we mean that $z_j = 1$ for all relevant j.

Theorem 4.4.4. *If* $\Pi(z) = E(\prod_j z_j^{X_j})$ *then, formally,*

$$\left[\left\{ \prod_j (\partial/\partial z_j)^{v_j} \right\} \Pi(z) \right]_{z=1} = E\left(\prod_j X_j^{(v_j)} \right). \tag{28}$$

Equivalently, one has the formal expansion

$$\Pi(1 + \theta_1, 1 + \theta_2, \ldots) = \sum_v E\left(\prod_j [\theta_j^{v_j} X_j^{(v_j)}/v_j!] \right). \tag{29}$$

So, for the multinomial distribution (7) we find that

$$E\left(\prod_j N_j^{(v_j)} \right) = N^{(\Sigma v_j)} \prod_j p_j^{v_j}.$$

We leave the reader to verify that this implies that

$$E(N_j) = Np_j, \qquad \text{cov}(N_j, N_k) = N(\delta_{jk}p_j - p_jp_k).$$

The fact that the covariance is negative for $j \neq k$ reflects the effect of the constraint $\sum N_j = N$; if a given molecule falls in a given subregion then there is one less molecule to fall in other regions.

EXERCISES AND COMMENTS

1. Consider the r.v. S_1 following the geometric distribution (19). Show that it has p.g.f.

$$E(z^{S_1}) = \frac{pz}{1 - qz},$$

and hence that

$$E(S_1) = p^{-1}, \qquad \text{var}(S_1) = qp^{-2}.$$

2. The *negative binomial distribution.* Consider a sequence of independent trials with success probability p, and let S_r be the number of the trial at which the rth success occurs. It follows from the IID character of the trials that $S_1, S_2 - S_1, S_3 - S_2, \ldots$ are IID, and hence that S_r has p.g.f.

$$E(z^{S_r}) = \left(\frac{pz}{1 - qz} \right)^r.$$

This is the p.g.f. of the *negative binomial distribution*

$$P(S_r = s) = \frac{(s-1)!}{(r-1)!(s-r)!} p^r q^{s-r} \qquad (s = r, r+1, \ldots).$$

One might imagine this as describing, for example, the distribution of the number of attempts needed to unlock an unknown r-digit combination lock, if each attempt on a given digit has success probability p and success can be recognized. Note that one could have obtained the expression for $P(S_r = s)$ by calculating the probability that a sequence of s trials yields r successes, with the constraint that the last trial yields a success.

3. Show that $E[(S_r - r)^{(v)}] = (r + v - 1)^{(v)}(q/p)^v$.

4. Consider the molecular distribution example of Sections 1 and 2 in the thermo-dynamic limit, with molecular density ρ. Suppose that A_1, A_2, ... are disjoint regions of respective volumes V_1, V_2, ... If N_i is the number of molecules in A_i show that the N_i are independent Poisson variables with respective parameters ρV_i.

5. Continuing Exercise 4, consider the situation in which the regions are not necessari-ly disjoint, and denote the volume of $A_1 \cap A_2$ by V_{12}. Show then that N_1 and N_2 have joint p.g.f. $\exp[\rho V_1(z_1 - 1) + \rho V_2(z_2 - 1) + \rho V_{12}(z_1 - 1)(z_2 - 1)]$. What is the covariance of N_1 and N_2?

6. Suppose that insurance claims can be for integral amounts $j = 1, 2, 3, \ldots$ of the currency. Let N_j be the number of claims of size j in a given time period, and suppose that the N_j are independent Poisson variables with $E(N_j) = \lambda_j$. Show that the total claim in the period, $X = \sum_j j N_j$, has p.g.f. $\exp[\sum_j \lambda_j(z^j - 1)]$.

   The distribution with this p.g.f. is known as the *compound Poisson distribution*. It gives the distribution of the total yield (e.g. of births) in a given time, when the yield comes in 'packets' which occur independently but can be multiple (e.g. twins, triplets). The distribution itself is not calculable, but the p.g.f. is immediate. Note that $E(X) = \sum_j j \lambda_j$, $\text{var}(X) = \sum_j j^2 \lambda_j$.

7. The same example could serve for molecules carrying an integral charge, N_j being the number of molecules in a region carrying charge j and X the total charge. However, j could now be negative. Suppose, for instance, that j can take the values ± 1 with $\lambda_1 = \lambda_{-1} = \lambda$. The p.g.f. of the total charge in the region is then $\exp[\lambda(z + z^{-1} - 2)]$. The expansion $\sum_x P(x)z^x$ must be valid on $|z| = 1$; one finds then that

$$P(x) = e^{-2\lambda_\lambda|x|} \sum_{j=0}^{\infty} \frac{\lambda^{2j}}{j!(j + |x|)!}.$$

   The distribution is not simple (it is expressible in terms of Bessel functions), but moments are easily calculated.

8. Note that the expectation values deduced for the multinomial distribution agree to first order with the maximizing values asserted in Exercise 4.2.1.

5. The St. Petersburg Paradox

Gaming has stimulated the study of probability and posed fundamental problems from the earliest days. The simplest situation is that in which two players, Alex and Bernard, say, play a sequence of statistically independent games, in each of which Alex can win with probability p. At each stage the loser pays unit amount to the winner, and the game continues until one of the players is 'ruined'; i.e. has exhausted his capital. One can ask various ques-tions. For example, for given values of the players' initial capitals: What is the probability that Alex wins the game? (i.e. ruins Bernard); What is the distribu-tion of the duration of the game? We shall return to these questions in Chapter 9.

The game is termed 'fair' if the expected net winnings from an individual round of the game are zero. In the case above, that would imply that $p = \frac{1}{2}$. That the game is fair does not mean that the players have equal chances of winning the whole sequence; we shall see that the richer player has an advantage.

One variant of the game is to allow the players to go into debt, so that some 'stopping rule' must be adopted other than that one of the players becomes ruined. Another variant is to allow the stake to be varied.

Suppose that Alex is allowed to go into debt, and may continue or withdraw from the game as he chooses. Then one might imagine that he could guarantee himself a net return of a, say, by continuing play until he has won (in net terms) exactly that amount. Whether such a moment ever comes depends sensitively on the value of p, as we shall see in Chapter 9. In the case of a fair game, such a moment will indeed come, but only after a time whose expectation is infinite.

However, there is a method of play (the 'martingale') which seems to offer the possibility of a guaranteed net return in a finite time. In this scheme Alex begins with a unit stake, continues to play with a doubled stake after every loss and retires at his first win. So, if he wins first at the nth round then his net return is

$$-(1 + 2 + 2^2 + \cdots + 2^{n-1}) + 2^n = -(2^n - 1) + 2^n = 1,$$

and so equals unity for all n. If the game is fair (as we assume) then the probability that play terminates at the nth round is $(\frac{1}{2})^n$. We interpret the relation $\sum_{n=1}^{\infty} (\frac{1}{2})^n = 1$ as implying that termination in a finite time is certain. Indeed, the expected time to termination is finite and equal to $\sum_{n=1}^{\infty} (\frac{1}{2})^n n = 2$. It seems, then, that Alex can guarantee himself a net return of one unit (and so of any amount, if he scales up his stakes) in a finite time.

However, the size of his gross win, when he makes it, has expectation $\sum_{n=1}^{\infty} (\frac{1}{2})^n 2^n = +\infty$. Since his win exceeds his previous outlay by exactly unity, it then appears that he must be prepared to make an infinite expected outlay before he finally recoups his guaranteed net profit. So, in the case of fixed stakes the certainty of a given return can be achieved only at the cost of an infinite expected playing time; in the case of the martingale policy, only at the cost of incurring a debt of infinite expected size before finally clearing it.

It is this conclusion which constitutes the St. Petersburg paradox, which has exercised gamblers and probabilists alike. As far as both gamblers and gaming houses go: Is the martingale policy a feasible one? Is it a 'legitimate' one? As far as probabilists go, the occurrence of infinite expectations was found disturbing, as we have noted in Section 3.4, and, as conjectured by David Stirzaker in his letter reproduced there, may have contributed to a subsequent distrust of expectation as a prime concept.

The problem lives on. There is now a substantial general theory, known indeed as 'martingale theory' (see Chapter 14). This is at least partly motivated by the question of whether advantageous gambling systems exist, and the St. Petersburg paradox supplies one of its simple pathological examples.

However, one can argue that if one is considering an optimization problem (and we are indeed considering the optimization of policy—the gambling system) then it should be properly framed. If the gambler is to finance his play he must borrow money at some rate of interest (or, equivalently, must sacrifice the possibility of earning interest on some of his own money). Suppose that we take the rounds of the game as units of time, and that an interest rate of $r\%$ is available over this unit time. A unit amount invested now will thus have grown to $(1 + r/100)^n$ after time n. Let us write this as β^{-n}, where $0 < \beta < 1$. Then one could equivalently say that a unit amount available n time units hence has a *present value* of β^n, since that sum invested now would produce unit amount after time n. The quantity β is known as the *discount factor*, since unit delay in acquiring a given sum will reduce the present value of that sum by a factor β.

If money sunk into the game must be borrowed at interest (or, equivalently, if interest on assets is thereby sacrificed), then all income and outlay for the game should be evaluated on its present value (at the beginning of play, say). This factor drastically clarifies conclusions. Let us assume, for definiteness, that discounting takes place between rounds of the game, so that the winnings in a round are not discounted relative to the stake in that round.

Theorem 4.5.1. *Suppose the martingale policy is followed, with an initial stake of unity. If discounting is strict (i.e. $\beta < 1$), then the expected discounted outlay and the expected discounted gross winnings are finite and equal, so that*

$$E(\text{net discounted return}) = \begin{cases} 1 & (\beta = 1), \\ 0 & (0 \le \beta < 1). \end{cases}$$

There is thus a discontinuity in expected return at $\beta = 1$; the game as a whole is fair if discounting is strict.

PROOF. This is a simple matter of verification. If play ends at the nth round then the discounted outlay is

$$\sum_{j=0}^{n-1} (2\beta)^j = \frac{(2\beta)^n - 1}{2\beta - 1}$$

and the discounted return is $2^n \beta^{n-1}$. The probability of this contingency is $(\frac{1}{2})^n$ $(n = 1, 2, 3, \dots)$. The expectations of the two components are finite if $\beta < 1$, and both equal to $1/(1 - \beta)$, as is easily verified. The expected net return is thus zero. We know already that the net return is $+1$ if there is no discounting. $\square$

So the introduction of a realistic discount factor resolves the St. Petersburg paradox: expectations exist and the total game is fair *identically in β for $\beta < 1$*.

We can say that the net value of the gain of one unit is exactly cancelled by the fact that the outlay occurs earlier in time than the win; this cancellation is exact for all positive interest rates.

The total game is fair in that Alex can expect to win nothing by the martingale policy, so there is indeed no point in the policy. It is also fair as far as his opponent Bernard (who may be the gaming house) is concerned. The certainty of an ultimate pay-out is balanced by the large bankable accumulation of winnings before that. However, one can also see why gaming houses do not like the system (which in general they indeed forbid). Despite the fact that there is zero loss on average, there is a constant succession of heavy transactions, in both directions. In fact, if transaction costs are taken into account, the policy is attractive to neither party.

EXERCISES AND COMMENTS

1. Suppose that discounting takes place *during* rather than between the rounds; i.e. between the placing of a stake and the collection of winnings. Show that the expected net discounted return is -1 if discounting is strict: the player loses his initial stake on average.

6. Matching, and Other Combinatorial Problems

There is a large number of problems in which one can plausibly start from equiprobability or independence assumptions, and for which calculation of the probability of an event then reduces to calculation of the number of realizations compatible with that event. These problems give great opportunity for combinatorial ingenuity, or, if one's taste runs that way, for ingenuity in the manipulation of generating functions.

The classic matching problem is one such. It takes many guises, one of which is this: n letters are written to different people, and envelopes correspondingly addressed. The letters are now mixed before being sealed in envelopes, the effect being to make all $n!$ allocations of letters to envelopes equally likely. What is the probability of the event A_j that just j letters are placed in their correct envelopes? For brevity, we shall say that a letter 'matches' if it is placed in its correct envelope.

Theorem 4.6.1. *The probability of exactly j matches for n letters is*

$$P(A_j) = \frac{1}{j!} \sum_{k=j}^{n} \frac{(-)^{k-j}}{(k-j)!}. \tag{30}$$

PROOF. 1. The direct combinatorial argument (which nevertheless relies heavily on expectation and indicator function techniques) goes as follows. Denote by P_j the probability that a given j letters match, and none of the others do.

Then, by symmetry,

$$P(A_j) = \binom{n}{j} P_j,$$

so the problem is solved if we can calculate the P_j. This is by no means immediate, however; the quantities which can be easily calculated are the probabilities

$$Q_j = \frac{(n-j)!}{n!}$$

that a given j letters match, no condition being placed on the remaining $n - j$. Now, if I_j is the indicator function of the event 'the jth letter matches', then

$$Q_j = E\left(\prod_{i=1}^{j} I_i\right)$$

(the expectation being the same for all choices of j distinct letters) and

$$P_j = E\left[\left(\prod_{i=1}^{j} I_i\right)\left(\prod_{h=j+1}^{n}(1 - I_h)\right)\right] = E\left[\left(\prod_{i=1}^{j} I_i\right)\sum_{m=0}^{n-j}(-)^m S_m\right]$$

$$= \sum_{m=0}^{n-j}(-)^m\binom{n-j}{m}Q_{m+j}.$$

Here S_m is the sum of all products of m distinct terms at a time chosen from $I_{j+1}, I_{j+2}, \ldots, I_n$. Since there are $\binom{n-j}{m}$ such terms, the final equality follows. Eliminating the P_j and Q_j from the last three relations, we obtain solution (30).

2. There is an argument in terms of p.g.f.s which is both faster and more powerful. Write A_j rather as A_{nj}, to emphasize its dependence also upon n, and define

$$\Pi_n(z) = \sum_{j=0}^{n} P(A_{nj})z^j,$$

the p.g.f. of the number of matches for n letters. Define also $M_{nj} = n!\,P(A_{nj})$, the number of the $n!$ permutations of n letters which lead to exactly j matches. Then

$$M_{nj} = \frac{j+1}{n+1}M_{n+1,j+1},$$

because the j-match case for n letters can always be regarded as the $(j + 1)$-match case for $n + 1$ letters, with the $(n + 1)$th letter always amongst the matched ones. However, by symmetry, only in a fraction $(j + 1)/(n + 1)$ of the $M_{n+1,j+1}$ cases will the $(n + 1)$th letter fall among the $j + 1$ which are matched. The last displayed relation implies that $P(A_{nj}) = (j + 1)P(A_{n+1,j+1})$ or

$$\frac{\partial}{\partial z}\Pi_{n+1}(z) = \Pi_n(z).$$

We can integrate this equation to

$$\Pi_{n+1}(z) = 1 + \int_1^z \Pi_n(w)\, dw, \tag{31}$$

since $\Pi_n(1) = 1$ for all n. Applying relation (31) repeatedly, starting from $\Pi_1(z) = z$, we deduce that

$$\Pi_n(z) = \sum_{k=0}^n \frac{(z-1)^k}{k!},$$

a truncated version of the Poisson p.g.f. $\exp(z-1)$. The coefficient of z^j in this expression is indeed equal to expression (30). □

EXERCISES AND COMMENTS

1. If J is the number of matches for n letters, note that $E(J^{(v)})$ equals 1 or 0 according as $v \le n$ or $v > n$.

2. Show that the probability of no matches for n letters is $[n!/e]/(n!)$, where $[x]$ is the integer nearest to x.

3. Suppose that one samples from a pack of m cards, with replacement, until each of the cards has occurred at least once. Assume the samples independent, with each of the m cards equally likely to be drawn. Show that if N is the size of this random sample, then

$$E(z^N) = \prod_{j=1}^m \left(\frac{p_j z}{1 - q_j z} \right),$$

where

$$p_j = \frac{m - j + 1}{m},$$

and hence that

$$E(N) = m \sum_1^m j^{-1},$$

which is asymptotic to $m \log m$ for large m.

The interest of this example is that it is just the 'picture card' problem. Suppose that a packet of cereal contains one card from a set of m; then N is the number of packets one must buy in order to complete the set. Note that $E(N)$ increases faster than m; the effort per card is greater for large sets than small ones.

7. Conditioning

The form of the multinomial distribution (7) suggests that the N_j would be independent distributed Poisson variables, but for the constraint

$$\sum_j N_j = N. \tag{32}$$

Suppose that they were indeed independent Poisson variables with respective parameters λ_j, so that

$$P(\mathbf{n}) = P(\mathbf{N} = \mathbf{n}) = \prod_j e^{-\lambda_j}(\lambda_j^{n_j}/n_j!). \tag{33}$$

However, suppose that we indeed also apply the constraint (32) for prescribed N, in that we consider distribution (33) confined to that part of $\mathbf{N}$-space for which (32) holds. Let us denote this set by A, so that A is just the event that (32) holds. However, if we were to consider the distribution on this reduced space A we should have to renormalize it, so that it integrates to unity on A. That is, we consider the constrained and renormalized distribution

$$P(\mathbf{n}|A) = P(\mathbf{n})/P(A) \tag{34}$$

confined to the set A. This is termed a *conditioning*, and expression (34) provides the distribution of $\mathbf{N}$ *conditional on the event* A. We shall consider the topic of conditioning systematically in the next chapter, but it is useful to introduce the idea already at this stage.

Theorem 4.7.1. *The distribution of the independent Poisson variables N_j conditional on the constraint (32) is multinomial, and given by expression (7) with*

$$p_j = \lambda_j \bigg/ \left(\sum_k \lambda_k \right).$$

PROOF. The r.v. $\sum_j N_j$ is Poisson with parameter $\sum_j \lambda_j$ (see Corollary 4.4.2) so that

$$P(A) = \exp\left(-\sum_j \lambda_j \right)\left[\left(\sum_j \lambda_j \right)^N \bigg/ N! \right]. \tag{35}$$

Substituting expressions (33) and (35) into (34) we confirm the assertion of the theorem. ☐

So, one might say that the multinomial distribution has a conditioning implicit in its structure, even though we have first now come upon the concept. For another example, consider a problem that we might have posed in Section 1, but could not then have handled. Let us determine the joint distribution of $(R_0, R_1, R_2, \ldots)$ where R_k is the number of cells containing k molecules. We could also interpret R_k as the number of people who have been sampled k times in the sampling-with-replacement model of Section 2.7.

Theorem 4.7.2. *Consider the spatial model of Section 1. Let R_k denote the number of cells containing k molecules $(k = 0, 1, 2, \ldots)$. Then the joint distribution of the R_k is*

$$P(\mathbf{r}) = P(R_k = r_k, k = 0, 1, 2, \ldots) = \frac{M!\,N!}{M^N} \prod_k (k!)^{-r_k}/r_k! \tag{36}$$

on the set of nonnegative integers R_k specified by

$$\sum_k R_k = M, \qquad \sum_k kR_k = N, \tag{37}$$

zero elsewhere.

PROOF. The occupation numbers N_j follow the special multinomial distribution (3). Let us suppose them independent Poisson variables with parameter λ and later apply the constraint (32). The probability that a given cell contains k molecules is then $p_k = e^{-\lambda}\lambda^k/k!$. Since the numbers in different cells are now independent, the distribution of the R_k is multinomial

$$P(\mathbf{r}) = M! \prod_k (p_k^{r_k}/r_k!)$$

and is to be regarded as already satisfying the first constraint of (37). However, we now have to apply constraint (32), which amounts to the second constraint of (37). The r.v. $\sum_k kR_k$ is just the total number of molecules in the system, which is Poisson with parameter $M\lambda$. The probability of the conditioning event is thus

$$P(A) = P\left(\sum_k kR_k = N\right) = e^{-M\lambda}(M\lambda)^N/N!.$$

The evaluation of the conditioned distribution $P(\mathbf{r})/P(A)$ derived from these last two expressions is just that asserted in (36).   □

Note that the irrelevant parameter λ has disappeared from the distribution. One could have deduced expression (36) for the distribution by the formal use of generating functions (see Exercise 3), but the concept of embedding the model in the more relaxed Poisson model and then conditioning is equivalent, and more 'probabilistic'. This technique also provides the natural way to calculate expectations (see Exercise 3); one can indeed say that the determination of the actual distribution (36) has only formal interest.

As a final example, consider the problem of capture/recapture sampling. We could see this in terms of the sampling model of Section 2.7, in which we sampled a town of N inhabitants. Suppose that one takes two successive samples, of sizes n_1 and n_2. Each sample is without replacement, but the two samples are independent. What is the distribution of the number of individuals the two samples have in common?

This double sampling is a technique of recognized value for wild-life surveys. Consider a geographically well-defined animal population, such as a lake containing N fish. In a first sample of the lake n_1 fish are caught, marked and returned to the lake. They are given time to mix with their unmarked brethren and recover from this experience, and then a second sample of n_2 is taken. It is assumed that marked and unmarked fish are equally likely to be caught in this second sample. What is then the distribution of the number X of marked fish in the second sample? This is of interest because $n_1 n_2/X$ supplies an estimate of the population size N. (We derive this estimate by equating the proportion X/n_2 of marked fish in the second sample to the proportion n_1/N of marked fish in the population.)

The distribution of X is determined as

$$P(x) = P(X = x) = \frac{\binom{n_1}{x}\binom{N - n_1}{n_2 - x}}{\binom{N}{n_2}} \qquad (0 \leq x \leq \min(n_1, n_2)), \qquad (38)$$

by a direct combinatorial argument. The denominator in (38) gives the number of ways in which the second sample can be chosen, all equally likely. The numerator gives the number of these which choose x of the marked fish and $n_2 - x$ of the unmarked ones.

Distribution (38) is the *hypergeometric distribution*, and from it one can make useful deductions. For example, the value of x maximizing this expression is approximately $n_1 n_2/N$, consistent with the estimate $\hat{N} = n_1 n_2/x$ of population size already suggested. This is also approximately the maximum likelihood estimate; the value of N maximizing expression (38).

However, the evaluation of expectations and moments from distribution (38) is impossible without some ingenuity, and this ingenuity is equivalent to an embedding and conditioning. Let us relax the problem in such a way that the sample sizes n_i themselves become r.v.s.

Suppose, indeed, that the first sample was taken, not with n_1 determined, but just by netting in a way that catches fish independently with probability $p_1 = 1 - q_1$. Suppose the second sample is taken in the same way, independently of the first and with catch probability p_2. The lake population then breaks into four classes: those not caught, those caught only in the first sample, those caught only in the second sample and those caught in both. The numbers in these categories are multinomially distributed with total N and respective probabilities $q_1 q_2, p_1 q_2, p_2 q_1$ and $p_1 p_2$. The observed numbers in these categories are $N - n_1 - n_2 + x, n_1 - x, n_2 - x$ and x. We can write down the multinomial expression for the probability of these values, and we can write down the expression

$$P(n_1, n_2) = \binom{N}{n_1} p_1^{n_1} q_1^{N - n_1} \binom{N}{n_2} p_2^{n_2} q_2^{N - n_2}$$

for the probability of the observed values of the independent binomial variables n_1 and n_2. We leave it to the reader to verify that the quotient of these two expressions, the conditional distribution of X for prescribed n_1 and n_2, reduces just to expression (38). Again, the irrelevant parameters p_1 and p_2 drop out.

This may seem a rather contorted way of deducing distribution (38), for which we already have a quick combinatorial derivation. However, for one thing, the second view may be the more realistic. That is, that one sets out with a fixed *probability* of catching fish rather than with a fixed *sample size* in mind. It is a question of a *stopping rule*: the rule that one fishes for a given time may be more realistic than the rule that one fishes until a required sample size has been attained.

The other point is that it is the second view which offers the natural path to the calculation of expectations; see Exercise 2.

EXERCISES AND COMMENTS

1. Suppose that an article has a lifetime S which is an integer-valued r.v. with distribution $P(S = s) = p_s$ $(s = 1, 2, \ldots)$. Suppose that the article has survived to age r without failing. The conditional probability that it survives for a further time s is then

$$P(S = r + s | S > r) = \frac{P(S = r + s)}{P(S > r)} = \frac{p_{r+s}}{\sum_{u > r} p_u} \qquad (s = 1, 2, 3, \ldots).$$

If this were equal to p_s for all relevant r, s, then the distribution would be 'memoryless' in that the distribution of residual lifetime at any age would be independent of age. Show that the only memoryless distribution is the geometric distribution (19).

2. Consider the capture–recapture example. We may as well suppose that $p_1 = p_2 = \frac{1}{2}$, since the values of the p's are irrelevant for the final conditioned X-distribution. The joint p.g.f. of the sample sizes N_1 and N_2 and the overlap X in the 'free' sampling case is then

$$E(w_1^{N_1} w_2^{N_2} z^X) = \left[\frac{1 + w_1 + w_2 + w_1 w_2 z}{4}\right]^N$$

and the p.g.f. of X conditional on prescribed values n_1 and n_2 of sample sizes is

$$\Pi(z) = E(z^X) \propto \text{coefficient of } w_1^{n_1} w_2^{n_2} \text{ in } (1 + w_1 + w_2 + w_1 w_2 z)^N.$$

Deduce then that

$$E(X^{(v)}) = \frac{n_1^{(v)} n_2^{(v)}}{N^{(v)}}.$$

3. Analogously, the p.g.f. of the numbers R_k of cells with k-fold occupation $(k = 0, 1, 2, \ldots)$ in the spatial model is

$$E\left(\prod_k z_k^{R_k}\right) \propto \text{coefficient of } w^N \text{ in } \left(\sum_k z_k w^k / k!\right)^M$$

and this relation is equivalent to the 'Poisson embedding and conditioning' of the text. By calculations analogous to those of Exercise 2, show that

$$E\left(\prod_k R_k^{(v_k)}\right) = M^{(a)} N^{(b)} \frac{(M - a)^{N-b}}{M^N} \prod_k (k!)^{-v_k},$$

where $a = \sum_k v_k$, $b = \sum_k k v_k$.

4. Consider the sampling example of Section 2.7. Let $S_n = \sum_1^n \xi_j$ be the sum of sample

X-values. Then we know that

$$E(e^{\alpha S_n}) = (A(e^{\alpha X}))^n$$

in the case of sampling with replacement (see Exercise 2.9.5). Show that, in the case of sampling without replacement,

$$\sum_{n=0}^{N} \binom{N}{n} w^n E(e^{\alpha S_n}) = \prod_{k=1}^{N} (1 + we^{\alpha x_k}) = \exp\{NA[\log(1 + we^{\alpha X})]\}, \qquad (39)$$

so that $E(e^{\alpha S_n})$ is proportional to the coefficient of w^n in the expansion of these expressions in powers of w. Hence confirm the formulae for the mean and variance of the sample mean S_n/n deduced in Section 2.7.

Use of identity (39) corresponds to an embedding in which one allows allows sample size n to be binomial, i.e. individuals are chosen independently for the sample, each with a fixed probability.

5. *Occupancy statistics.* Almost a last gasp for the spatial model. Suppose that the p.g.f. of the occupation numbers N_j of the cells is proportional, for prescribed N, to the coefficient of w^N in the expansion of a generating function $\Phi(wz_1, wz_2, \ldots, wz_M)$. Thus Φ is a generating function for a more relaxed model in which N is allowed to vary.

The case we have been working with is effectively that for which $\Phi(z_1, z_2, \ldots, z_M) = \exp(\sum_j z_j)$, and the N_j are independent Poisson variables in the 'free' model. This assumption embodies what are known as *Boltzmann statistics*; the molecules are identifiable and placed independently.

The choice $\Phi = \prod_j (1 - z_j)^{-1}$ corresponds to what is known as *Bose–Einstein statistics*; molecules are not identifiable in that configurations which differ only by a permutation of the molecules are counted as a single realization. (That is, they cannot be distinguished at any level of description.) In the 'free' model the N_j are independent geometric variables (although with a distribution starting from 0 rather than from 1). We shall encounter this case in some of the queueing and statistical–mechanical models of Chapter 10.

The choice $\Phi = \prod_j (1 + z_j)$ corresponds to what is known as *Fermi–Dirac statistics*. A cell can be occupied by one molecule at most. In the 'free' model, the N_j are IID variables taking only the values 0 or 1.

8. Variables on the Continuum: the Exponential and Gamma Distributions

If we take our spatial model to both the continuum limit and the thermodynamic limit then we have a model in which molecules are distributed uniformly and independently throughout infinite space at density ρ. Such a model is termed a *Poisson process*. It can serve as a model for many situations other than the distribution of molecules (e.g. the distribution of plants in space, or of events in time) if only the assumptions of uniformity and independence are satisfied. Let us then use the more neutral term 'particles' rather than 'molecules'.

The immediate property of a Poisson process is that which we have already stated in Exercise 4.4.4, and which is easily proved by the methods of Theorem 4.2.2.

Theorem 4.8.1. *Consider a Poisson process of density ρ. Consider disjoint regions A_j of respective volumes V_j, and let N_j be the number of particles in A_j. Then the N_j are independent Poisson variables with respective expectations ρV_j.*

We have not considered the dimensionality of the space in which the particles are distributed. Up to now it has not mattered, and we have even done some violence to dimensionality by supposing, in the discussions of Exercises 4.1.2 and 4.2.5, that the cells could be linearly ordered. However, we could indeed consider a one-dimensional version which represents the distribution of events in time rather than of particles in space. In this case, one tends to denote the 'density of events' by λ rather than ρ, and usually refers to this as the *rate* or the *intensity* of the process. One uses the term 'event' in this context to denote simply something which happens at a point in time, such as a birth, an accident or a click on a Geiger counter—this is not quite the use of the term in the technical sense of Section 3.1.

Let us, in particular, consider events on the nonnegative time axis $t \geq 0$. Let S_r denote the time of occurrence of the rth event after time zero.

Theorem 4.8.2. *Consider a Poisson process in time of rate λ. The time elapsing until the occurrence of the first event after some prescribed time is a r.v. with probability density*

$$f(s) = \lambda e^{-\lambda s} \qquad (s \geq 0). \tag{40}$$

PROOF. Let $N(s)$ be the number of events in the time interval $0 < t \leq s$. Then $N(s)$ is Poisson distributed with parameter λs. Further,

$$P(S_1 > s) = P(N(s) = 0) = e^{-\lambda s}. \tag{41}$$

Differentiating relation (41) with respect to s we obtain (40). (We appeal to the fact that $f(s) = -(d/ds)P(S_1 > s)$, see Exercise 2.4.4. It is understood that $f(s)$ is a density with respect to Lebesgue measure on the line.) $\qquad\square$

The distribution with density (40) is the *exponential distribution*. It is the continuous-time equivalent of the geometric distribution (19).

The distribution of S_r (which must be the continuous-time equivalent of the negative binomial distribution of Exercise 4.4.2) is obtained almost as easily.

Theorem 4.8.3. *Consider a Poisson process in time of rate λ. The time elapsing until the occurrence of the rth event after some prescribed time is a r.v. with probability density*

$$f(s) = \frac{\lambda^r e^{-\lambda s} s^{r-1}}{(r-1)!} \qquad (s \geq 0). \tag{42}$$

PROOF. The analogue of relation (41) is

$$P(S_r > s) = P(N(s) < r) = \sum_{j=0}^{r-1} e^{-\lambda s}(\lambda s)^j/j!.$$

On differentiating this relation with respect to s we obtain an expression for $f(s)$ which reduces gratifyingly to (42).   □

The distribution with density (42) is the *gamma distribution*, parameterized by the scale parameter λ and the shape parameter r. It is a unimodal distribution with a maximum at $s = (r - 1)/\lambda$. As one might have expected from the discussion of negative binomial distribution in Exercise 4.2.2 it is the distribution of the sum of r independent exponential variables; see Exercises 2 and 3.

We shall return to the Poisson process in Section 10.3, tackling it directly as a continuous-time process rather than as the limit of the discrete cellular process. One can regard it as the continuous-time version of a sequence of independent Bernoulli trials (i.e. trials with only two outcomes). In Section 10.10 we shall discuss the general continuous-time analogue of a sequence of IID trials.

EXERCISES AND COMMENTS

1. The statement in p.g.f. form of Theorem 4.8.1 is

$$E\left(\prod_j z_j^{N_j}\right) = \exp\left(\rho \sum_j V_j(z_j - 1)\right),$$

or

$$E\left\{\exp\left(\sum_j \alpha_j N_j\right)\right\} = \exp\left(\rho \sum_j V_j(e^{\alpha_j} - 1)\right).$$

A formal limit version of this is

$$E\left\{\exp\left(\int \alpha(x)N(dx)\right)\right\} = \exp\left(\rho \int (e^{\alpha(x)} - 1)\, dx\right), \tag{43}$$

where $N(A)$ is the number of particles in a region A and x is the coordinate and dx the infinitesimal volume element in the space considered. Relation (43) might be considered as characterizing the Poisson process.

2. Prove that, for the Poisson process in time, the r.v.s $S_1, S_2 - S_1, S_3 - S_2, \ldots$ are IID.

3. Confirm from expression (42) that

$$E(e^{\alpha S_r}) = \left(\frac{\lambda}{\lambda - \alpha}\right)^r, \tag{44}$$

and note the consistency of this result with Exercise 2.

4. Consider the spatial Poisson process with density ρ and in d dimensions, say. Let $A(r)$ denote a ball of radius r centred on the origin. The probability that this is empty is $\exp(-\rho K_d r^d)$, where K_d is the volume of the unit ball in d dimensions.

Show that the distance S_1 from the origin to the nearest particle has density

$$f(s) = (\rho d K_d) s^{d-1} \exp(-\rho K_d s^d).$$

Show that this implies that the volume of the ball with radius S_1 is exponentially distributed with parameter ρ (and so expectation $1/\rho$).

5. Continue with the assumptions of Exercise 4. Let V_j be the volume of the spherical shell centered on the origin and with inner and outer radii S_{j-1} and S_j, where $S_0 = 0$ and S_j is the distance from the origin to the jth nearest particle to the origin. Show that V_1, V_2, V_3, ... are distributed independently as exponential variables with parameter ρ.

CHAPTER 5

Conditioning

1. Conditional Expectation

The idea behind *conditioning* is that a partial observation can give one some idea of where in sample space the ω specifying the realization must lie: in a set A, say. Then, in effect, the sample space has contracted: instead of taking expectations over the full space Ω, one does so only over A.

For example, suppose that a doctor is to examine a patient, and that ω represents 'the patient's physical condition', as revealed by a standard medical examination. We suppose that the doctor has no foreknowledge of the patient, so that, if Ω is the sample space representing the whole range of bodily state he encounters in his practice, then expectations must be taken over Ω before examination.

However, suppose that as soon as the patient enters the room the doctor notices that he suffers from shortness of breath, with its implications of damage to lungs or heart. In effect, the doctor has performed a partial observation, and the patient is no longer 'any patient' (with ω in Ω) but 'a patient with shortness of breath' (with ω in A, if A is the event 'suffers from shortness of breath'). The doctor's expectations will now be changed, because he is dealing with a more specific case. For example, his expectation of the r.v. 'extent of lung damage' will certainly have increased, because patients in set A are known to have greater lung damage on average than patients in general.

So, the doctor would carry out further tests, narrowing down the potential sample space every time, until at last a confident diagnosis is possible, and ω has been sufficiently determined for the particular patient.

We shall now define the *expectation of the r.v. X conditional on the event* A as

$$E(X|A) = \frac{E[XI(A)]}{E[I(A)]} = \frac{E[XI(A)]}{P(A)}, \tag{1}$$

where $I(A)$ is, as ever, the indicator function of the set A. Definition (1) is reasonable. The inclusion of $I(A)$ in the numerator means that X is averaged only over A, the more specific set to which ω is known to belong. However, in contracting the sample space we must renormalize the expectation, in order to retain the property of an average, $E(1|A) = 1$. We therefore divide through by $P(A)$.

Operation (1) corresponds exactly to what one does when one calculates a partial average; for the census data of section 1.3, for example. So, if X is income, then the average income for the community is $\sum_k n_k X(\omega_k)/(\sum_k n_k)$ where k runs over the census categories. However, suppose that we are interested in 'the average income for males', so that the conditioning event A is 'male sex'. Then the sums in numerator and denominator would be restricted to male categories, which is exactly what we are doing in (1).

The example makes it clear that the conditioning event must be defined as a subset of Ω. For the census example one could not calculate 'average male income' if sex had not been recorded in the census, because the number of males in each category would then be indeterminate. Also, it is necessary that there should be *some* males, i.e. $\sum_{\text{male}} n_k > 0$, otherwise the partial average in meaningless, and, indeed, takes the indeterminate form 0/0. The corresponding condition in (1) is $P(A) > 0$, and we shall retain this condition for the next section or two. However, one does in fact run quite quickly into situations where it is not only reasonable, but also necessary, to consider conditioning events of zero probability. This will force us to a more generally valid characterization of a conditional expectation. However, for the moment the obvious definition (1) will do.

Let us consider some examples. One conditional expectation that interests us all is 'expectation of life'. Suppose the length of a human life can be regarded as a r.v. with probability density $f(x)$. A person has survived to age a: how much longer can he expect to live? The conditioning event is that the person has survived to age a, equivalent to $X > a$. The time he has yet to live is $X - a$, and the conditional expectation of this quantity is the 'expected residual life at age a', equal to

$$E(X - a|X > a) = \frac{\int_a^\infty (x - a)f(x)\,dx}{\int_a^\infty f(x)\,dx}$$

by definition (1).

Suppose that R and S are two r.v.s with joint p.g.f. $\Pi(w, z) = E(w^R z^S)$. Then one sees from (1), with a little thought, that the p.g.f. of S conditional on a value r of R is

$$E(z^S|R = r) = \frac{\text{coefficient of } w^r \text{ in } \Pi(w, z)}{\text{coefficient of } w^r \text{ in } \Pi(w, 1)}. \tag{2}$$

A conditional p.g.f. implies a conditional distribution, and we have already evaluated such conditional distributions. For example, in Section 4.7 we found that the distribution of independent Poisson variables conditional on the value of their total was multinomial. In that section, we also considered the composition of a pair of independent samples of a fish population, drawn with a fixed catch probability per fish, and found that the distribution of the overlap in the two samples conditional on sample sizes was hypergeometric.

The principal formal properties of a conditional expectation are summed up in

Theorem 5.1.1.

(i) *If* $P(A) > 0$ *then the conditional expectation operator* $E(\cdot|A)$ *has the properties demanded of an expectation operator in the axioms of Section 2.2.*

(ii) *If* $\{A_k\}$ *is a decomposition of* Ω *then*

$$E(X) = \sum_k P(A_k)E(X|A_k). \tag{3}$$

The proof of the first part is direct. The second assertion follows from the fact that the right-hand member of equation (3) equals

$$\sum_k E[XI(A_k)] = E\left[X\sum_k I(A_k)\right] = E(X),$$

the last relation following because $\sum_k I(A_k) = 1$.

Relation (3) is basic, and can be regarded as at least a partial inverse to (1). Whereas (1) gives the rule for the transformation of E when the sample space is *contracted* from Ω to A, relation (3) gives the transformation when a sample space is *expanded* from individual spaces A_k to an over-all space Ω by averaging with respect to an extra variable. It can be regarded as a formula which contructs an expectation in two stages: first by averaging within an event A_k and then by averaging over a r.v. k which labels these disjoint events. This is an aspect which will recur.

For example, suppose electric lamps may be of various manufactures, and that the lifetime X for lamps of manufacture k has probability density

$$f_k(x) = \mu_k e^{-\mu_k x},$$

where we shall later see μ_k as a failure rate for the lamp (Exercise 4). Lamps of manufacture k thus have expected lifetime $E(X|A_k) = \mu_k^{-1}$. We have written this expectation as conditional on the event A_k that the lamp is of manufacture k.

Suppose now that one buys a lamp at random, and that this has probability $P(A) = \pi_k$ of being of manufacture k. Its expected lifetime is then, by relation (3),

$$E(X) = \sum_k \pi_k \mu_k^{-1}.$$

This expectation has been obtained in two stages: by averaging over the fluctuations in lifetime of a lamp of given manufacture and then by averaging over k, the different sources of manufacture.

There are various special cases of a conditional expectation. For example, the probability of an event B conditional on the occurrence of an event A is

$$P(B|A) = E[I(B)|A] = \frac{E[I(A)I(B)]}{E[I(A)]} = \frac{P(AB)}{P(A)}. \tag{4}$$

If for a scalar r.v. X there exists a function $f(x|A)$ for which

$$E[H(X)|A] = \int H(x)f(x|A)\, dx$$

for a sufficiently wide class of functions H, then $f(x|A)$ is the probability density function of X condition on A.

EXERCISES AND COMMENTS

1. Show for the example of Exercise 4.4.5 that

$$E(N_2|N_1 = n_1) = \rho V_2 + \frac{V_{12}}{V_1}(n_1 - \rho V_1).$$

2. What is the probability density of residual lifetime for an individual who has reached age a?

3. Consider an article whose lifetime X takes integral values 0, 1, 2, ... and define $p_j = P(X = j)$. If we define also

$$b_j = P(X = j + 1 | X > j) = \frac{p_{j+1}}{\sum_{k>j} p_k}, \tag{5}$$

then this is interpretable as the *failure probability* at age j: the probability that an article of age j fails in the next unit of time. Correspondingly, $a_j = 1 - b_j$ is interpretable as the *survival probability* at age j. Show from relation (5) that

$$P(X > j) = a_0 a_1 a_2 \ldots a_j \quad \text{and} \quad P(X = j) = a_0 a_1 \ldots a_{j-1} b_j. \tag{6}$$

The geometric distribution is characterized by a constant failure rate.

4. Consider the continuous-time analogue of Exercise 3; suppose that lifetime has a probability density $f(x)$. Then the probability of failure by age b given that the article has survived to age a ($<b$) is

$$P(a < X \le b | X > a) = \frac{\int_a^b f(x)\, dx}{\int_a^\infty f(x)\, dx}.$$

For $b - a$ small this will be of order $(b - a)$, say $(b - a)\mu(a) + o(b - a)$, where

$$\mu(a) = \frac{f(a)}{1 - F(a)}$$

could be regarded as the *failure rate* of the article at age a. (The term *hazard rate* is also used, and is of more general application.) Demonstrate the following relations, analogous to (6):

$$P(X > a) = \exp\left[-\int_0^a \mu(s)\,ds\right],\qquad(7)$$

$$f(x) = \mu(x)\exp\left[-\int_0^x \mu(s)\,ds\right].\qquad(8)$$

The exponential distribution is thus characterized by a constant hazard rate.

5. Suppose that lamps of manufacture k have constant failure rate μ_k, and that π_k is the probability that a randomly chosen lamp is of manufacture k. Show that for a randomly chosen lamp the failure rate at age x is

$$\mu(x) = \frac{\sum \pi_k \mu_k e^{-\mu_k x}}{\sum \pi_k e^{-\mu_k x}},$$

and the expected residual life at age x is

$$L(x) = \frac{\sum \pi_k \mu_k^{-1} e^{-\mu_k x}}{\sum \pi_k e^{-\mu_k x}}.$$

Show that these quantities are, respectively, decreasing and increasing functions of x. The summands for which the failure rate is small dominate for large x. This is 'survival of the fittest' (by attrition rather than by competition): those lamps which have survived long are most probably of good manufacture.

2. Conditional Probability

The conditional probability (4) is analogous to a partial proportion, just as a conditional expectation is analogous to a partial average. So, to return to the census example of Section 1.3, if A is the event 'male' and B the event 'employed', then $P(B)$ is analogous to 'the proportion of the population which is employed', but $P(B|A)$ is analogous to 'the proportion of the *male* population which is employed'.

The classic card-playing and dice-throwing examples abound in aspects which are usefully viewed conditionally. For example, suppose A is the event 'the first n draws from a pack of m_1 red and m_2 black cards ($m_1 + m_2 = m$) are all red' and B is the event 'the $(n + 1)$th draw is red'. Then $P(B|A)$ equals m_1/m or $(m_1 - n)/(m - n)$ according as sampling is with or without replacement. One can see this directly, or formally from (4), as we leave the reader to verify.

Note that we can rewrite formula (4) as

$$P(AB) = P(A)P(B|A).\qquad(9)$$

This can be read: 'The probability that A and B both occur equals the probability that A occurs times the probability of B conditional on the occurrence of A'. From the symmetry of $P(AB)$ with respect to A and B we see that formula (9) implies that

$$P(A|B) = \frac{P(A)P(B|A)}{P(B)}. \qquad (10)$$

A particular case of relation (3) is

$$P(B) = \sum_k P(A_k)P(B|A_k), \qquad (11)$$

where $\{A_k\}$ is a decomposition of Ω. This is sometimes known as the 'generalized addition law of probability', and in many cases provides the only way to calculate $P(B)$, in that it is the quantities in the right-hand member of (11) which are given. For, instance, take the lamp example with which we concluded the previous section, and consider the event B: that a randomly chosen lamp survives at least to age x. By (11) this has probability

$$P(B) = \sum_k \pi_k \int_x^\infty f_k(s)\, ds = \sum_k \pi_k e^{-\mu_k x}.$$

Combining (10) and (11) we see that

$$P(A_j|B) = \frac{P(A_j)P(B|A_j)}{\sum_k P(A_k)P(B|A_k)}. \qquad (12)$$

This result is known as *Bayes' Theorem*. Although mathematically trivial, it has become celebrated for its quasi-philosophical implications. We shall discuss these in a moment, but first we consider some of its straightforward implications.

Returning to the lamp example, we see that Bayes' formula (12) yields

$$P(A_j|B) = \frac{\pi_j e^{-\mu_j x}}{\sum_k \pi_k e^{-\mu_k x}}$$

as the probability that the lamp is of manufacture j, given that it has not yet failed at age x. This distribution will assign increasing weight as x increases to those values of j for which failure rate μ_j is small. A lamp that has lasted well is probably of good manufacture.

The quantities $P(A_j)$ and $P(A_j|B)$ are, respectively, known as the *prior* and *posterior* probabilities of A_j; prior and posterior, that is, to the observation of the event B.

One can equally well find use for Bayes' formula in the medical diagnosis problem with which we began this chapter. Suppose a patient may be suffering from various conditions, which we shall label A_j ($j = 1, 2, \ldots$). For simplicity, we shall assume these to be exhaustive and exclusive, so that $\{A_j\}$ is a decomposition of Ω. Suppose that for purposes of diagnosis the doctor carries out a test, with result B. Then formula (12) gives the posterior probability (i.e. the probability conditional on the test result B) that the patient is suffering from condition A_j. Here $P(A_j)$ is the prior probability of the same event, i.e.

the proportion of patients 'in general' who suffer from condition A_j. The probability $P(B|A_j)$ is the proportion of patients suffering from condition A_j for whom the test will give result B. One would like a completely decisive test: i.e. one for which $P(B|A_j)$ is unity for one condition and zero for all others. However, in practice the test will be less conclusive, and one can only hope that it will lead to a substantial sharpening of one's inferences: i.e. that the posterior distribution will be sensibly more concentrated than the prior.

Bayes' theorem has more generally been invoked in cases where one has a number of exhaustive and mutually exclusive hypotheses H_j, the event A_j is defined as 'H_j is true' and B is a piece of experimental evidence. Formula (12) then indicates how the probabilities of truth of the various hypotheses are changed when one takes the experimental evidence into account.

Other question is, whether one can meaningfully attach a probability to the event 'H_j is true'. In the medical case one could: one could interpret $P(A_j)$ as the proportion of people attending a doctor's surgery who suffer from the jth condition. This was also possible in the lamp: one interprets $P(A_j)$ as the proportion of lamps on sale which are of manufacture j. But suppose the hypotheses H_j are scientific hypotheses such as, e.g. 'the quark is an elementary particle'. In our universe the hypothesis is either true or false (or, possibly, meaningless), and there is no conceivable 'sample space' of universes in which such a statement would be sometimes true, sometimes false, and over which one could average. However, there is a second difficulty, which may be more fundamental: Can one list an exhaustive set of hypotheses? The actual formulation of a class of hypotheses that certainly includes the true one may be beyond current thinking, and will intrinsically be so for a research problem of any depth. That is, one may not even be able to conceive the hypotheses that will be needed, let alone form an exhaustive list. In such a case, discussion in these terms is meaningless.

EXERCISES AND COMMENTS

1. Show that $P(ABC) = P(A)P(B|A)P(C|AB)$ and generalize.

2. A family with two children can have any of the four constitutions bb, bg, gb and gg, where b = boy, g = girl, and account is taken of order. Suppose that these four possibilities are equally probable. Show that $P(bb|\text{elder child a boy}) = \frac{1}{2}$, but that $P(bb|\text{at least one boy}) = \frac{1}{3}$. One is included to think that these probabilities should be equal, since in both cases one is given the information that the family contains at least one boy, and the extra information given in the first case (that this boy can be labelled as the elder) seems irrelevant. However, the conditioning event contains two realizations in the first case and three in the second. If the conditioning and conditioned events are denoted A and B, then note that in both cases $AB = B$, since $B \subset A$.

3. Cards are dealt to four players, all 52! orders of dealing being equally likely. Show

that

$$P(\text{the first player holds all four aces}|\text{he holds the ace of hearts})$$

$$= \binom{48}{9} \bigg/ \binom{51}{12} = \frac{132}{12,495},$$

$$P(\text{the first player holds all aces}|\text{he holds at least one ace})$$

$$= \binom{48}{9} \bigg/ \left[\binom{52}{13} - \binom{48}{13} \right] = \frac{5}{1318}.$$

4. Consider the serial number example of Exercise 4.1.3. Suppose that the event B is the set of registration numbers observed (repetitions allowed) in a sample of fixed size n from a town in which cars are registered serially from 1 to M, if the town contains M cars. Suppose that M is a r.v. (i.e. one is calculating expectations over towns of varying sizes) and that the distribution $\pi_m = P(M = m)$ of town size is known. Show that

$$P(M = m|B) = \begin{cases} \pi_m m^{-n} \bigg/ \left(\sum_{k \geq u} \pi_k k^{-n} \right) & (m \geq u), \\ 0 & (m < u), \end{cases}$$

where u is the largest registration number observed in the sample.

5. *Detection problems.* Suppose that the event 'radiation leakage' occurs at a random time σ and that $P(\sigma = j) = \pi_j$ $(j = 0, 1, 2, \ldots)$. Suppose that at each instant j one makes an observation Y_j on radiation, where these observations are independent conditional on the value of σ, and have probability density $f_0(y)$ or $f_1(y)$ according as leakage has not or has occurred. Let $B_t(y)$ denote the event that the observation history up to time t is $(y_0, y_1, \ldots, y_t)$ i.e. that $Y_j = y_j$ $(0 \leq j \leq t)$. Show then that the distribution of σ conditional on this information is

$$P[\sigma = j|B_t(y)] = \begin{cases} c\pi_j \left(\prod_{i=0}^{j-1} f_0(y_i) \right) \left(\prod_{i=j}^{t} f_1(y_i) \right) & (j \leq t), \\ c\pi_j \prod_{i=0}^{t} f_0(y_i) & (j > t), \end{cases}$$

where c is a normalizing constant.

Suppose that π_j has the particular form

$$\pi_j = \begin{cases} \rho_0 & (j = 0), \\ (1 - \rho_0)(1 - p)(1 - p)^{j-1}p & (j > 0). \end{cases}$$

That is, conditional on $\sigma > 0$, the r.v. σ is geometrically distributed. Show then that this generalizes in that

$$P[\sigma = j|B_t(y)] = (1 - \rho_t)(1 - p)^{j-t-1}p \qquad (j > t),$$

where $\rho_t = P[\sigma \leq t|B_t(y)]$. That is, the posterior distribution of *future* values of σ always remains geometric. The quantity ρ_t is the probability, conditional on the observations at time t, that leakage has already occurred at time t. Show that it obeys the updating equation

$$\rho_{t+1} = \frac{[\rho_t + (1 - \rho_t)p]f_1(y_{t+1})}{[\rho_t + (1 - \rho_t)p]f_1(y_{t+1}) + (1 - \rho_t)(1 - p)f_0(y_{t+1})}.$$

3. A Conditional Expectation as a Random Variable

Very often the alternative events by which a r.v. may be conditioned correspond to the possible alternative values y of another r.v. Y. For example, X might be the income of a randomly chosen individual and Y the district in which the lives, the event $Y = y$ being an expression of the fact that he belongs to the district y.

In such a case it is natural to speak of the expectation of X conditional on Y, denoted $E(X|Y)$, this being regarded as the r.v. which takes the value $E(X|Y = y)$ with probability $P(Y = y)$.

The notation is perhaps confusing, in that $E(X|A)$ and $E(X|Y)$ are different objects. However, the two usages are distinguished by the fact that the conditioning argument is a set in one case and a r.v. in the other. Note, however, that X must be numerical-valued, whereas Y is unrestricted (or, rather, subject as yet to the condition of discreteness, but no other).

So, suppose indeed that we are taking an individual at random from a given finite population so that $P(Y = y)$ is in fact the proportion of the population living in district y, and $E(X|Y = y)$ is equal to $A(X|Y = y)$, the average income in district y.

If the sampled individual's income is recorded, then X is observable and known. However, if only the district y in which he lives is recorded, then the best guess one could make of his income would be $A(X|Y = y) = E(X|Y = y)$, the average income of that district. Thus $E(X|Y)$, as a random variable, is a kind of coarsened version of X itself: it is the best estimate one can make of X from knowledge of Y. In other words, it is a function of Y which approximates X in a sense we yet have to quantify.

Note that relation (3) can now be given the more pleasing form

$$E(X) = E[E(X|Y)]. \tag{13}$$

That is, the expectation over X can be seen as taken in two stages: first as an expectation over X conditional on Y, and then as an expectation over Y.

However, as soon as one introduces the ideal of conditioning by the value of a random variable Y, then one runs into problems. If Y is continuously distributed then the probability $P(Y = y)$ is zero for any prescribed y, and the conditional expectation defined by (1) takes the indeterminate form $0/0$. One encounters this situation, if, for example, one wishes to investigate the distribution of wind-speed for winds in a given direction, or of the maximum height reached by a sounding rocket (which means the distribution of rocket height X given that vertical velocity Y is zero).

One's natural approach is to consider the conditioning event $Y = y$ as the limit of the event $|Y - y| < h$ as h tends to zero. One's next discovery is then that, in this sense, the event $Y = y$ is not the same as the event $Y/X = y/X$, for example (see Exercise 4). So, there seems to be a severe ambiguity in the definition of $E(X|Y)$ in general cases, and an alternative view is needed to resolve it.

We shall adopt the following definition, which is wholly in terms of expectations, but which has become accepted as the standard and natural characterization. The conditional expectation $E(X|Y)$ of a scalar r.v. X is defined as that scalar function of Y which satisfies

$$E\{[X - E(X|Y)]H(Y)\} = 0 \tag{14}$$

for any scalar function $H(Y)$ of Y. This is a way of saying that, as a r.v., $E(X|Y)$ is indistinguishable from X in its interaction with Y. A firmer interpretation follows.

Theorem 5.3.1.

(i) *The characterization (14) agrees with definition (1) in the case when Y is discrete-valued.*

(ii) *If $E(X^2) < \infty$ then $E(X|Y)$ can be interpreted as the LS approximant to X in terms of Y: i.e. the function $\psi(Y)$ of Y which minimizes*

$$D = E\{[X - \psi(Y)]^2\}. \tag{15}$$

(iii) *At least in the case $E(X^2) < \infty$, condition (14) is self-consistent in that it possesses a solution, and this solution is unique in that all solutions are m.s. equivalent.*

Thus, we have the very definite characterization of $E(X|Y)$ as the function of Y which approximates X best in mean square (m.s.). Note, however, that it is in general a *nonlinear* least square approximant; the class of candidate functions ψ in (15) is unrestricted. Nevertheless, the ideas of LLS approximation developed in Sections 2.8 and 2.9 largely carry over.

PROOF. Consider first the case of discrete-valued Y. Taking $H(Y)$ as the indicator function $I(Y = y)$ in (14) we deduce that $E(X|Y)$ takes the value

$$E(X|Y) = \frac{E[XI(Y = y)]}{E[I(Y = y)]} = E(X|Y = y)$$

when $Y = y$. Definitions (1) and (14) are thus mutually consistent.

Functions $\psi(Y)$ minimizing the criterion (15) certainly exist; let us denote any one of them by $E(X|Y)$. By requiring that criterion (15) be stationary under a perturbation of this solution to $E(X|Y) + \varepsilon H(Y)$ for small ε we deduce the necessity of relation (14). Conversely, relation (14) is sufficient to establish the m.s. minimizing character of $E(X|Y)$, for, by setting $H(Y) = \psi(Y) - E(X|Y)$ in (14), we deduce that

$$E\{[X - \psi(Y)]^2\} = E\{[X - E(X|Y)]^2\} + E\{[\psi(Y) - E(X|Y)]^2\}. \tag{16}$$

Finally, by setting $\psi(Y)$ equal in (16) to any other evaluation of $E(X|Y)$, i.e. to any other function minimizing D, we deduce that all such evaluations have zero m.s. difference, and so are m.s. equivalent. $\square$

Even in the case of discrete-valued Y there may be some indeterminacy in the form of $E(X|Y)$; we could assign $E(X|Y)$ any value at $Y = y$ if $P(Y = y) = 0$. However, this indeterminacy has no significance, since the value y almost certainly never occurs.

Condition (14) is a nonlinear version of the orthogonality relation (Exercise 2.8.1) that characterized LLS estimates. One can also state this condition and a generalization of Theorem 5.3.1 for vector X: see Exercise 4. In a sense this adds little, in that the conditional expectation of the vector is just the vector of conditional expectations of the components. However, it will generally be understood henceforth that the conditioned variable may be vector-valued.

The conditional expectation has a number of important properties; these must all be consequences of the characterizing relation (14) alone.

Theorem 5.3.2. *Let* X, Y_1, Y_2, ... *be random variables, and suppose that* $E(X^2) < \infty$. *Then all the following assertions (of equality, etc.) hold in a m.s. sense:*

(i) *The conditional expectation has all the properties required of an expectation by the axioms of Section 2.2.*

(ii) *The following relation holds for iterated conditional expectations:*

$$E[E(X|Y_1, Y_2)|Y_1] = E(X|Y_1). \tag{17}$$

(iii) *If* $E(X|Y_1, Y_2, Y_3)$ *turns out to be a function* $\psi(Y_1)$ *of* Y_1 *alone, then* $\psi(Y_1)$ *can also be interpreted as* $E(X|Y_1)$ *or* $E(X|Y_1, Y_2)$.

PROOF. The verification of fulfilment of the axioms is not completely straightforward, because the properties hold in a m.s. sense rather than unqualifiedly. Suppose $X \geq 0$. Let $\psi(Y)$ be any possible evaluation of $E(Y|X)$ and let $\psi_-(Y)$ be the negative part of $\psi(Y)$, i.e. the function that is equal to $\psi(Y)$ or zero according as $\psi(Y)$ is negative or positive. Set $H(Y)$ equal to $\psi_-(Y)$. We deduce then from (14) that $E[\psi_-(Y)^2] \leq 0$. Thus equality must hold, so that ψ_- is zero in m.s., or $\psi(Y) \geq 0$ in m.s., confirming fulfilment of Axiom 1.

One sees readily from (14) that $\sum_j c_j E(X_j|Y)$ is a possible evaluation of $E(\sum_j c_j X_j|Y)$, and it follows from Theorem 5.3.1(iii) that any other solution is m.s. equivalent to it. Correspondingly, 1 is a possible evaluation of $E(1|Y)$ and any other evaluation is m.s. equivalent to it. Axioms 2–4 are thus confirmed.

For a version of Axiom 5, let $\{X_j\}$ be a sequence of r.v.s increasing to X. It follows then from the equivalent of Axiom 1 that $E(X_j|Y)$ is increasing in j and is not greater than $E(X|Y)$. Thus

$$\Delta(Y) = E(X|Y) - \lim E(X_j|Y)$$

is nonnegative (all assertions being understood in a m.s. sense). It follows from (14) that

$$E\{[X_j - E(X_j|Y)]\Delta(Y)\} = 0,$$
$$E\{[X - E(X|Y)]\Delta(Y)\} = 0.$$

But $X_j \Delta(Y)$ is increasing in j, whence it follows from Axiom 5 that $E[X\Delta(Y)] = \lim E[X_j \Delta(Y)]$. Taking the difference of the two equations above and letting j tend to infinity we deduce that $E[\Delta(Y)^2] = 0$, so that $E(X|Y)$ equal $\lim E(X_j|Y)$ in m.s. Assertion (i) is thus proved.

To prove assertion (ii), denote possible evaluations of $E(X|Y_1)$, $E(X|Y_1, Y_2)$ and $E[E(X|Y_1, Y_2|Y_1])$ by $\psi_1(Y_1)$, $\psi_2(Y_1, Y_2)$ and $\phi(Y_1)$, respectively. Condition (14) then yields the relations

$$E[(X - \psi_1)H_1(Y_1)] = 0, \qquad E[(\psi_2 - X)H_2(Y_1, Y_2)] = 0,$$

$$E[(\phi - \psi_2)H_3(Y_1)] = 0.$$

Set $H_1 = H_2 = H_3 = \phi - \psi_1$ and add these relationships. We then obtain $E[(\phi - \psi_1)^2] = 0$, which implies the m.s. validity of (17).

In assertion (iii) the hypothesis is that there is a function $\psi(Y_1)$ which satisfies

$$E\{[X - \psi(Y_1)]H(Y_1, Y_2, Y_3)\} = 0$$

By characterization (14) we see $\psi(Y_1)$ as $E(X|Y_1, Y_2, Y_3)$. But since we could take H as an arbitrary function of Y_1 and Y_2 alone, or of Y_1 alone, the other two identifications follow.   □

EXERCISES AND COMMENTS

1. Note that in the present notation we could write the assertion of Exercise 5.1.1 simply as

$$E(N_2|N_1) = \rho V_2 + \frac{V_{12}}{V_1}(N_1 - \rho V_1),$$

and that this is indeed a r.v. and a function of N_1.

2. To show that the elementary definition of conditional expectation can gives ambiguous results in the limit of a conditioning event of zero probability, supose that X and Y are scalars with joint probability density $f(x, y)$. Suppose we interpret the event $Y = y$ as the limit as h tends to zero of the event $A(h)$ that $|Y - y| < ha(X)$ for some prescribed nonnegative function $a(X)$. The conditional expectation

$$E[X|A(h)] = \frac{E\{XI[A(h)]\}}{E\{I[A(h)]\}}$$

then has the evaluation

$$E[X|A(0)] = \frac{\int xa(x)f(x, y)\, dx}{\int a(x)f(x, y)\, dx}$$

in the limit of small h. But this evaluation then depends upon $a(x)$, so that the manner of approach to the event $Y = y$ radically affects the limit value of the conditional expectation.

3. Consider a distribution uniform on the surface of the unit sphere, in that the probability that a sample point lies in a given region is proportional to the area of that region. The sample point will have Cartesian coordinates $x = \cos \phi \cos \theta$, $y = \cos \phi \sin \theta$ and $z = \sin \phi$ ($-\frac{1}{2}\pi \leq \phi \leq \frac{1}{2}\pi; 0 \leq \theta \leq 2\pi$), if ϕ and θ are its angles

of latitude and longitude. Show that if the point is constrained to lie on the meridian $\theta = 0$, then its ϕ coordinate has conditional density $\frac{1}{2} \cos \phi$ or $1/\pi$ according as the meridian is specified by 'θ arbitrarily small' or 'y arbitrarily small'.

4. Suppose that X is a vector with elements X_j, and define $E(X|Y)$ as the vector of conditional expectations $E(X_j|Y)$. Show that $E(X|Y)$ satisfies (14), and also the matrix equivalent of (16):

$$E\{[(X - \psi(Y)][X - \psi(Y)]^T\} = E\{[X - E(X|Y)][X - E(X|Y)^T\}$$
$$+ E\{[\psi(Y) - E(X|Y)][\psi(Y) - E(X|Y)]^T\}.$$

This vector extension is useful, but introduces no new element, in that the evaluation of $E(X|Y_1)$ is unaffected by the fact that $E(X|Y_2)$ is also being evaluated.

4. Conditioning on a σ-Field

The measure-theoretic approach to probability leads one to speak of conditioning with respect to a σ-field of sets (events) rather than to values of a random variable. In this vector we shall merely indicate how our language would translate into this.

The sets in Ω that one can consider are those belonging to some basic σ-field $\mathscr{F}$. The r.v. Y is $\mathscr{F}$-measurable if the ω-sets corresponding to sets of constant Y belong to $\mathscr{F}$. The σ-field generated from these sets is then a σ-field $\mathscr{F}_1$ which is a coarsened version of $\mathscr{F}$; we express this by writing $\mathscr{F}_1 \subset \mathscr{F}$.

What we spoke of as 'a function of Y' will now be spoken of as 'an $\mathscr{F}_1$-measurable r.v.', and the conditional expectation which we wrote as $E(X|Y)$ will now be written $E^{\mathscr{F}_1}(X)$. The definition of $E(X|Y)$ associated with (14) will now rather be expressed: $E^{\mathscr{F}_1}(X)$ is the $\mathscr{F}_1$-measurable r.v. for which

$$E\{[X - E^{\mathscr{F}_1}(X)]Z\} = 0$$

for any $\mathscr{F}_1$-measurable Z.

Assertion (ii) of Theorem 5.4.2 will now be phrased as

$$E^{\mathscr{F}_1}[E^{\mathscr{F}_2}(X)] = E^{\mathscr{F}_1}(X)$$

if $\mathscr{F}_1 \subset \mathscr{F}_2$. Assertion (iii) would be restated as saying that, if $E^{\mathscr{F}_2}(X)$ turns out to be $\mathscr{F}_1$-measurable, where $\mathscr{F}_1 \subset \mathscr{F}_2$, then it can also be interpreted as $E^{\mathscr{G}}(X)$ for any $\mathscr{G}$ such that $\mathscr{F}_1 \subset \mathscr{G} \subset \mathscr{F}_2$.

5. Independence

We have already introduced the idea of independence in Sections 4.1 and 4.3: r.v.s $X_1, X_2, \ldots X_N$ are independent if

$$E\left[\prod_k H_k(X_k)\right] = \prod_k E[H_k(X_k)] \tag{18}$$

for all functions H_k for which the right-hand member of (18) can be given a meaning. This definition seems natural enough, but we can now justify it in terms of conditional expectations.

Theorem 5.5.1. *If X and Y are independent then $E[H(X)]$ is a possible evaluation of $E[H(X)|Y]$. If also $E[H(X)^2] < \infty$, then $E[H(X)|Y] \overset{m.s.}{=} E[H(X)]$.*

PROOF. We characterize $E[H(X)|Y]$ as the function of Y which satisfies

$$E\{(H(X) - E[H(X)|Y])G(Y)\} = 0$$

for any G. Under the assumption of independence this is satisfied by $E[H(X)|Y] = E[H(X)]$. If $E(H^2) < \infty$ then we know from Theorem 5.4.1 (iii) that any other evaluation is m.s. equivalent to this.   □

That is, distribution of X conditional on Y is essentially the same as its unconditioned distribution, which is indeed what one would understand by independence. The original characterization (18) has the merit that it treats the variables symmetrically, however, and also makes no appeal to the concept of conditioning.

The further examples of Chapter 4 will have sufficiently emphasized the importance of the independence concept. As well as formulating a natural and essential idea, it provides a powerful tool for building up interesting processes from simple elements, e.g. the sequences of IID trials considered in Chapter 4.

With the formalization of the independence concept we can now attack problems of genuine scientific interest. The theory of Mendelian inheritance provides examples of models which can be simple without being idealized to the point where they have no practical value.

The gene is the unit of heredity, and in the simplest cases, genes occur in pairs: each gene of a particular pair can assume two forms (*alleles*), A and a. There are then, with respect to this gene, three types of individual (*genotype*): AA, Aa and aa. The pure genotypes AA and aa are termed *homozygotes*, the mixed one Aa, a *heterozygote*. If A *dominates* a then the heterozygote Aa will be outwardly indistinguishable from the homozygote. AA; there are then only two outwardly distinguishable types (*phenotypes*) (AA or Aa) and aa. For example, brown eyes are dominant over blue, so that an individual with a 'blue' and a 'brown' gene will have brown eyes. (The situation is actually more complicated, but such a model is a first approximation to it.)

An individual receives one gene at random from each of its parents, i.e. it receives a given maternal gene and a given paternal gene with probability $\frac{1}{4}$ for each of the four possibilities. Thus the mating $AA \times Aa$ would yield progeny of types AA or Aa, each with probability $\frac{1}{2}$; the mating $Aa \times Aa$ would yield AA, Aa or aa with respective probabilities $\frac{1}{4}$, $\frac{1}{2}$ and $\frac{1}{4}$.

Suppose now that we have a large population, and that the proportions of genotypes AA, Aa and aa in the nth generation are p_n, q_n and r_n, respectively. The point of assuming the population large is that p_n can then be equated with the probability that a randomly chosen individual is an AA and p_n^2 with the probability that a randomly chosen *pair* are both AA, etc.

Assume now that mating takes place at random, i.e. the two parents are chosen randomly and independently from the population. Thus, the probability of an $AA \times Aa$ mating in the nth generation would be $2p_n q_n$, etc. The probability that the offspring from such a mating is an AA is then

$$p_{n+1} = p_n^2 + 2p_n q_n(\tfrac{1}{2}) + q_n^2(\tfrac{1}{4}) = (p_n + \tfrac{1}{2}q_n)^2 = \theta_n^2,$$

say, where θ_n is the proportion of A genes in the nth generation. We leave it to the reader to verify similarly that

$$q_{n+1} = 2\theta_n(1 - \theta_n), \qquad r_{n+1} = (1 - \theta_n)^2,$$

so that

$$\theta_{n+1} = p_{n+1} + \tfrac{1}{2}q_{n+1} = \theta_n.$$

That is, the gene frequency θ_n stays constant from generation to generation (at θ, say), and, after one generation of random mating, the genotype frequencies become fixed at θ^2, $2\theta(1 - \theta)$ and $(1 - \theta)^2$. Hence

$$4pr = q^2$$

(the *Hardy–Weinberg law*).

EXERCISES AND COMMENTS

1. Consider the genetic example of the text. If A is dominant and aa is regarded as 'abnormal', show that the probability that the first child of normal parents is abnormal is $[(1 - \theta)/(2 - \theta)]^2$. (We assume that mating has been random in previous generations.)

2. *Controlled overbooking.* Suppose that an airline allows n passengers to book on a flight with a maximum capacity of m ($n \geq m$). Each passenger who flies brings the airline revenue a; each booked passenger for whom there is no place at flight costs the airline an amount b. Passengers have a probability $1 - p$ of cancelling their flight, independently of one another, so that X, the number of passengers finally wishing to fly, is binomially distributed with parameters n and p. Show that the value of n maximizing the expected revenue is the smallest value for which

$$P(X < m) \leq \frac{b}{a + b}.$$

This looks similar to the solutions of Section 2.6, but the model requires an appeal to independence.

3. The number of printing errors in the proofs of an article is Poisson with parameter λ. (We suppose the article long enough and the incidence of errors low enough that this assumption is realistic.) A proofreader corrects the text, but has a probability p of missing any given error, independently of other events. Show that the number of residual errors is Poisson with parameter λp. If the proofreader discovers n_1 and n_2 errors on a first and second reading, show that a reasonable estimate of the number of errors remaining after the second reading is $n_2^2/(n_1 - n_2)$.

4. Continuing the example of Exercise 3, suppose that the number of original errors had p.g.f. $\Pi(z)$. Show that, under the detection assumptions stated, the number of errors remaining after one reading has p.g.f. $\Pi(pz + q)$, where $q = 1 - p$.

5. Continuing yet further, suppose that the proofreader and setter actually introduce new errors after every revision, the number r of new errors being independent of earlier events and with p.g.f. $\psi(z)$. Show that the p.g.f. of the number of errors remaining after one revision is $\psi(z)\Pi(pz + q)$. It then follows formally that, if repeated revisions are carried out, the distribution of the number of residual errors R will settle down to a limit form with p.g.f.

$$E(z^R) = \prod_{j=0}^{\infty} \psi[1 + p^j(z - 1)].$$

One would expect this formal result to be valid if $p < 1$, i.e. if the reader manages to correct some positive proportion of existing errors.

6. Statistical Decision Theory

Suppose that a situation depends upon a number of random variables which we can partition into (X, Y), where Y denotes just those variables which have been observed, and so whose values are known. Suppose also that one can take an action or decision, denoted by u, and that one suffers a cost $L(u, x, y)$ if one takes action u and the actual values of X and Y are x and y. The aim is to choose one's action in such a way as to minimize expected cost.

For example, X might denote what the weather would be on the following day and Y the current meteorological information available. The actions that one could take might be, for an individual: whether or not to prepare for an outing the following day, whether or not to take waterproof clothes, etc. For an army general they might be: whether to launch an attack, whether to postpone but maintain readiness, or whether to defer attack altogether for the moment.

What is clear is that the decision can depend only upon Y: the information that one possesses. The determination of an action rule (or policy) means the determination of a function $u(y)$ specifying what action one would take for all possible values y of the information, and we wish to determine the *optimal action rule $u^*(y)$* which minimizes the expected cost $E\{L[X, Y, u(Y)]\}$.

Theorem 5.6.1. *The optimal action $u^*(Y)$ is determined as the value of u minimizing $E[L(X, Y, u)| Y]$.*

PROOF. If an action rule $u(Y)$ is followed then the expected cost is

$$E\{L[X, Y, u(Y)]\} = E(E\{L[X, Y, u(Y)]| Y\}) \geq E\left(\inf_{u} \{E[L(X, Y, u)| Y]\}\right),$$

$$(19)$$

and the lower bound is attained by the rule suggested in the theorem.   □

The theorem may seem trivial, but the reader should understand its point: the reduction of a constrained minimization problem to a free one. The initial problem is that of minimizing $E\{L[X, Y, u(Y)]\}$ with respect to the *function* $u(Y)$, so that the minimizing u is constrained to be a function of Y at most. This is reduced to the problem of minimizing $E[L(X, Y, u)|Y]$ *freely* with respect to the parameter u.

The conclusion is formally immediate, but the whole formulation has implications which have provoked considerable discussion over the years. That observations Y should be regarded as r.v.s is not controversial. However, X represents what is sometimes called a 'state of nature', the state of the world in some important respect which is not revealed to us, and there are cases where this can scarcely be regarded as a r.v.; see the discussion at the end of Section 2.

We shall merely remark that there is a whole gamut of cases, for which at one extreme (weather prediction, medical prognosis, signal extraction) the approach is clearly acceptable and at the other (theories of the universe, or of an afterlife) it is inappropriate, for the reasons indicated in Section 2.

The notion that a cost function $L(x, y, u)$ must be specified has also been criticized. However, one scarcely has a basis for decision unless one can quantify costs. Moreover, it has been shown by Savage (1954) and others that any decision rule obeying certain axioms of rationality can be regarded as optimal on the basis of some hypothetical cost function and prior distribution (of the 'states of nature' X).

EXERCISES AND COMMENTS

1. Suppose, for simplicity, that X can take only a discrete set of values, with $\pi(x) = P(X = x)$ being the prior probability of a state of nature x. Suppose also that Y has a probability density $f(y|x)$ conditional on the event $X = x$, this being a density relative to some measure independent of x. Show that if Y has been observed to be equal to y, then the optimal choice of u is that minimizing

$$\sum_x \pi(x) f(y|x) L(x, y, u).$$

2. *The two-hypothesis two-action case.* Continuing with the last example, let us suppose that x and u can each take only the two values 0 and 1, say, and that the cost function $L(x, u)$ depends only on x and u. Suppose also that $u = x$ would be the better decision if x were known (so that $L(x, x) < L(x, u)$ for $u \neq x$). Show that it is optimal to take decision 1 if

$$\frac{f(y|1)}{f(y|0)} \geq \frac{\pi(0)[L(0, 1) - L(0, 0)]}{\pi(1)[L(1, 0) - L(1, 1)]}. \tag{20}$$

That is, if the *likelihood ratio* $f(y|1)/f(y|0)$ exceeds the critical value given by the right-hand member.

For example, states $x = 0$ and 1 might correspond to 'all quiet' and 'intruder present' and actions $u = 0$ and 1 might correspond to 'do nothing' and 'give the alarm', respectively. Relation (20) determines the set of values y of the observations for which one should give the alarm.

3. *Least square approximation.* Suppose x and u scalar, for simplicity. Then the choice of cost function $L(x, y, u) = (u - x)^2$ leads exactly to formulation of the problem of least square approximation which we have already considered in Sections 2.8, 2.9 and 5.3. That is, to the problem of choosing a function of Y to approximate X as well as possible in mean square. If u is unrestricted as a function of Y then we know from Section 3 that the optimal estimate $u^*(Y)$ is just the conditional expectation $E(X|Y)$. If u is further restricted to being linear (a constraint of a type not envisaged in Theorem 5.6.1) then u^* is the LLS estimate of Sections 2.8 and 2.9.

7. Information Transmission

Some of the nicest examples of statistical decision theory, and also of asymptotic probabilistic arguments, occur in statistical communication theory. Suppose that a sender wishes to transmit one of M possible messages along a communication channel. He does this by sending a signal X which can take values $x_1, x_2, x_3, \ldots, x_M$. That is, the signal x_j is the codeword representing message j. The code is one that is adopted, not for reasons of security, but simply because, if the message is to be transmitted, it must be given some physical representation.

However, the channel is 'noisy', so that the signal Y which emerges at the receiver end of the channel is only imperfectly related to the signal X which was transmitted. In fact, it is related by a conditional probability $P(Y|X)$ whose form (i.e. whose functional dependence on X and Y) reflects the statistical characteristics of the channel.

How should the receiver decide how to decode Y? In other words, how should he decide which message was intended on the basis of the observation Y? We assume that he knows the prior probabilities π_j of the various messages, the codewords x_j which represent them, and the statistical characteristics of the channel, as specified by the functional form of $P(Y|X)$.

Of course, Y may not be discrete-valued; let us suppose that $f(y|X = x_j)$ is the conditional density of Y relative to a measure independent of j, and let us for brevity denote this simply $f_j(y)$. Suppose that the recipient of the message incurs a cost L_{jk} if he infers that message k was intended when message j was in fact intended. Then, by Exercise 5.6.1, he should, on the basis of received signal y, infer that message $k(y)$ has been sent, where $k(y)$ is the value of k minimizing $\sum_j \pi_j f_j(y) L_{jk}$.

Suppose that all errors are regarded as equal serious, so that

$$L_{jk} = \begin{cases} 0 & (j = k), \\ 1 & (j \neq k), \end{cases}$$

say. Then the rule must be to choose the value of k minimizing $\sum_{j \neq k} \pi_j f_j(y)$, i.e. the value maximizing $\pi_k f_k(y)$.

As the final simplification, suppose that all M message values are equally likely, so that $\pi_j = 1/M$ for all j. (This is, in fact, the least favourable case;

see Exercise 1.) Then the optimal decision rule reduces simply to the *maximum likelihood rule*: infer the value of k which maximizes $f_k(y)$.

In this case one can derive a simple and useful bound for the probability error.

Theorem 5.7.1. *If all M message values are equally likely then an upper bound to the probability of error in inferring the intended message is*

$$P(\text{error}) \leq \frac{1}{M} \sum_{j} \sum_{\neq k} \rho_{jk}, \qquad (21)$$

where

$$\rho_{jk} = P[f_k(Y) \geq f_j(Y)|X = x_j].$$

PROOF. Under the assumptions stated one will use maximum likelihood inference. Thus, if x_j has been sent then one will certainly make an error if $f_k > f_j$ and may make an error if $f_k = f_j$ for some $k \neq j$. If A_{jk} is the event $f_k(Y) \geq f_j(Y)$ we thus have

$$P(\text{error}|X = x_j) \leq P\left(\bigcup_{k \neq j} A_{jk}|X = x_j\right) \leq \sum_{k \neq j} P(A_{jk}|X = x_j) = \sum_{k \neq j} \rho_{jk}, \quad (22)$$

where the second inequality follows from Boole's inequality (3.13). Averaging inequality (22) over j we deduce the bound (21). □

Having solved the decision aspects of the problem, one should now turn to the design aspects. How should one choose the codewords x_j in the space of possible signals so as to minimize the probability of error? How large can M be for a given channel, consistent with an acceptable error probability? These questions stimulate the development of the concepts of channel coding and of channel capacity. We sketch some ideas in the exercises.

EXERCISES AND COMMENTS

1. Suppose that, for a given procedure (i.e. given coding and decoding rules), the probability of error if message j is sent is P_j (error). The overall probability of error is then $P(\text{error}) = \sum_j \pi_j P_j$ (error). Show that one can always find a reallocation of codewords to messages such that

$$P(\text{error}) \leq \overline{P}(\text{error}) = \frac{1}{M} \sum_j P_j(\text{error}). \qquad (23)$$

In this sense, the case of equiprobable codewords is least favourable.

[Obviously, one should allocate error-prone words to less frequent messages. However, the easy proof is to note that, if one chose a coding randomly and equiprobably from the $M!$ codings which employ the M given words, then the expected error probability would be just $\overline{P}$(error). At least one of the codings must thus satisfy (23).]

Exercises 2–5 all concern the *binary symmetric memoryless channel.*

2. Suppose that both X and Y consist of a signal extended over n discrete-time steps, so that $X = (X_1, X_2, \ldots, X_n)$ and $Y = (Y_1, Y_2, \ldots, Y_n)$, where X_t and Y_t represent the 'letters' transmitted and received at time t. Suppose that $x_j = (x_{j1}, x_{j2}, \ldots, x_{jn})$. The channel is *binary* if X_t and Y_t can take only two values, say 0 and 1. It is *memoryless* if

$$P(Y|X) = \prod_{t=1}^{n} P(Y_t|X_t),$$

so that the Y_t are independent conditional on the X_t. Under these suppositions it is *symmetric* if

$$P(Y_t = b|X_t = a) = \begin{cases} q & (b = a), \\ p = 1 - q & (b \neq a), \end{cases}$$

for all t. Thus p is the probability of error in the transmission of a letter. Define the *Hamming distance* $d(X, Y)$ between X and Y as the number of letters in which X and Y differ. Then show that the maximum likelihood inference rule is to infer the value of k minimizing or maximizing $d(x_k, Y)$, according as $p < \frac{1}{2}$ or $p > \frac{1}{2}$.

3. Note that $f_k(Y)/f_j(Y) = (p/q)^S$ where $S = \sum_1^n \Delta_t$ and

$$\Delta_t = |x_{kt} - Y_t| - |x_{jt} - Y_t| = \begin{cases} 0 & (x_{jt} = x_{kt}), \\ 1 & (Y_t = x_{jt} \neq x_{kt}), \\ -1 & (Y_t = x_{kt} \neq x_{jt}). \end{cases}$$

Hence show that if z is any number in $(0, 1]$ then

$$\rho_{jk} = P(S \leq 0|X = x_j) \leq E(z^S|X = x_j) = (pz^{-1} + qz)^{d(x_j, x_k)},$$

whence

$$\rho_{jk} \leq \theta^{d(x_j, x_k)}, \tag{24}$$

where $\theta = \sqrt{4pq} \leq 1$.

4. Let E_x denote an averaging over a *random coding* in which the codewords are constructed by giving the letters x_{jt} values 0 and 1 independently and with equal probability. Show then from (21) and (24) that

$$E_x[P(\text{error})] \leq M E_x[\theta^{d(x_1, x_2)}] = M \left[\frac{1 + \sqrt{4pq}}{2} \right]^n. \tag{25}$$

5. Suppose that $M = 2^{nR + o(n)}$, indicating that one is attempting to transmit the equivalent of $nR + o(n)$ binary digits in time n, or R bits per unit time. We see from (25) that $E_x[P(\text{error})]$ tends to zero with increasing n provided that

$$R < \log_2 \left[\frac{1 + \sqrt{4pq}}{2} \right] = \underline{C},$$

say. That is, for the binary symmetric memoryless channel there exists a coding which is *reliable* if $R < \underline{C}$, in that the probability of error tends to zero as one codes in longer time blocks. The supremum of reliable transmission rates is termed the *capacity* of the channel, denoted C, and we have effectively shown that $C \geq \underline{C}$. The actual evaluation is

$$C = 1 + p \log_2 p + q \log_2 q.$$

To prove this requires more careful argument, but C and $\underline{C}$ are similar in their dependence upon p. If we write capacity as a function $C(p)$ of p then $C(p) = C(1 - p)$, and $C(p)$ decreases from 1 to 0 as p increases from 0 to $\frac{1}{2}$. Interpret the extreme cases $p = 0, \frac{1}{2}$ and 1.

8. Acceptance Sampling

Suppose that a contractor is supplied by a subcontractor with a batch of N items (castings, say) and that he must decide whether these are up to specification before he accepts them. The contractor takes a sample of n castings for test, and finds that y of them are substandard. He must then make his decision on the basis of this observation. Let us suppose that there are only two possible decisions: to accept or to reject the whole batch. (There would be other possibilities: e.g. to sample further, or to examine the whole batch, accepting only the satisfactory items.)

Let us suppose that the cost of rejecting the batch has a flat value $L = aN$, say, and that the cost of acceptance has a value $L = bX$, where X is the number in the batch which are defective. The value of X is of course unknown; in order to be able to proceed we must make some assumptions. Let us assume that the individual castings are defective with probability p, independent of each other, so that if Y is the r.v. 'number of defective items observed' then

$$P(Y = y|p) = \binom{n}{y}p^y(1 - p)^{n-y}. \tag{26}$$

We have written this as conditional on the value of p, because p itself is to be regarded as a r.v., with a distribution reflecting the inspector's experience of the subcontractor.

We have then

$$E(\text{cost of acceptance}|Y = y) = b[y + (N - n)E(p|Y = y)] \approx bNE(p|Y = y).$$

This last approximation holds if the sampling fraction n/N is small, so that the principal concern is the number of defectives so far undetected.

The criterion for acceptance of a batch is then that

$$bE(p|Y = y) < a, \tag{27}$$

and we must determine $E(p|Y = y)$ if we are to determine the decision rule. A natural prior distribution for p is the B-distribution (read this as 'beta-distribution') with density

$$f(p) = \frac{p^{\alpha-1}(1 - p)^{\beta-1}}{B(\alpha, \beta)} \quad (0 \le p \le 1) \tag{28}$$

and parameters α and β (see Exercise 1). The distribution of p conditional on

$Y = y$ then has density

$$f(p|Y = y) = \frac{f(p)P(Y = y|p)}{\int f(p)P(Y = y|p)\, dp} = \frac{p^{\alpha+y-1}(1-p)^{\beta+n-y-1}}{B(\alpha + y, \beta + n - y)}$$

and so still has the B-form (which is what makes the assumption of a B-prior natural in this context). The expectation of p on the basis of this distribution is

$$E(p|Y = y) = \frac{B(\alpha + y + 1, \beta + n - y)}{B(\alpha + y, \beta + n - y)} = \frac{\alpha + y}{\alpha + \beta + n}. \tag{29}$$

We see from (27) and (29) that one would accept or reject the batch according as $y < y(n)$ or $y > y(n)$, where

$$y(n) = (\alpha + \beta + n)\frac{a}{b} - \alpha. \tag{30}$$

The decision in the case $y = y(n)$ is immaterial.

We have thus solved the decision problem for the contractor. We can now pose a design problem: how large should n, the sample size, be? In deciding this one has to balance the two components of cost: the cost incurred when a decision is made and the cost of sampling. We should also consider whether we are making decisions only for this single batch or for a whole sequence of batches. In this latter case, increased intensity of inspection now might lead to economies later. However, this consideration of a sequential version of the problem would take us too far for present purposes. Moreover, if one is guarding against variability in manufacture rather than trying to assess the subcontractor, then p might be assumed to have the distribution (28) afresh for each batch, regardless of experience with earlier batches.

Suppose that the cost of sampling n items is $c(n)$, an increasing function of n. At the time of determination of the sample size p and Y both must be regarded as r.v.s, with joint distribution $f(p)P(y|p)$ specified by (26) and (28).

If one takes n observations then the expected cost of the sampling/decision exercise is

$$C(n) = c(n) + \int dp \sum_y f(p)P(y|p)L(y), \tag{31}$$

where

$$L(y) = \begin{cases} aN & (y > y(n)), \\ bNp & (y < y(n)), \end{cases}$$

and $y(n)$ is the threshold value determined by (30). The sample size n must be chosen to minimize expression (31). We can perform the p-integration in (31); we have

$$\int p^s f(p)P(y|p)\, dp = \binom{n}{y}\frac{B(\alpha + y + s, \beta + n - y)}{B(\alpha, \beta)} = \mu_{ns}(y),$$

say, so that

$$C(n) = c(n) + aN \sum_{y \geq y(n)} \mu_{n0}(y) + bN \sum_{y < y(n)} \mu_{n1}(y).$$

At this point the minimization must probably be performed numerically.

EXERCISES AND COMMENTS

1. The B-distribution is natural when observations Y are binomially distributed, because it retains its form under Bayesian updating, as we observed in the text. The normalizing constant in (28) in just the reciprocal of the B-function

$$B(\alpha, \beta) = \frac{\Gamma(\alpha)\Gamma(\beta)}{\Gamma(\alpha + \beta)},$$

where the Γ-function

$$\Gamma(\alpha) = \int_0^\infty e^{-x} x^{\alpha-1} \, dx$$

has the evaluation $\Gamma(\alpha) = (\alpha - 1)!$ if $\alpha = 1, 2, 3, \ldots$, and for $\alpha > 1$ obeys the recursion $\Gamma(\alpha) = (\alpha - 1)\Gamma(\alpha - 1)$.

2. Suppose that Y is Poisson distributed with parameter λ, and that λ itself follows a Γ-distribution. Show that the distribution λ conditional on the value of Y still has the Γ-form.

CHAPTER 6

Applications of the
Independence Concept

The reader may now realize in hindsight that the study of basic models in Chapter 4 was essentially a sustained appeal to the concept of independence; an appeal made explicit already in Section 4.1. Now that the concept has been set in the wider context of conditioning we can continue with the description of some further models which, while still standard, are of a more advanced character. These models are all somewhat individual. In Chapter 7 we shall follow one of the major general consequences of independence: the road to the various limit theorems.

1. Renewal Processes

A *Bernoulli* trial is one which has only two outcomes: conventionally, success or failure. An economical specification of what we mean by an infinite sequence of independent Bernoulli trials would be to state that, for such a sequence,

$$E\left(\prod_{j=1}^{\infty} z_j^{\xi_j}\right) = \prod_{j=1}^{\infty} (pz_j + q), \tag{1}$$

where ξ_j is the indicator variable of success at the jth trial, and we regard (1) as valid for all $\{z_j\}$ for which the product is convergent.

In this way we can obtain the binomial distribution of the number of successes in the first N trials immediately (by setting $z_j = z$ for $j \leq N$ and $z_j = 1$ for $j > N$). However, we have established the geometric distribution of the number of trials to first success only by first deducing from (1) the probability of a given sequence of trial outcomes and working on from that.

Brief and meaningful as this calculation is, one still wonders whether it is possible to deduce the result directly from (1) without the intermediate step of determining probabilities of individual sequences. In attempting to do so, we shall find that we can treat a much more general class of processes: *renewal processes* and *recurrent events*.

The practical context of a renewal process is a situation in which an article, such as a machine tool, is replaced by a new article as soon as it wears out. The interest is in the probability that replacement (or *renewal*) will take place at a definite instant, and in the distribution of the number of renewals made in a given time. The situation also has a kind of converse, the idea of recurrence, which we shall consider in the next section.

We suppose that the lifetimes of consecutive tools, denoted $T_1, T_2, \ldots$, are IID r.v.s. Then the total lifetime of the first j tools

$$S_j = \sum_{k=1}^{j} T_j$$

is just such a sum of IID r.v.s as we have considered a number of times. If the first article (the one of lifetime T_1) was installed at time zero then S_j is also the instant at which the jth renewal takes place.

We shall assume that the lifetimes T_k are integral-valued, so that we can work in integral time, t. The assumption makes for simplicity without losing much realism; we are in effect rounding off lifetimes to the nearest number of whole time units. The r.v. 'lifetime' will then have p.g.f.

$$\Pi(z) = \sum_{t=0}^{\infty} p_t z^t,$$

where p_t is the probability of a lifetime $T = t$. It is usual to set $p_0 = 0$; that is, to exclude the possibility that the article has zero lifetime, and so needs replacement the very moment it is installed. We shall not require this, but merely require that $p_0 < 1$; that is, that the article does not *always* fail immediately! The p.g.f. $\Pi(z)$ is always convergent in $|z| \leq 1$, and, because of the assumption $p_0 < 1$, we shall have $|\Pi(z)| < 1$ if $|z| < 1$.

Let R_t be the number of renewals made *at* time t, and define

$$u_t = E(R_t). \tag{2}$$

The installation of the original article is counted in R_0. If $p_0 = 0$ then R_t can only take the values 0 or 1, and u_t then has the interpretation of the *probability of renewal at time t*.

Theorem 6.1.1. *If $|z| < 1$ then*

$$\sum_{t=0}^{\infty} R_t z^t = \sum_{j=0}^{\infty} z^{S_j}, \tag{3}$$

(with the convention $S_0 = 0$), and

$$\sum_{t=0}^{\infty} u_t z^t = \frac{1}{1 - \Pi(z)}.$$ (4)

PROOF. Relation (3) is the key identity. Its formal validity is evident, since a renewal at time t will contribute z^t to each side. The question is whether these infinite series with random coefficients converge. If $p_0 = 0$ then R_t can only take the values 0 or 1, so the series are dominated by $\sum |z|^t$, and necessarily converge. In the case $0 < p_0 < 1$ a slightly more careful argument is needed, which we give in Chapter 13. This yields the conclusion that the series converge in the conventional sense, with probability one.

The second result (4) follows from (3) by the taking of expectations. If we define the generating function of the u_t by $U(z)$ then

$$U(z) = \sum_{t=0}^{\infty} u_t z^t = \sum_{j=0}^{\infty} E(z^{S_j}) = \sum_{j=0}^{\infty} \Pi(z)^j = \frac{1}{1 - \Pi(z)}. \qquad \square$$

Relation (3) essentially determines the u_t in terms of the lifetime distribution.

Let us consider a pair of simple examples. Suppose that lifetime is fixed and equal to L, so that $\Pi(z) = z^L$. Then

$$U(z) = (1 - z^L)^{-1} = \sum_{j=0}^{\infty} z^{jL},$$

with the obvious interpretation that there is a single renewal when t is a multiple of L, and none at any other time.

If we consider a geometric lifetime distribution, so that

$$\Pi(z) = \frac{pz}{1 - qz},$$ (5)

then (3) yields

$$U(z) = 1 + \frac{pz}{1 - z},$$ (6)

or $u_0 = 1$, $u_t = p \ (t > 0)$. That is, the renewal probability is constant after the initial installation, a fact whose significance will be brought out in the next section.

A r.v. of interest is

$$N_t = \sum_{k=0}^{t} R_k.$$

the *total number of renewals* made up to time t. The key fact for this variable is that $N_t < j$ is the same event as $S_j > t$; essentially the fact embodied in (3). One can use this relation to obtain exact results; we shall use it to obtain the asymptotic assertions of

Theorem 6.1.2. *For large t*

$$E(N_t) = \frac{t}{\mu} + o(t), \tag{7}$$

$$\mathrm{var}(N_t) = \frac{t\sigma^2}{\mu^3} + o(t), \tag{8}$$

where μ and σ^2 are, respectively, the mean and variance of lifetime.

PROOF. Let us denote N_t simply by N, to ease notation. Let A_j denote the event $N < j$. Then $\overline{A}_j$ implies that $S_j \leq t$, and so that

$$0 \leq E(S_j | \overline{A}_j) \leq t. \tag{9}$$

Furthermore, if $j \geq t$ then

$$P(\overline{A}_j) \leq p_0^{j-t}, \tag{10}$$

since there must be at least $j - t$ zero lifetimes.

We have now

$$j\mu = E(S_j) = P(A_j)E(S_j | A_j) + P(\overline{A}_j)E(S_j | \overline{A}_j). \tag{11}$$

But

$$E(S_j | A_j) = E(S_j - S_N + S_N | A_j) = E[(j - N)\mu + S_N | A_j], \tag{12}$$

since S_N is the instant of first renewal after time t, and, conditional on the value of N, the r.v. $S_j - S_N$ is freely distributed as the sum of $j - N$ unconditioned lifetimes. From (11) and (12) we thus deduce that

$$P(A_j)E(S_N - \mu N | A_j) + P(\overline{A}_j)[E(S_j | \overline{A}_j) - j\mu] = 0. \tag{13}$$

Letting j tend to infinity in (13) we obtain

$$E(S_N - \mu N) = 0. \tag{14}$$

We have appealed to the fact that the expectation conditional on A_j in (13) tends to the unconditional expectation, by virtue of (10), and that the term involving $\overline{A}_j$ tends to zero, by virtue of (9) and (10).

Relation (14) implies that

$$E(N) = \frac{E(S_N)}{\mu}. \tag{15}$$

Now S_N, the moment of the first renewal after time t, will be greater than t, but only by an amount of the order of a single lifetime. It is plausible that the 'expected overshoot' $E(S_N - t)$ is uniformly bounded in t, whence (7) follows from (15).

By applying the argument starting at (11) to $E(S_j^2)$ we obtain, analogously to (14),

$$E[(S_N - \mu N)^2 - \sigma^2 N] = 0, \tag{16}$$

which leads to (8), by the same overshoot argument. $\square$

Since $E(N_t) = \sum_{k=0}^{t} u_k$, we see that (7) implies that

$$\frac{1}{t} \sum_{k=0}^{t} u_k \to \frac{1}{\mu}$$

with increasing t. Under wide conditions one can prove the much stronger result

$$u_t \to \frac{1}{\mu}. \tag{17}$$

That is, that the renewal rate u_t tends to a constant with increasing t, which must necessarily be the reciprocal of expected lifetime. A general proof is not straightforward, although the result is fairly immediate in simple cases (see Exercise 8).

EXERCISES AND COMMENTS

1. Calculate p_t and u_t for the case $\Pi(z) = [p/(1 - qz)]^m$ ($m = 1, 2$). What is the limit value of u_t for large t?

2. Show that $u_t \le 1/(1 - p_0)$.

3. Relation (4) is equivalent to the relations

$$u_0 = 1 + p_0 u_0,$$

$$u_t = \sum_{k=0}^{t} p_k u_{t-k} \qquad (t = 1, 2, \ldots).$$

   Deduce these directly.

4. Interpret the coefficient of $s^j z^t$ in the expansion of $[1 - s\Pi(z)]^{-1}$ in nonnegative powers of s and z.

5. Show from (3) that

$$\sum_{j=0}^{\infty} \sum_{k=0}^{\infty} w^j z^k \operatorname{cov}(R_j, R_k) = \frac{\Pi(wz) - \Pi(w)\Pi(z)}{[1 - \Pi(w)][1 - \Pi(z)][1 - \Pi(wz)]}.$$

6. Confirm (17) for the cases of Exercise 1.

7. Show that the equation $\Pi(z) = 1$ has no roots inside the unit circle. Show also that if we require that lifetime distribution be *aperiodic* (i.e. we exclude the case where the distribution is confined to multiples of some fixed integer greater than unity), then the only root on the unit circle is a simple one at $z = 1$.

8. Consider the case when lifetime is bounded (so that $\Pi(z)$ is a polynomial) and the lifetime distribution is aperiodic. Show from the partial fraction expansion of $[1 - \Pi(z)]^{-1}$ that, for large t,

$$u_t = \mu^{-1} + O(z_0^{-t}),$$

   where z_0 is the root of $\Pi(z) = 1$ smallest in modulus after $z = 1$.

9. *Wald's identity.* By applying the argument beginning at (11) to $E(z^{S_j})$, show that

$$E[z^{S_N}\Pi(z)^{-N}] = 1 \tag{18}$$

for all z such that $\Pi(z)$ exists and $|\Pi(z)| > p_0$. Note that (14) and (16) follow from differentiation of the identity with respect to z at $z = 1$. There are two r.v.s in the bracket of (18): N and S_N. That is, the number of renewals by time t and the time of the first renewal after time t. The 'no overshoot' approximation is to set $S_N \approx t$.

2. Recurrent Events: Regeneration Points

There are cases where it is the renewal probabilities u_t, rather than the probabilities p_t of failure at age t, which are known, so that in this case one would invert relation (4) to obtain

$$\Pi(z) = \frac{U(z) - 1}{U(z)}. \tag{19}$$

For example, this is the situation when one is considering occurrence of successes in a sequence of Bernoulli trials. One knows that there is a constant probability p of success, so that $u_t = p \, (t > 0)$, with $u_0 = 1$ by convention. The generating function $U(z)$ thus has the evaluation (6), and substitution into (19) gives the p.g.f. $\Pi(z)$ of (5). We thus deduce the geometric character of the lifetime distribution by a direct generating function argument.

Of course, for this converse argument to hold, it is necessary that the basic hypothesis of the renewal process be fulfilled: that the intervals between consecutive occurrences of a success be IID r.v.s. This is evident in the case of a Bernoulli sequence.

In general, suppose that a situation $\mathscr{A}$ can occur from time to time in a sequence of trials (not necessarily independent), and that the process is such that the numbers of trials between consecutive occurrences of $\mathscr{A}$ are IID random variables. Then $\mathscr{A}$ is termed a *recurrent event*. This term is a double misnomer, but is established. We shall clarify the point and formalize the definition after examining a few more examples.

Thus, 'renewal' was a recurrent event in the last section; 'success' is a recurrent event for a Bernoulli sequence. In many cases the u_t are quickly determinable and are interpretable as $P(\mathscr{A}$ occurs at time $t)$, because multiple occurrences are excluded. Equation (19) then determines the p.g.f. $\Pi(z)$ of *recurrence times* of $\mathscr{A}$. In effect, it determines the probability that $\mathscr{A}$ *first occurs* at time t (after the initial occurrence at $t = 0$) in terms of the probability that $\mathscr{A}$ occurs at time t.

For example, consider again a sequence of Bernoulli trials and let S_t be the number of successes less the number of failures at time t (both reckoned from time zero). This can be regarded as the net winnings of a gambler playing for unit stakes. Let us say that $\mathcal{A}$ occurs at time t if $S_t = 0$, i.e. if the player breaks even at time t. Then $\mathcal{A}$ certainly occurs at $t = 0$, and is a recurrent state.

Then u_t is the probability of $\frac{1}{2}t$ successes in t trials; this is zero if t is odd, and has the binomial expression

$$u_{2r} = \binom{2r}{r}(pq)^r \qquad (r = 0, 1, 2, \ldots)$$

if t is even. The series $U(z)$ can be summed: we have

$$U(z) = \sum_{r=0}^{\infty} \binom{2r}{r}(pq)^r z^{2r} = (1 - 4pqz^2)^{-1/2}, \qquad (20)$$

as the reader can verify directly (see also Exercise 5). That root is taken in (20) which tends to unity as z tends to zero. The recurrence time p.g.f. is thus, by (19),

$$\Pi(z) = 1 - (1 - 4pqz^2)^{1/2}, \qquad (21)$$

so the probability that the recurrence time τ has value t is

$$p_t = \begin{cases} 0 & (t \text{ odd or zero}), \\ \binom{2r}{r}\dfrac{(pq)^r}{2r - 1} & (t = 2r; r = 1, 2, 3, \ldots). \end{cases} \qquad (22)$$

Note that the distribution is not an aperiodic one, but is restricted to even values of t, as it obviously must be.

A very interesting point is that

$$\Pi(1) = 1 - (1 - 4pq)^{1/2} = 1 - |p - q|, \qquad (23)$$

so that $\Pi(1) < 1$ unless $p = q$, that is, unless the game is a fair one. Now, $\Pi(1) = \sum_t p_t$, and, for all the distributions we have encountered, $\sum_t p_t = 1$. How then are we to interpret (21)?

The actual interpretation of the infinite sum $\sum_t p_t$ is as a limit such as $\lim_{n \to \infty} \sum_0^n p_t$ or $\lim_{z \uparrow 1} \sum_0^\infty p_t z^t$. The first interpretation makes it plain that the sum is to be regarded as the *probability that the recurrence time τ is finite*. The inequalities

$$\sum_0^n p_t \le \lim_{z \uparrow 1} \sum_0^\infty p_t z^t \le \lim_{n \to \infty} \sum_0^n p_t$$

show that these two limits must be equal, so that $\Pi(1)$ can also be interpreted as the probability of a finite recurrence time. All the r.v.s considered hitherto have been finite with probability one, and here we have the first exception, and in a natural problem. If there is a deficit, $\Pi(1) < 1$, we say that recurrence is *uncertain*. The reason for uncertain recurrence is intuitively clear in the present case. If the game is an unfair one, the player with the advantage has a good chance of building up a substantial lead; once he has this he is unlikely to lose it.

However, even in the case $p = q = \frac{1}{2}$, when the game is a fair one, the time to recurrence is extraordinarily long. The mean recurrence time, if it exists, should be given by $\Pi'(1)$; we see from (21) that this is infinite. In fact, an application of Stirling's formula to expression (22) shows that

$$p_t \approx \sqrt{\frac{2(4pq)^t}{\pi t^3}}$$

for t large and even. Thus, if $p = q$, the distribution tails away at the slow rate of $t^{-3/2}$, implying that large recurrence times are relatively probable.

The objection to the term 'recurrent event' is that $\mathscr{A}$ is, in the technical sense, neither an event nor necessarily recurrent. An event is something which, for a given realization, occurs either once or not at all (the realization being, in this case, the course of the whole sequence of trials). It is then not something that can happen repeatedly in the sequence. For the other point, we shall later (Chapter 9) follow convention by using the description 'recurrent' only in cases when recurrence is certain. However, we have seen from the break-even example that definition of the break-even event $\mathscr{A}$ is useful even when its recurrence is uncertain. The reader is required to flinch and make these mental reservations every time he uses the term 'recurrent event'.

More fundamental than the concept of a recurrent event is that of a regeneration point, by which is meant a situation that effectively brings about a fresh start of the process whenever it occurs. In fact, it is usually by establishing the existence of a regeneration point that one establishes the existence of a recurrent event.

Formally, suppose that one considers a sequence of random variables in time $\{X_t\}$; what we shall later term a stochastic process. Suppose that occurrence of $\mathscr{A}$ at time t corresponds to $X_t \in A$, where A is a fixed set of x-values. Then occurrence of $\mathscr{A}$ constitutes a *regeneration point* for the process $\{X_t\}$ if the future and the past of the process (at arbitrary t) are independent conditional on occurrence of $\mathscr{A}$ (at t). That is, we require that

$$E(Y_t Z_t \mid X_t \in A) = E(Y_t \mid X_t \in A) E(Z_t \mid X_t \in A), \tag{24}$$

where Y_t is any function of past-and-present $\{X_s; s \le t\}$ and Z_t is any function of present-and-future $\{X_s; s \ge t\}$. This relation would imply, for example, that

$$E(Z_t \mid X_t \in A, Y_t \in B) = E(Z_t \mid X_t \in A)$$

for arbitrary B, which is a way of saying that the process makes a fresh start after the occurrence of $\mathscr{A}$.

Theorem 6.2.1. *The intervals of time between consecutive regeneration points are independent random variables.*

This follows immediately from the definition (24); the intervals before and after a given regeneration point lie in past and future, respectively. One requires some assumption of constancy of structure if one is to establish that the intervals are identically distributed as well as independent.

Such a regeneration point occurred when the two players in the game broke even; it occurs in a queueing process (for most models considered) when the queue becomes empty.

EXERCISES AND COMMENTS

1. A rephrasing of the proof of Theorem 6.1.1, which makes no assumption that recurrence (renewal) time is finite, goes as follows. As in that theorem, let R_t be the number of recurrences at time t, and let τ be the time of first recurrence. Then

$$U(z) = E\left(\sum_{t=0}^{\infty} R_t z^t\right) = 1 + E\left(\sum_{t=0}^{\infty} R_{t+\tau} z^{t+\tau}\right) = 1 + E(z^\tau)E\left(\sum_{t=0}^{\infty} R_{t+\tau} z^t\right)$$

$$= 1 + \Pi(z)U(z),$$

whence (4) and (19) follow. The 1 in the third expression corresponds to the initial occurrence of $\mathcal{A}$ counted in R_0 and the factorization in the fourth expression follows from the independence assumption.

2. Note that $\Pi(1) = [U(1) - 1]/U(1)$, so that the probability of recurrence, $\Pi(1)$, is unity if and only if the expected number of recurrences, $U(1)$, is infinite. If recurrence is certain then the expected recurrence time is

$$E(\tau) = \Pi'(1) = \lim_{z \uparrow 1} \frac{1 - \Pi(z)}{1 - z} = \lim_{z \uparrow 1} [(1 - z)U(z)]^{-1}. \tag{25}$$

The quantity $\lim_{z \uparrow 1} (1 - z)U(z)$ defines a type of average limit value of u_t, the Abel mean. Relation (25) thus implies an averaged version of (17).

3. Show that if τ is the first break-even point in a sequence of Bernoulli trials, then

$$E(\tau | \tau < \infty) = 1 + |p - q|^{-1}.$$

4. Show that for the break-even problem u_t converges to zero with increasing t, which, in the case of certain recurrence ($p = q$), is consistent with (17) and $\mu = \infty$.

5. Show that the infinite sum in (20) is the absolute term in the Laurent expansion of $[1 - \frac{1}{2}z(pw + qw^{-1})]^{-1}$ in powers of w on the unit circle. Interpret and derive the final expression in (20).

6. Let S_t again denote the capital at time t of a player waging unit stakes in a sequence of Bernoulli trials. Is $S_t = k$ a recurrent state? Is $S_t > k$ a recurrent state? (In both cases assume the event to have occurred at time zero, for consistency with our convention.)

7. Suppose that occurrence of $\mathcal{A}$ constitutes a regeneration point, as does occurrence of $\mathcal{B}$, and that the recurrence time to $\mathcal{B}$ cannot be zero. Define

$$U_{\mathcal{AB}}(z) = \sum_{t=0}^{\infty} z^t P(\mathcal{B} \text{ at } t | \mathcal{A} \text{ at } 0), \qquad U_{\mathcal{BB}}(z) = \sum_{t=0}^{\infty} z^t P(\mathcal{B} \text{ at } t | \mathcal{B} \text{ at } 0),$$

and let $\Pi_{\mathcal{BB}}(z)$ and $\Pi_{\mathcal{AB}}(z)$ be correspondingly the p.g.f.s recurrence time to $\mathcal{B}$ and of *first passage time* from $\mathcal{A}$ to $\mathcal{B}$, so that $U_{\mathcal{BB}} = (1 - \Pi_{\mathcal{BB}})^{-1}$. Show that $U_{\mathcal{AB}} = \Pi_{\mathcal{AB}}U_{\mathcal{BB}}$.

8. Suppose that the recurrent events $\mathscr{A}$ and $\mathscr{B}$ correspond to $S = 0$ and $S = b \,(>0)$ in a Bernoulli process (see Exercise 6). Show (see Exercise 5) that $U_{\mathscr{A}\mathscr{B}} = \phi^b/\Delta$, where $\Delta = \sqrt{(1 - 4pqz^2)}$ and $\phi = (1 - \Delta)/(2qz)$. Hence show that $\Pi_{\mathscr{A}\mathscr{B}} = \phi^b$, and that the probability that S ever equals b (given $S_0 = 0$) is 1 if $p \geq q$ and $(p/q)^b$ otherwise.

3. A Result in Statistical Mechanics: the Gibbs Distribution

A model for a system such as a perfect gas would be somewhat as follows. Imagine a collection of N molecules, each possessing an energy which must take one of the values $\varepsilon_j \,(j = 0, 1, 2, \ldots)$. It is assumed that there is no energy of interaction, so that the total energy of the system is

$$\mathscr{E} = \sum_j n_j \varepsilon_j, \tag{26}$$

where n_j is the number of molecules of energy ε_j.

Suppose that the total energy $\mathscr{E}$ of the system is prescribed, as is the total number of molecules

$$N = \sum_j n_j. \tag{27}$$

Then a basic axiom of statistical mechanics states that all possible allocations of the total energy $\mathscr{E}$ to the N (distinguishable) molecules are equally likely. The distribution of molecules over energy levels is then just like the distribution of molecules over cells in Section 4.1, and the probability of any given vector of occupation numbers $\mathbf{n} = \{n_j\}$ is

$$P(\mathbf{n}) \propto \prod_j \frac{1}{n_j!} \tag{28}$$

as in (4.2), except that distribution (28) is now subject to the energy constraint (26) as well as to the numbers constraint (27).

The basic assumption is subject to much qualification, modification and refinement, but for present purposes we shall take it in the form stated. In Chapter 10 we shall consider a dynamic model which would yield it as a consequence.

Suppose we now go to the thermodynamic limit, with prescription of both particle density and energy density. That is, we suppose the particles distributed in a region of volume V, and let V, N and $\mathscr{E}$ all tend to infinity in such a way that N/N and $\mathscr{E}/V$ have prescribed limit values. Then a basic theorem of statistical mechanics, a consequence of the axiom just stated, asserts that the proportion of molecules in the jth energy level, n_j/N, tends to

$$\pi_j \propto e^{-\beta \varepsilon_j}, \tag{29}$$

where β is a constant adjusted to give the correct energy per molecule. This is *Gibbs' distribution*. Actually, since we are considering the molecules as distributed also in space, a better way to put this might be to assert that the spatial density (in the thermodynamic limit) of molecules of energy ε_j is

$$\rho_j = e^{-\alpha - \beta \varepsilon_j}, \tag{30}$$

where the constants α and β are adjusted to give the correct molecular and energy densities.

The usual 'proof' of (30) goes as follows. We look for the most probable values of the n_j, i.e. the values maximizing expression (28) subject to constraints (26) and (27) for large $\mathscr{E}$ and N. The n_j will then be correspondingly large, so if we use the Stirling approximation for $\log(n_j!)$ in the expression for $\log P(\mathbf{n})$ and take account of constraints (26) and (27) by Lagrangian multipliers α and β we find ourselves maximizing an expression

$$\sum_j (n_j - n_j \log n_j) + \alpha \left(N - \sum_j n_j \right) + \beta \left(\mathscr{E} - \sum_j n_j \varepsilon_j \right) + \log V \sum_j n_j,$$

where the last term has been added for convenience. If we neglect the fact that the n_j must be integer-valued then we find that the maximum is attained for the value of n_j/V asserted in (30).

This proof leaves many gaps. One is the question of whether the most probable values locate limit values in any sense. Another is the propriety of the Lagrangian calculations. However, the neglect of the integral character of the n_j raises an even more delicate point: for given $\mathscr{E}$ and ε_j, do equations (26) and (27) possess sufficiently many solutions in integers that these methods work at all? Suppose, for example, that $\varepsilon_j = \log p_j$, where p_j is the jth prime. Then, since the decomposition of an integer into prime factors, $e^{\mathscr{E}} = \prod_j p_j^{n_j}$, is unique, there is at most one set of integers solving (26) and (27), and there is no reason at all why these values should be compatible with (30).

These difficulties vanish if the solution is given a little latitude, by allowing $\mathscr{E}$ to take values in a narrow band rather than one prescribed value. (Note how frequently it is easier for a result to hold in a 'blurred' rather than a 'hard' version.) However, we shall avoid them completely by taking a case which permits exact treatment by elementary methods. This is the case $\varepsilon_j = hj$, for which a molecule of energy ε_j can be regarded as possessing j energy quanta of magnitude h ($j = 0, 1, 2, \ldots$). In this case we can easily prove a considerably stronger result than (30).

Theorem 6.3.1. *Consider the process with energy levels $\varepsilon_j = hj$ ($j = 0, 1, 2, \ldots$) in the thermodynamic limit. Let N_j be the number of molecules of energy ε_j in a specimen region of unit volume. Then the r.v.s N_j are independently distributed as Poisson variables with respective expectations ρ_j given by (30), where the parameters α and β have the values which assure the prescribed molecular and energy densities.*

PROOF. Denote the specimen region by A, and its complement (in the case of finite V) by $\bar{A}$. Also set $\mathscr{E} = Mh$, so that M is the prescribed number of energy quanta in the system. Let the number of molecules of energy hj in $\bar{A}$ be denoted $\bar{N}_j$. Then the joint distribution of the energy occupation numbers in A and $\bar{A}$ is given by

$$P(\mathbf{n}, \bar{\mathbf{n}}) \propto \prod_j \frac{(V-1)^{\bar{n}_j}}{n_j! \, \bar{n}_j!}, \tag{31}$$

this being subject to

$$\sum_j (n_j + \bar{n}_j) = N, \qquad \sum_j j(n_j + \bar{n}_j) = M. \tag{32}$$

Let us now write

$$n = \sum_j n_j, \qquad m = \sum_j jn_j,$$

for the numbers of molecules and quanta in A. Summing expression (31) over $\bar{\mathbf{n}}$ consistent with (32) we find that

$$P(\mathbf{n}) \propto c(m, n) \prod_j \frac{1}{n_j!}, \tag{33}$$

where

$$c(m, n) = \text{coefficient of } z^{M-m} \text{ in } \left((V-1) \sum_j z^j \right)^{N-n} \Big/ (N-n)!$$

$$= (V-1)^{N-n} \times \text{coefficient of } z^{M-m} \text{ in } (1-z)^{n-N}/(N-n)!$$

$$= \frac{(V-1)^{N-n}(M+N-m-n-1)^{(M-m)}}{(M-m)! \, (N-n)!}$$

$$\propto \frac{M^{(m)} N^{(n)} (N-1)^{(n)}}{(V-1)^n (M+N-1)^{(m+n)}}$$

$$\to e^{-\alpha n - \beta h m} = \prod_j e^{-(\alpha + \beta h j) n_j}, \tag{34}$$

and

$$e^{-\alpha} = \frac{\rho^2}{\rho + \sigma}, \qquad e^{-\beta h} = \frac{\sigma}{\rho + \sigma}. \tag{35}$$

Here ρ and σ are the prescribed spatial densities of molecules and quanta.

Relations (33) and (34) imply the distributional assertions of the theorem. The reader can check that the evaluations (35) of α and β indeed give ρ and σ as the expected numbers of molecules and quanta in A. □

EXERCISES AND COMMENTS

1. In the case when energy levels take the values $\varepsilon_j = hj$ the Gibbs distribution (29) effectively asserts that the number of energy quanta associated with a randomly chosen molecule is geometrically distributed (although with a distribution starting from $j = 0$ rather than $j = 1$). This is a consequence of the basic axiom; distribution (28) essentially corresponds to the assumption that molecules are distinguishable

but quanta are not. If quanta had been distinguishable then (28) would have been replaced by

$$P(\mathbf{n}) \propto \prod_j \frac{1}{n_j!} \left(\frac{1}{j!}\right)^{n_j}$$

and the number of quanta associated with a molecule would have been Poisson distributed.

2. The distribution specified by (26)–(28) has a p.g.f. $E(\prod_j z_j^{n_j})$ which is proportional to the coefficient of $w^N \exp(\theta \mathscr{E})$ in the expansion of

$$\phi(z, w, \theta) = \prod_j \exp(w z_j e^{\theta \varepsilon_j}). \tag{36}$$

This expression of the p.g.f is another example of the 'embedding and conditioning' technique of Section 4.7, in which one regards the problem as a conditioned version of a freer problem. In expression (36) the n_j are indeed seen as independent Poisson variables, which are then conditioned by the constraints (26) and (27). This view of the situation is also to be found in the statistical–mechanical literature. If N and $\mathscr{E}$ are both prescribed then it is said that one is dealing with the micro-canonical ensemble; if $\mathscr{E}$ is allowed to take an appropriate random distribution then it is said that one is dealing with the canonical ensemble; if $\mathscr{E}$ and N are both allowed to be random then it said that one is dealing with the grand canonical ensemble. Note that we essentially moved from the first case to the last when we went to the thermodynamic limit and examined the statistics of a specimen finite region.

3. Expression (36) is that valid for Boltzmann statistics, when molecules are regarded as distinguishable. For Bose–Einstein statistics (when molecules are regarded as indistinguishable) and Fermi–Dirac statistics (when at most one molecule can occupy a given energy a given level) we have

$$\phi = \prod_j (1 \pm w z_j e^{n \varepsilon_j})^{\pm 1},$$

if we take the lower and upper option of sign, respectively. Note the similarity of the Fermi–Dirac case to the problem of Exercise 4.7.4: that of determining the distribution of total income in a sample chosen without replacement.

4. In a generalized version of the problem of the text one considers the 'joint Poisson' distribution (28) subject to r linear constraints

$$\sum_j n_j a_{jk} = M_k \qquad (k = 1, 2, \ldots, r),$$

where the a_{jk} and the M_k are integral. These would be the appropriate hypotheses if n_j were the number of molecules of type j, where such a molecule contains a_{jk} atoms of element k, this element being present in abundance M_k ($k = 1, 2, \ldots, r$). If one goes to the thermodynamic limit (so that the M_k become infinite, but in fixed ratios), then the analogue of the Gibbs distribution is that the density of molecules of type j has the limiting form

$$\rho_j = \exp\left(-\sum_k a_{jk} \beta_k\right),$$

where the β_k are Lagrange multipliers ('chemical potentials') associated with the abundance constraints.

4. Branching Processes

In 1873–74 Galton and de Candolle remarked on the many instances of family names that had become extinct, and, prompted by this observation, raised the general question: What is the probability that a natural population dies out in the course of time? Such a question cannot be answered by appeal to a deterministic model; the model must have a probabilistic element. For example, suppose that we measure time in generations, and that X_t is the number of population members in the tth generation. Then the simplest deterministic model would be

$$X_{t+1} = \alpha X_t, \tag{37}$$

where α is the multiplication rate from one generation to the next. The population will then increase indefinitely, decrease to zero or just maintain itself according as to whether α is greater than, less than or equal to unity. But these statements are too crude, particularly in small populations. They take no account of fluctuations, or even of the fact that X must be integer-valued. One would like to be able to determine, for instance, the probability that the line descending from a single initial ancestor ultimately becomes extinct.

A model for this situation was proposed and partially analysed by a clergy-man, H.W. Watson, in 1874; the analysis was completed by J.F. Steffensen in 1930. The model is the simplest example of what we now know as a *branching process*. It serves as a model, not only for population growth, but for other multiplicative phenomena, such as the spread of an infection or the progress of a nuclear fission reaction. The event 'indefinite survival of a population' corresponds in these two cases to the occurrence of an epidemic or of a nuclear explosion, respectively.

For concreteness we shall formulate our model in terms of the surname example and make the following idealizing assumptions: that the number of sons of different individuals (in whatever generation) are IID r.v.s. We restrict attention to male members of a line simply because it is through these that the family name is inherited. (The study of the statistical consequences of the fact that two sexes are necessary for reproduction is interesting; the study of the evolutionary advantages of sexual reproduction is fascinating; we have space to discuss neither.) We suppose that the probability that a man has j sons is p_j ($j = 0, 1, 2, \ldots$) independently of the numbers of individuals in his or previous generations or of the number of sons sired by other members of his generation. We can thus define a *progeny p.g.f.*

$$G(z) = \sum_0^\infty p_j z^j$$

with the property

$$E(z^{X_{t+1}} \mid X_t = 1) = G(z), \tag{38}$$

where X is now understood to be integral.

The model is thus a very idealized one: descendants are rather arbitrarily grouped into generations (which may ultimately overlap in time); effects such as environmental limitations or variation of birth-rate with population size are neglected, as is also sexual structure of the population. However, the model is still a valuable generalization of (37), and produces some interesting new effects and ideas.

If X_t has the value k then X_{t+1} is the sum of k independent variables each with p.g.f. $G(z)$, whence we see that relation (38) generalizes to

$$E(z^{X_{t+1}} \mid X_t) = G(z)^{X_t}. \tag{39}$$

Taking expectations in (39) with respect to X_t we then find that

$$\Pi_{t+1}(z) = \Pi_t[G(z)], \tag{40}$$

where $\Pi_t(z) = E(z^{X_t})$ is the p.g.f. of X_t. Relation (40) is the fundamental relation for this problem. It is a generalization of the deterministic recursion (37) and is, of course, very much more informative, since it relates distributions rather than simple numbers. Let us formalize our conclusion.

Theorem 6.4.1. *For the simple branching process the p.g.f. $\Pi_t(z)$ of the number of individuals in the tth generation obeys recursion* (40), *where $G(z)$ is the progeny p.g.f. If $X_0 = 1$, so that $\Pi_0(z) = z$, then*

$$\Pi_t(z) = G^{(t)}(z), \tag{41}$$

where $G^{(t)}(z)$ is the tth iterate of $G(z)$.

The second statement, for the case of a single ancestor in the 'zeroth' generation, follows by repeated application of (40). We then have $\Pi_1(z) = G(z)$, $\Pi_2(z) = G[G(z)]$, and, in general, (41) holds, where $G^{(t)}(z)$ is the function obtained by applying the transformation $z \to G(z)$ to z t times.

Equation (41) solves the problem in the same sense that $X_t = \alpha^t X_0$ solves the deterministic relation (37); it determines the distribution of X_t as well as is possible from knowledge of X_0. The probabilistic problem is thus reduced to the analytic one of calculating the tth iterate of a function $G(z)$. The problem is classic and difficult, and one can solve it explicitly only in a few cases (see Exercises 4–6). However, one can extract a certain amount of useful information from relations (40) and (41) without actually evaluating the iterate.

For example, we can use (40) to obtain recursions for the moments of X_t. Differentiating once and setting $z = 1$ we obtain

$$E(X_{t+1}) = \alpha E(X_t), \tag{42}$$

where

$$\alpha = G'(1) = \sum_0^\infty jp_j \tag{43}$$

is the expected number of sons born to a man.

Relation (42) corresponds nicely to the deterministic relation (37), and has the general solution $E(X_t) = \alpha^t E(X_0)$. Differentiating relation (40) twice at $z = 1$ we obtain, with some reduction

$$\text{Var}(X_{t+1}) = \alpha^2 \text{ var}(X_t) + \beta E(X_t),$$

where β is the variance of the number of sons born to a man. This difference equation has the solution in terms of values at $t = 0$

$$\text{var}(X_t) = \alpha^2 \text{ var}(X_0) + \frac{\beta \alpha^{t-1}(\alpha^t - 1)}{\alpha - 1} E(X_0).$$

One can continue in this way, and calculate the moments of X_t as far as one has patience. However, it is more illuminating to calculate the *extinction probability* of which we spoke at the very beginning of the section. Define

$$\rho_t = P(X_t = 0) = \Pi_t(0), \tag{44}$$

so that ρ_t is the probability of extinction by the tth generation. This is not to be confused with the event 'extinction *at* the tth generation', which would have probability $\rho_t - \rho_{t-1}$. The sequence $\{\rho_t\}$ is nondecreasing (as shown by this last remark). Since it is also bounded, it must then have a limit value $\rho = \lim_{t \to \infty} \rho_t$. This is to be interpreted as 'the probability of extinction in finite time' or 'the probability of ultimate extinction'.

Theorem 6.4.2. *Let ρ_t be the probability of extinction by generation t and let ρ be the probability of ultimate extinction, both for the case of a single initial ancestor. Then:*

(i) *The recursion*

$$\rho_{t+1} = G(\rho_t) \qquad (t = 0, 1, 2, \ldots) \tag{45}$$

holds, with initial condition $\rho_0 = 0$, and ρ is a root of the equation

$$z = G(z). \tag{46}$$

(ii) *If $G(z)$ is not identically equal to z (i.e. if a man does not have exactly one son with probability one), then equation (46) has just two positive real roots, of which $z = 1$ is always one. The extinction probability ρ is the smaller of the two roots, and is less than or equal to unity according as $\alpha > 1$ or $\alpha \leq 1$.*

This last result tallies with conclusions from the deterministic case, although it is interesting to note that extinction is also certain in the transitional case $\alpha = 1$, when a man replaces himself exactly on average.

PROOF. Both the assertions of (i) follow from

$$\rho_{t+1} = G^{(t+1)}(0) = G[G^{(t)}(0)] = G(\rho_t). \tag{47}$$

We shall give only an outline of the proof of (ii); the neglected details are relegated to Exercise 13.

Note first that, since $G(z)$ is a power series with positive coefficients, its derivatives exist on $z \geq 0$ in any open interval for which $G(z)$ converges (certainly for $0 \leq z < 1$) and are themselves positive. In particular, since z has an increasing first derivative (i.e. it is *convex*) the graph of $G(z)$ is intersected by any straight line in at most two points. (The condition that $G(z)$ be not identically equal to z, i.e. that there is some spread in the distribution of progeny, implies that G is *strictly* convex.) Thus equation (46) has at most two roots, and one of these is certainly $z = 1$.

We leave it to the reader to convince himself that the cobweb construction in Fig. 6.1 generates the sequence $\{\rho_t\}$. Graphically, it is obvious that the sequence converges to the first intersection of the two curves, i.e. to the smaller positive root of (46). The point is proved analytically in Exercise 13.

Now, since at the smaller root the graph of $G(z)$ crosses the graph of z from above, we must have $G'(\rho) < 1$ if the two roots of (46) (say ρ and ρ') are distinct. It is then also true that $G'(\rho') > 1$. Now, since one of ρ and ρ' equals unity and $\alpha = G'(1)$, we must have $\rho = 1$ if $\alpha < 1$, and $\rho' = 1$ (i.e. $\rho < 1$) if $\alpha > 1$, as asserted. In the transitional case $\alpha = 1$ equation (46) will have a double root, so that ρ and ρ' coincide and both equal unity.   □

As an example, consider the progeny p.g.f.

$$G(z) = \frac{pz + q}{1 + r - rz}. \tag{48}$$

This is indeed a p.g.f. if $p + q = 1$ and $p, q, r \geq 0$. It is the p.g.f. of a *modified geometric distribution*, in which p_j falls off geometrically from $j = 1$ onwards.

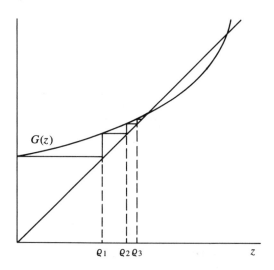

Figure 6.1. A construction for the extinction probabilities of a branching process; iteration of the function $G(z)$.

For this case, $\alpha = p + r$ and equation (46) has the two solutions $z = 1$ and $z = q/r$, so that $\rho = \min(1, q/r)$. Indeed, q/r is less than unity just when $\alpha = p + r$ exceeds unity.

Lotka found that the distribution of sons for U.S. males in 1920 was very well represented by $p_0 = 0.4981$, $p_j = 0.4099(0.5586)^j$ ($j > 0$), which corresponds to the p.g.f. (48) with $q/(1 + r) = 0.4981$ and $r/(1 + r) = 0.5586$. Thus $\alpha = p + r = 1.14$ and $\rho = q/r = 0.89$, so that, on our idealizing assumptions, a surname carried by just one man will ultimately disappear with probability 0.89. If it is known to be held by k men, then the probability reduces to $(0.89)^k$. So, the Smiths of this world are unlikely to become extinct, despite the fact that a probability of 0.89 seems high.

EXERCISES AND COMMENTS

1. Show that the assumptions made imply the stronger form of (39):
$$E(z^{X_{t+1}} | X_0, X_1, X_2, \ldots, X_t) = G(z)^{X_t}.$$

2. Show that the general solution of equation (40) is $\Pi_t(z) = \Pi_0[G^{(t)}(z)]$.

3. Consider the case of one initial ancestor. Show from first principles, without appealing to the solution (41), that $\Pi_{s+t}(z) = \Pi_s[\Pi_t(z)]$ for $s, t = 0, 1, 2, \ldots$.

4. Consider the p.g.f. (48). Show that its tth iterate has the form $(a_t z + b_t)/(c_t z + d_t)$, where
$$\begin{bmatrix} a_t & b_t \\ c_t & d_t \end{bmatrix} = \begin{bmatrix} p & q \\ -r & 1+r \end{bmatrix}^t.$$

5. Show that $1 - \gamma(1 - z)^\delta$ is a p.g.f. if $0 \le \gamma, \delta \le 1$. Evaluate its tth iterate.

6. Show that if $G(z)$ is the p.g.f. of a nonnegative r.v. and can be represented $G(z) = J^{-1}[1 + J(z)]$ then $G^{(t)}(z) = J^{-1}[t + J(z)]$, and that this is also a p.g.f.

7. Use relation (45) to show that ρ_t is nondecreasing.

8. Complete the derivation of the expression for $\text{var}(X_t)$.

9. Show that, for the case of Exercise 5, $\alpha = +\infty$ and $\rho = 1 - \gamma^{1/(1-\delta)}$.

10. Suppose we modify the process by introducing immigration, so that $X_{t+1} = X'_{t+1} + Y_{t+1}$, where X'_{t+1} is the number of progeny of the tth generation and Y_{t+1} is the number of immigrants into generation $t + 1$. Suppose that Y_{t+1} is independent of all previous r.v.s, including X'_{t+1}, and has p.g.f. $H(z)$. Show that recursion (40) is modified to
$$\Pi_{t+1}(z) = H(z)\Pi_t[G(z)],$$
and deduce the form of the limit p.g.f. (if any) in the case $G(z) = pz + q$, $H(z) = \exp[\lambda(z - 1)]$.

11. Suppose that there are several types of individual in each generation. Find a model of this situation for which the treatment of this section generalizes, as least as far as deduction of an analogue to (40).

12. Suppose $X_0 = 1$, and define $S = \sum_{t=0}^{\infty} X_t$, the total size of the line. This will be a finite r.v. only in the case $\rho = 1$. Show that its p.g.f. $\Phi(z)$ obeys the relation $\Phi(z) = zG[\Phi(z)]$ and evaluate Φ in the case (48). This calculation is of interest in the theory of polymers—organic substances whose molecules are formed by the association of many smaller units. In some cases one can imagine that the molecule grows by branching, and S is then the total size of the molecule. If conditions change so that ρ becomes less than unity, then infinite values of S become possible. This corresponds to a change in state of the polymer substance: from the 'sol' state to the 'gel' state.

13. Let the smaller real root of (46) be denoted by ξ. Then $G'(\xi)$ will exist and $0 \le G'(\xi) < 1$. Show that $\xi \ge p_0$. Show also that, in the range $0 \le z \le \xi$,

$$\xi + G'(\xi)(z - \xi) \le G(z) < \xi + \frac{\xi - p_0}{\xi}(z - \xi).$$

Using these inequalities and (47) show that

$$\left(\frac{\xi - p_0}{\xi}\right)(\xi - \rho_t) \le (\xi - \rho_{t+1}) \le G'(\xi)(\xi - \rho_t).$$

The first inequality demonstrates that if ρ_t is not larger than ξ, then neither is ρ_{t+1}, so that ξ is the only possible limit point for the sequence $\{\rho_t\}$, and $\rho = \xi$. From the second inequality we conclude that this limit is reached exponentially fast:

$$|\rho_t - \rho| \le \rho[G'(\rho)]^t.$$

CHAPTER 7

The Two Basic Limit Theorems

Suppose that X_1, X_2, X_3, ... are scalar-valued random variables, independently and identically distributed (IID). There is an enormous literature on the behaviour of the sums $S_n = \sum_{j=0}^{n} X_j$ and averages $\overline{X}_n = S_n/n$ of such variables as n becomes large. The motivation is clear from the discussion of Section 1.2. Sample averages of actual data are observed to 'converge' with increasing sample size, and it was this 'limit' which we idealized to provide the concept of an expectation. There is then interest in confirming whether such behaviour can be reproduced within the theory, in that one can demonstrate convergence of $\overline{X}_n$ in some well-defined sense, under appropriate assumptions.

Convergence can indeed be demonstrated, and the techniques for doing so turn out to be of considerable interest in themselves.

In this chapter we shall prove two basic limit results: the law of large numbers and the central limit theorem, both stated as theorems on *convergence in distribution*. The methods used do not yield the strongest possible results, but are natural, direct and powerful.

1. Convergence in Distribution (Weak Convergence)

Let ξ_1, ξ_2, ξ_3, ... be a sequence of r.v.s and ξ yet another r.v., all of these taking values in the same space. Then we shall say that the sequence $\{\xi_n\}$ *converges to ξ in distribution* if

$$E[H(\xi_n)] \to E[H(\xi)] \tag{1}$$

as $n \to \infty$, for any scalar-valued, bounded, continuous H. This is expressed in brief as $\xi_n \xrightarrow{D} \xi$. Convergence in distribution does not imply actual convergence of ξ_n to ξ, in that the difference between the two r.v.s would approach zero in some sense, but rather that the difference between the *probability distributions* of ξ_n and ξ approaches zero. The requirement (1) of the convergence of expectations for test functions H in the class indicated provides the operational touchstone.

What one means by 'continuous' depends upon the space Ξ in which the r.v.s take values, but it is clear enough what is meant if $\Xi = \mathbb{R}^d$, say. For an example, let ξ_n be the r.v. which takes the values j/n ($j = 1, 2, 3, \ldots, n$) each with probability $1/n$. Then

$$E[H(\xi_n)] = \frac{1}{n} \sum_{j=1}^{n} H(j/n) \to \int_0^1 H(x)\, dx,$$

which implies that $\xi_n \overset{D}{\to} \xi$, where ξ is a r.v. which is *rectangularly* distributed on $(0, 1)$. That is, it is a r.v. whose probability density is constant on this interval and zero elsewhere.

So, the distribution of ξ_n does not posses a density for any n, but, as far as the expectations of bounded continuous functions are concerned, its distribution is essentially uniform for large n. It is only because we assume the test functions H continuous that the increasing density of possible values of the discrete-valued r.v. ξ_n ultimately gives the effect of a true probability density. The following example brings out the point further.

Suppose that ξ_n takes the sole value $1/n$, so that, as n increases, ξ_n approaches the value 0 without ever actually reaching it. Then, for any bounded continuous H,

$$E[H(\xi_n)] \to H(0), \tag{2}$$

corresponding to a limit distribution concentrated on $\xi = 0$. On the other hand, suppose we take H as the indicator function of the set $\xi \le 0$, and so discontinuous. Then $H(0) = 1$, but $H(1/n) = 0$, so that

$$E[H(\xi_n)] = 0 \quad \to \quad 0 \ne H(0). \tag{3}$$

The discrepancy between (3) and (2) stems from the fact that, in case (3), one has chosen a test function which implies that one can distinguish between the values $\xi = 0$ and $\xi = 0+$. Such a distinction can be held to be meaningless in that it is physically unverifiable. The fact that we consider only continuous test functions is a reflection of this view.

The reasons for restricting one's self to bounded test functions are similar, see Exercise 1, although less compelling. The point is that bounded continuous functions always possess an expectations, so that this class of test functions has the useful property of not being specific to the problem. There are good reasons for wishing to consider the expectations of unbounded functions as well, but the functions of interest will indeed be specific to the problem.

One may ask: In order to establish convergence in distribution, is it sufficient to establish the convergence (1) for some subclass of the bounded continuous functions, or must one consider the full set? In fact, there are

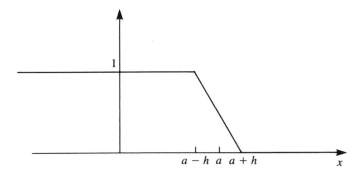

Figure 7.1. A continuous approximation to the indicator function $I(x \leq a)$.

subclasses (the so-called separating classes) which provide a sufficient set of test functions.

In the special case when the r.v.s ξ_n take values x on the real line there are two important separating classes. One consists of the continuous approximations to the indicator functions $I(x \leq a)$:

$$H(x; a, h) = \begin{cases} 1 & (x \leq a - h), \\ \frac{1}{2}[1 + (a - x)/h] & (a - h \leq x \leq a + h), \\ 0 & (x \geq a + h); \end{cases} \qquad (4)$$

see Fig. 7.1. The parameters a and h are real, with h nonnegative. The other class consists of the complex exponentials

$$H(x; 0) = e^{i\theta x}, \qquad (5)$$

with the parameter θ taking real values.

We shall not prove the separating character of the class (4), although it is easily seen that convergence (1) for members of this class has the implications for which one would hope (see Exercise 2). Consideration of class (5) leads us to define the *characteristic functions*

$$\phi_n(\theta) = E(e^{i\theta\xi_n}), \qquad \phi(\theta) = e(e^{i\theta\xi}),$$

in terms of which convergence (1) becomes

$$\phi_n(\theta) \to \phi(\theta). \qquad (6)$$

The characteristic function (henceforth often abbreviated to c.f.) is the generalization of the p.g.f. to the case of continuous r.v.s, and supplies the natural tool for the study of sums of independent r.v.s, just as did the p.g.f. That the convergence of c.f.s (6) is sufficient for convergence in distribution (i.e. that the class (5) is separating) is something we shall prove first in Section 15.6, although the material of the next section almost implies it.

An alternative to the term 'convergence in distribution' which is much used is 'weak convergence'. Correspondingly, one says that ξ_n 'converges weakly' to ξ if $\xi_n \overset{D}{\to} \xi$. The term is acceptable in that it is in wide use, and a good one in that it marks a break from the rather narrow idea of distribution on the real line. However, one must be careful not to confuse it with the concept of 'weak convergence in probability' (see Section 13.2) which is something quite different.

EXERCISES AND COMMENTS

1. Suppose the r.v. ξ_n takes the values j/n ($j = 1, 2, 3, \ldots, n-1$) and n, each with probability $1/n$. Then $\xi_n \overset{D}{\to} \xi$, where ξ is uniformly distributed on $(0, 1)$. Note, however, that $E(\xi_n^\nu)$ converges to $E(\xi^\nu)$ if $0 \le \nu < 1$, but converges to $+\infty$ if $\nu > 1$. The discrepancy is caused by the probability $1/n$ assigned to the value n. Whether the escape of this shrinking atom of probability to infinity has real significance is something that, in a physical problem, would be resolved on physical grounds.

2. Consider the case of variables on the real line, and suppose that ξ_n and ξ have distribution functions $F_n(x)$ and $F(x)$. (That is, $F(x) = P(\xi \le x)$.) Show that convergence (1) for the function (4) implies that

$$\int_{a-h}^{a+h} F_n(x)\, dx \to \int_{a-h}^{a+h} F(x)\, dx,$$

and hence that $F_n(a) \to F(a)$ for all a at which F is continuous. As argued in the text, absence (or ambiguity) of convergence at the discontinuity points is immaterial.

3. Suppose that $\xi_n \overset{D}{\to} \xi$ and the limit r.v. ξ is a constant: a, say. Then this is the one case where convergence in distribution implies a convergence of ξ_n to ξ itself. The convergence holds in the sense that, if $d(\xi, \xi')$ is any bounded continuous distance function between the elements of Ξ then, in this case, $E[d(\xi_n, \xi)] = E[d(\xi_n, a)] \to d(a, a)$.

   Suppose that the ξ_n take values in a finite-dimensional Euclidean space. Then show that the assumption above implies *convergence in probability*, i.e. that $P(|\xi_n - a| > \varepsilon) \to 0$ for any prescribed positive ε. Show that convergence in probability to ξ (not necessarily constant) also implies convergence in distribution to ξ. In fact, all modes of stochastic convergence imply convergence in probability (see p. 238); the weakest convergence concept.

2. Properties of the Characteristic Function

The characteristic function (c.f.) will prove a consistent and valuable tool; we should derive its principal properties. If X is a scalar r.v. then we define

$$\phi(\theta) = E(e^{i\theta X}) \tag{7}$$

as the c.f. of X. We shall write it as $\phi_X(\theta)$ if there is need to emphasize its relation to X.

If X is integer-valued then ϕ is related to the p.g.f. $\Pi(z)$ by

$$\phi(\theta) = \Pi(e^{i\theta}),$$

and so differs from it only by a transformation of the argument.

We note a number of elementary properties of c.f.s.

Theorem 7.2.1.

(i) $\phi(\theta)$ exists for θ real, when $|\phi(\theta)| \le 1$, with $\phi(0) = 1$.
(ii) If θ is real then $\overline{\phi(\theta)} = \phi(-\theta)$.
(iii) If a and b are constants, then

$$\phi_{a+bX}(t) = e^{i\theta a}\phi(b\theta).$$

In particular, the c.f. of a r.v. taking the value a with probability one is $e^{i\theta a}$.

Assertion (i) follows from the fact that

$$|E(e^{i\theta X})| \le E(|e^{i\theta X}|) = E(1) = 1.$$

The others are immediate consequences of the definition.

However, the key property of a c.f., as it is of a p.g.f., is the following.

Theorem 7.2.2. If X and Y are independent r.v.s then

$$\phi_{X+Y}(\theta) = \phi_X(X)\phi_Y(\theta). \tag{8}$$

This follows, as ever, from the factorization consequent on independence:

$$E(e^{i\theta(X+Y)}) = E(e^{i\theta X})E(e^{i\theta Y}).$$

Again, the result extends to the sum of several variables.

Another characteristic which is useful, although secondary, is the moment generating property. By formally expanding the exponential under the expectation sign in (7) we obtain

$$\phi(\theta) = \sum_{j=0}^{\infty} \frac{(i\theta)^j}{j!} E(X^j), \tag{9}$$

$$E(X^j) = i^{-j}\phi^{(j)}(0) \qquad (j = 0, 1, 2, \ldots), \tag{10}$$

where $\phi^{(j)}$ is the jth differential of ϕ at $\theta = 0$ (see Theorem 4.4.3). However, these results cannot be generally valid, because $E(X^j)$ may not exist. We shall give an exact statement in Theorem 7.2.4.

One may ask whether knowledge of the c.f. for real θ determine the distribution. Essentially it does. We have

$$\phi(\theta) = \int_{-\infty}^{\infty} e^{i\theta x}\, dF(x) = \int_{-\infty}^{\infty} e^{i\theta x} f(x)\, dx, \tag{11}$$

the first relation holding generally, and the second if a density f exists. The *inversion theorem*, which we state without proof (although see Exercise 7.5.6) then states that (11) has the inversion

$$f(x) = \frac{1}{2\pi} \int_{-\infty}^{\infty} e^{-i\theta x} \phi(\theta) \, d\theta \tag{12}$$

if a density exists, and that, in general,

$$F(x) - F(y) = \frac{1}{2\pi} \int_{-\infty}^{\infty} \frac{e^{-i\theta x} - e^{-i\theta y}}{-i\theta} \phi(\theta) \, d\theta \tag{13}$$

if x and y are continuity points of F. (Now one can see why convergence of c.f.s should be sufficient for convergence in distribution: the c.f. essentially determines the distribution function, and this in turn determines the expectation of any bounded continuous function.)

One often defines the expansion

$$\log \phi(\theta) = \sum_{j=0}^{\infty} \frac{\kappa_j (i\theta)^j}{j!}$$

(in so far as it is valid) where the κ_j are the *cumulants* of the distribution. Note that $\kappa_1 = \mu_1 = E(X)$ and $\kappa_2 = \mu_2 - \mu_1^2 = \text{var}(X)$. We give an interpretation of other cumulants in Section 5. A key property of cumulants follows from Theorem 7.2.2: the cumulant of order j for a sum of independent r.v.s is the sum of the cumulants of order j of the summands.

The c.f.s for the standard distributions we have encountered hitherto are most easily presented in a table.

Distribution	Probability or density	Characteristic function
Degenerate	$X = a$ with probability one	$e^{i\theta a}$
Binomial	$P(X = r) = \binom{n}{r} p^r q^{n-r}$	$(pe^{i\theta} + q)^n$
Poisson	$P(X = r) = e^{-\lambda}\lambda^r/r!$	$\exp[\lambda(e^{i\theta} - 1)]$
Negative binomial	$P(X = r) = \binom{n-1}{r-1} p^r q^{n-r}$	$[pe^{i\theta}/(1 - qe^{i\theta})]^r$
Rectangular	$f(x) = (b-a)^{-1}$ on (a, b)	$(e^{i\theta b} - e^{i\theta a})/[i\theta(b-a)]$
Exponential	$f(x) = \lambda e^{-\lambda x}$ $(x \geq 0)$	$(1 - i\theta/\lambda)^{-1}$
Gamma	$f(x) = \lambda^\nu e^{-\lambda x} x^{\nu-1}/\Gamma(\nu)$	$(1 - i\theta/\lambda)^{-\nu}$

The following theorem clears the way for a proper treatment of the moment generating property.

Theorem 7.2.3. *If X is finite with probability one, then $\phi(\theta)$ is uniformly continuous. If $E(|X|^\nu) < \infty$ then $\phi_\nu(\theta) = E(X^\nu e^{i\theta X})$ is uniformly continuous in θ.*

PROOF. Note that for real θ and θ^*

$$|e^{i\theta} - e^{i\theta^*}| \leq 2$$

and also that

$$|e^{i\theta} - e^{i\theta*}| \leq \left|\int_{\theta*}^{\theta} \frac{de^{i\zeta}}{d\zeta} d\zeta\right| \leq \left|\int_{\theta*}^{\theta} \left|\frac{de^{i\zeta}}{d\zeta}\right| d\zeta\right| = |\theta - \theta*|.$$

Thus

$$|\phi(\theta) - \phi(\theta*)| \leq E(|e^{i\theta X} - e^{i\theta*X}|) \leq 2P(|X| \geq A) + A|\theta - \theta*|,$$

as we see by using the first bound above for $|X| \geq A$ and the second for $|X| < A$. By virtue of the finiteness assumption we can choose A so large that $2P(|X| \geq A) \leq \frac{1}{2}\varepsilon$, and then $|\theta - \theta*|$ so small that $A|\theta - \theta*| \leq \frac{1}{2}\varepsilon$, and so obtain $|\phi(\theta) - \phi(\theta*)| \leq \varepsilon$, where ε is an arbitrarily small positive quantity. The first assertion is thus established, and the second is proved analogously. $\square$

Theorem 7.2.4 (The Limited Expansion Theorem). *If $E(|X|^{\nu}) < \infty$ for a given integer ν then*

$$\phi(\theta) = \sum_{j=0}^{\nu} \frac{(i\theta)^j}{j!} E(X^j) + o(\theta^{\nu}) \tag{14}$$

and (10) holds for $j = 0, 1, 2, \ldots, \nu$.

PROOF. We have the limited Taylor expansion

$$e^{i\theta X} = \sum_{j=0}^{\nu} \frac{(i\theta X)^j}{j!} + \frac{(i\theta)^{\nu}}{(\nu - 1)!} \int_0^1 (e^{i\theta Xt}X^{\nu} - X^{\nu})(1 - t)^{\nu-1} \, dt.$$

Taking expectations in this relation, we find that we have expansion (14), but with the remainder term evaluated as

$$\frac{(i\theta)^{\nu}}{(\nu - 1)!} \int_0^1 [\phi_{\nu}(\theta t) - \phi_{\nu}(0)](1 - t)^{\nu-1} \, dt.$$

But the integral is $o(1)$ for small θ, because of the uniform continuity of ϕ_{ν}. The whole expression is thus $o(\theta^{\nu})$. The validity of (14) thus proved, and the second assertion of the theorem follows. $\square$

One can ask whether the converse result holds: that the existence of $\phi^{(\nu)}(0)$ implies the existence of $E(X^{\nu})$. The statement is true for ν even, but not for ν odd; see Section 15.6.

One can also consider the joint c.f. of several r.v.s. If X is a random vector with elements X_j then we can define the c.f. with vector argument θ

$$\phi(\theta) = E\left[\exp\left(i \sum_j \theta_j X_j\right)\right] = E(e^{i\theta^T X}). \tag{15}$$

All the theorems above have fairly obvious vector versions, especially since expression (15) can be regarded as the univariate c.f. of a linear function of the X_j.

EXERCISES AND COMMENTS

1. One can also define the *moment generating function* (m.g.f.) $M(\alpha) = E(e^{\alpha X})$, which is indeed just the c.f. if we set $\alpha = i\theta$. This may or may not exist if α is not purely imaginary, but is often convenient to work with. Note that we have already appealed to the notion of a c.f or of a m.g.f. in the income distribution example of Exercises 2.9.5 and 4.7.4 and in the energy distribution example of Exercises 6.3.2 and 6.3.3.

2. It is not obvious that the function $\phi(\theta) = e^{-|\theta|}$ is a c.f., but in fact it is. Appealing to the inversion formula (12) we have

$$f(x) = \frac{1}{2\pi}\left(\int_0^\infty e^{-\theta - i\theta x}\, dx + \int_{-\infty}^0 e^{\theta - i\theta x}\, dx\right) = \frac{1}{2\pi}\left(\frac{1}{1 + ix} + \frac{1}{1 - ix}\right) = \frac{1}{\pi(1 + x^2)},$$
(16)

which is indeed a density, the density of the *Cauchy distribution*. We shall find a physical origin for this distribution in Chapter 8, and shall see that it has other features before then. Note from (16) that the integral representing $E(X)$ is divergent; the distribution does not posses a mean. This is reflected in the fact that the c.f. does not have a Taylor expansion about the origin which goes as far even as the term linear in θ.

3. If $\phi(\theta)$ is a c.f., then of what r.v. is $|\phi(\theta)|^2$ the c.f.?

4. Consider a sum $S_N = \sum_{j=0}^N X_j$, where the X_j are IID with c.f. $\phi(\theta)$ and N is also a r.v., independent of the X_j, with p.g.f. $\Pi(z)$. Show that S has c.f. $\Pi[\phi(\theta)]$, and hence that $E(S) = E(N)E(X)$ and $\text{var}(S) = E(N)\,\text{var}(X) + \text{var}(N)[E(X)]^2$. For example, S might be the total claim paid by an insurance company in a year, on a random number of random claims.

5. Consider the renewal problem of Section 6.1 in the case when lifetimes may be continuously distributed. Show that, under the assumptions of that section (independent lifetimes with c.f. $\phi(\theta)$ and an initial renewal at $t = 0$), we have

$$\int_0^\infty e^{i\theta t}\, dM(t) = \frac{1}{1 - \phi(\theta)},$$

where $M(t)$ is the expected number of renewals in $[0, t]$. (Actually, the formula is valid only for θ such that $\text{Im}(\theta) > 0$, just as (6.1.4) is valid only for $|z| < 1$.)

6. Suppose the vector r.v. X has expectation μ and covariance matrix V, and let $\phi(\theta)$ be its c.f. Show that

$$\log \phi(\theta) = i\theta^T\mu - \tfrac{1}{2}\theta^T V\theta + \cdots,$$

where $+ \cdots$ indicates terms of smaller order than the second in θ.

3. The Law of Large Numbers

Theorem 7.3.1. *Suppose that scalar r.v.s X_1, X_2, X_3, ... are IID, and possess mean μ. Define $S_n = \sum_{j=1}^n X_j$. Then $\bar{X} = S_n/n$ converges in distribution to the constant μ as $n \to \infty$.*

PROOF. By Theorem 7.2.4 the c.f. of the X_j has the limited expansion

$$\phi(\theta) = 1 + i\mu\theta + o(\theta) = e^{i\mu\theta + o(\theta)},$$

and so $\overline{X}_n$ has c.f.

$$E(e^{i\theta\overline{X}_n}) = [\phi(\theta/n)]^n = e^{i\mu\theta + no(\theta/n)} \to e^{i\mu\theta}.$$

The final limit relation holds because the term $no(\theta/n)$ tends to zero as n increases for fixed θ. The limit expression is the c.f. of a distribution degenerate at μ, whence the assertion of the theorem. ☐

As noted in Exercise 7.1.3, convergence in distribution to a constant is equivalent to convergence to that constant in probability, so that

$$P(|\overline{X}_n - \mu| > \varepsilon) \to 0. \tag{17}$$

This is the result often referred to as 'the weak law of large numbers'. However, the conclusion is in some respects weaker than the result already obtained by simple means in Exercises 2.8.6 and 2.9.14, where it was shown that if the X_j are uncorrelated, with common mean μ and common variance, then $\overline{X}_n$ converges to μ in mean square, and so also in probability. Theorem 7.3.1 thus yields a weaker conclusion, under hypotheses which are in one sense stronger (independence is assumed, rather than lack of correlation) and in one sense weaker (existence of a first moment is assumed, but not of a second).

However, the techniques used in the proof of the theorem lead on directly to stronger results: the central limit theorem of the next section. For results which are stronger in other respects (the strong law of large numbers: that $\overline{X}_n$ converges almost surely to μ) we will have to wait until Chapter 13.

EXERCISES AND COMMENTS

1. Suppose the X_j follow the Cauchy distribution of Exercise 7.2.2. Show, by evaluation of its c.f., that $\overline{X}_n$ follows the same Cauchy distribution. The sample average thus converges in distribution, but not to a constant. Why does Theorem 7.3.1 not apply in this case?

2. Continuing Exercise 1, show that the more general linear function $\sum_j c_j X_j / \sum_j |c_j|$ also follows the same Cauchy distribution.

4. Normal Convergence (the Central Limit Theorem)

We are now about to encounter the normal distribution, a distribution which occupies a central and unique place in probability theory. However, for all that it pervades both theory and application, it is not a distribution which arises as immediately from a physical context as do those which we have encountered up to now and listed in Section 2. It arises as a limit distribution, and we meet it first through its c.f.

$$\phi(\theta) = e^{-\theta^2/2}. \tag{18}$$

This is the c.f. of the *standard normal distribution*. That of the normal distribution with mean μ and variance σ^2 is

$$\phi(\theta) = e^{i\mu\theta - (1/2)\sigma^2\theta^2}. \tag{19}$$

We shall examine the distribution itself in the next section.

Theorem 7.4.1 (The Central Limit Theorem). *Suppose that scalar r.v.s* X_1, X_2, X_3, ... *are IID and possess a mean* μ *and variance* σ^2. *Define* $S_n = \sum_{j=1}^n X_j$. *Then*

$$u_n = \frac{S_n - n\mu}{\sigma\sqrt{n}} \tag{20}$$

converges in distribution to a standard normal variable as $n \to \infty$.

One usually telescopes the statement by simply saying that S_n (or u_n) *converges to normality*.

The theorem may be regarded as a refinement of Theorem 7.3.1, which stated that

$$\overline{X}_n = \mu + v_n,$$

where v_n has a limit distribution concentrated on zero. Theorem 7.4.1 refines this by asserting that

$$\overline{X}_n = \mu + \frac{\sigma u_n}{\sqrt{n}},$$

where u_n has a limit distribution: the standard normal distribution.

PROOF. The c.f. of the X_j has the limited expansion

$$\phi(\theta) = \exp[i\mu\theta - \tfrac{1}{2}\sigma^2\theta^2 + o(\theta^2)].$$

Then S_n has c.f. $\phi(\theta)^n$ and u_n has c.f.

$$E(e^{i\theta u_n}) = \phi\left(\frac{\theta}{\sigma\sqrt{n}}\right)^n \exp\left(-\frac{i\mu\theta\sqrt{n}}{\sigma}\right) = \exp[-\tfrac{1}{2}\theta^2 + no(\theta^2/n)] \to e^{-(1/2)\theta^2}. \quad \square$$

The result thus follows by a reapplication of the argument of Theorem 7.3.1.

The first proofs of the central limit theorem were restricted to the binomial distribution, and involved direct evaluation of an effective density function (de Moivre (1733), Laplace (1812)); see Exercise 5.1. Laplace gave a heuristic proof for the more general IID case, rigorized by Liapunov in 1901. Gauss (1809) derived the normal distribution, not as a limit distribution, but by demanding a characterizing property (that the sample average, should always provide the maximum likelihood estimate of the mean). The attempts during the nineteenth century to establish the central limit theorem generated a rich theory (e.g. of generalized Chebyshev inequalities, see Section 12.5, and of the 'moment problem', concerning the extent to which distributions are determined by moments). This work is of interest and value in its own right, but, as far as the central limit theorem is concerned, has been largely and economically by-passed by appeals to weak convergence and the characteristic function.

We have now to determine what manner of thing we have evoked in the normal distribution.

EXERCISES AND COMMENTS

1. Note that the transformation (20) from S_n to u_n is just the standardizing linear transformation bringing the variable to zero mean and unit variance; see Exercise 2.5.2.

2. We have implicitly assumed that expression (18) *is* a c.f., since it is a limit of c.f.s, and this indeed it will prove to be. Some caution is needed however. If we calculate the c.f. $\phi_n(\theta)$ of $S_n/\sqrt{n}$ in the case when the X_j are standard Cauchy variables then we find indeed a limit value:

$$\phi_n(\theta) \rightarrow \begin{cases} 1 & (\theta = 0), \\ 0 & (\theta \neq 0). \end{cases} \tag{21}$$

However, this limit does not have the continuity properties asserted for a c.f. in Theorem 7.2.3. It is the condition of finiteness required there which is contravened; expression (21) is in fact the c.f. of a r.v. which is infinite with probability one.

5. The Normal Distribution

The standard normal distribution has a density

$$f(x) = \frac{1}{\sqrt{2\pi}} e^{-x^2/2} \tag{22}$$

over the whole x-axis. The distribution thus has the striking feature that its density (22) and its c.f. (18) have the same functional form, apart from the normalization constant.

One can deduce expression (1) from the inversion formula

$$f(x) = \frac{1}{2\pi} \int_{-\infty}^{\infty} e^{-i\theta x - (1/2)\theta^2} \, d\theta$$

if one completes the square in the exponent by setting $\theta^2 + 2i\theta x = (\theta + ix)^2 + x^2$ and then integrates with respect to the new variable $\theta + ix$. However, to justify the change of integration path needs some care; we indicate an elementary derivation of (22) in Exercises 1 and 2.

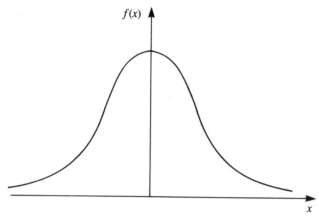

Figure 7.2. A graph of the standard normal distribution.

The density (1) has the symmetric unimodal form illustrated in Fig. 7.2 and often referred to as 'bell-shaped'. The names of Laplace and Gauss are both frequently and rightly attached to the distribution. So, if one reads of 'Gaussian r.v.s' or a 'Gaussian process' then one knows that the r.v.s concerned are jointly normally distributed. (We discuss multivariable distributions later in the section.)

The term 'normal' comes from the more extended description 'normal law of errors'. The belief was that the limit property demonstrated in Theorem 7.4.1 implied that experimental errors would be normally distributed, being made up of many small, independent additive effects. Normality is indeed often assumed in such cases, but certainly cannot be asserted as a universal truth. However, the limit property of the distribution indeed gives it a unique and universal character.

The next thing we notice from the form of the c.f. (18), or its non-standardized form (19), is that the cumulants κ_j are zero for $j > 2$, so that the distribution is fully specified by its first and second moments. Other distributions then tend to be compared to the normal distribution in this respect. The scale-standardized third cumulant $\kappa_3/\kappa_2^{3/2} = \kappa_3/\sigma^3$ is a measure of *skewness*; of the degree of asymmetry of the distribution about its mean. The scale-standardized fourth cumulant $\kappa_4/\kappa_2^2 = \kappa_4/\sigma^4$ is a measure of *kurtosis*: the degree to which, for a given value of $\sigma^2 = \text{var}(X)$, the distribution is weighted towards its tails ($\kappa_4 > 0$) or its centre ($\kappa_4 < 0$) relative to the normal distribution. If the distribution has been standardized to zero mean, so that $\mu_1 = 0$, then these measures become μ_3/σ^3 and $(\mu_4/\sigma^4) - 3$, respectively, in terms of the moments.

The density function of the normal distribution of mean μ and variance $v = \sigma^2$ is

$$f(x) = \frac{1}{\sqrt{2\pi v}} \exp\left(-\frac{(x - \mu)^2}{2v}\right) \tag{23}$$

as we see from (22) by a change of variable (see Exercise 3).

Other properties of the distribution emerge most strikingly in the vector case. Let us say simply that the random vector X is normally distributed (or its components $X_1, X_2, \ldots, X_m$ are jointly normally distributed) if all cumulants of higher order than the second vanish. This implies the form

$$\phi(\theta) = E(e^{i\theta^T X}) = \exp(i\theta^T \mu - \tfrac{1}{2}\theta^T V \theta) \qquad (24)$$

for the multivariable c.f., with $\mu = E(X)$ and $V = \text{cov}(X)$ (see Exercise 2.6).

Theorem 7.5.1. *Expression* (24) *is indeed a c.f. if* $V \geq 0$. *If* $V > 0$ *it corresponds to a probability density*

$$f(x) = \frac{1}{\sqrt{(2\pi)^m |V|}} exp(-\tfrac{1}{2}(x-\mu)^T V^{-1}(x-\mu)) \qquad (25)$$

in $\mathbb{R}^m$.

PROOF. Let $u = (u_1, u_2, \ldots, u_m)$ be a vector of independent standard normal variables, so that

$$\phi_u(\theta) = E(e^{i\theta^T u}) = \exp\left(-\frac{1}{2}\sum_j \theta_j^2\right) = \exp(-\tfrac{1}{2}\theta^T \theta).$$

Now, if expression (24) is to be the c.f. of a random vector X, then V can be identified as $\text{cov}(X)$, and so necessarily $V \geq 0$. One can then find a square matrix M such that $MM^T = V$. If we then define a random vector X by

$$X = \mu + Mu, \qquad (26)$$

then indeed $E(X) = \mu$ and $\text{cov}(X) = MM^T = V$, and X has c.f.

$$\phi_X(\theta) = e^{i\theta^T \mu}\phi_u(M\theta) = \exp(i\theta^T \mu - \tfrac{1}{2}\theta^T V \theta),$$

which is just expression (24). Hence expression (24) indeed defines a c.f. if $V \geq 0$.

The vector u has probability density in $\mathbb{R}^m$

$$f(u) = \frac{1}{\sqrt{(2\pi)^m}}\exp\left(-\frac{1}{2}\sum_j u_j^2\right) = \frac{1}{\sqrt{(2\pi)^m}}\exp(-\tfrac{1}{2}u^T u). \qquad (27)$$

If V is nonsingular, then so is M, and the form of the density (25) for X follows from that of u, (27), by the transformation of variable (26); see Exercise 3. $\square$

The proof of the theorem indicates an immediate corollary.

Theorem 7.5.2. *Linear functions of jointly normal r.v.s are jointly normal.*

PROOF. If X has c.f. (24) then $Y = A + BX$ (where A is a vector and B a matrix of appropriate dimensions) has c.f. $\phi_Y(\theta) = e^{i\theta^T A}\phi_X(B^T \theta) = \exp[i\theta^T(A + B\mu) - \tfrac{1}{2}\theta^T(BVB^T)\theta]$. This is the c.f. of a normal distribution with the correct moments. $\square$

We note that quadratic forms play a considerable role for the normal distribution, both in its c.f. (24) and its density function (25). We shall see, in Chapter 11, that there is indeed a natural linkage with LLS ideas. Related to this fact is the following.

Theorem 7.5.3. *Suppose the random vectors X and Y are jointly normally distributed. Then X and Y are independent if and only if they are uncorrelated.*

That is, for the normal distribution, independence and orthogonality (lack of mutual correlation) are equivalent.

PROOF. If we write $E(X) = \mu_X$ and $\mathrm{cov}(X, Y) = V_{XY}$, etc., then the logarithm of the joint c.f. has the expression

$$\log \phi(\theta_X, \theta_Y) = i\theta_X^T \mu_X + i\theta_Y^T \mu_Y - \frac{1}{2}\begin{bmatrix} \theta_X \\ \theta_Y \end{bmatrix}^T \begin{bmatrix} V_{XX} & V_{XY} \\ V_{YX} & V_{YY} \end{bmatrix} \begin{bmatrix} \theta_X \\ \theta_Y \end{bmatrix}. \qquad (28)$$

This decomposes into the sum of a function of θ_X and a function of θ_Y (such a decomposition being equivalent to independence) if and only if $V_{XY} = 0$ (which is a statement of orthogonality). □

Finally, a couple of widely accepted conventions on notation. We shall often abbreviate the statement that a random vector X is normally distributed with mean μ and covariance matrix V to the statement that X is $N(\mu, V)$. The distribution function of a $N(0, 1)$ variable

$$\Phi(X) = \frac{1}{\sqrt{2\pi}} \int_{-\infty}^{x} e^{-u^2/2} \, du$$

is a nonelementary function known as the *normal integral*. It is important, not only in probability, but also in diffusion theory, and is extensively tabulated.

EXERCISES AND COMMENTS

1. We can derive the density (22) by evaluating the effective asymptotic density for a model which we know shows normal convergence. This is essentially de Moivre's calculation. Let S be the sum of $2N$ independent r.v.s taking the values ± 1 with equal probability. Then the number of 1's is binomially distributed with parameters $2N$ and $\frac{1}{2}$, so that

$$P(S = 2j) = \binom{2N}{N+j} 2^{-2N} = \frac{(2N)!}{(N+j)!\,(N-j)!} 2^{-2N} \qquad (-N \le j \le N). \quad (29)$$

Since S has mean 0 and variance $2N$ we should consider the standardized variable $X = S/\sqrt{2N}$. If we write expression (29) as $g_N(j)$ then the r.v. X will have an effective probability density $f_N(x) = \sqrt{N/2}\, g(N, [x\sqrt{N/2}])$. Here the factor $\sqrt{N/2}$ takes account of the differing j and x scales, and $[a]$ is the integer nearest to a.

We could determine the asymptotic form of $f_N(x)$ by appealing to Stirling's approximation for the factorials in expression (29). However, we can make a direct calculation which avoids this appeal. If we regard $f_N(x)$ as a function of x then we see from (29) that

$$\log f_N(x) = \text{const.} + \sum_{j=1}^{[x\sqrt{N/2}]} \{\log(N-j) - \log(N+j)\}$$

$$\sim \text{const.} + N \int_0^{x/\sqrt{2N}} \{\log(1-u) - \log(1+u)\}\, du$$

$$\sim \text{const.} - N \int_0^{x/\sqrt{2N}} 2u\, du$$

$$= \text{const.} - \tfrac{1}{2}x^2,$$

where the $\sim$ sign indicates neglect of lower-order terms in N for fixed x. We have thus deduced expression (22) for the standard normal density, to within evaluation of the normalizing constant.

2. The normalizing constant in (22) can also be evaluated by elementary methods. Suppose it to be C, so that $C \exp(-\tfrac{1}{2}x^2)$ is a standard normal density. Then

$$1 = C^2 \iint \exp[-\tfrac{1}{2}(x^2 + y^2)]\, dx\, dy.$$

Transforming to polar coordinates $x = r\cos\alpha$ and $y = r\sin\alpha$ we then have

$$1 = C^2 \left(\int_0^{2\pi} d\alpha\right)\left(\int_0^\infty r \exp(-\tfrac{1}{2}r^2)\, dr\right) = C^2(2\pi)(1),$$

whence the evaluation $C = 1/\sqrt{2\pi}$ follows.

3. The calculation of Exercise 2 has more than manipulative significance. It is plainly a characterizing property of the normal distribution that the joint density of independent $N(0, v)$ variables X_j depends on $x = \{x_j\}$ only through $\sum_j x_j^2$. Let us express this better. Suppose that X is a random vector whose elements X_j are IID, and that the distribution of X is rotation-invariant in that $E[H(UX)] = E[H(X)]$ for any H and for any orthogonal matrix U. Show that the X_j are normally distributed with zero mean. (This is Maxwell's characterization, based on the assumptions or conclusions that the Cartesian velocity components of a molecule in a perfect gas are independent and that the gas is isotropic (rotationally invariant) in its properties.)

4. Suppose that a scalar r.v. X has probability density $f(x)$. Consider the linearly transformed r.v. $Y = a + bX$ and suppose that has density $g(y)$. We must then have an equivalent representation of expectations

$$\int H(a + bx)f(x)\, dx = \int H(y)g(y)\, dy$$

for all H. Show, by a change of variable in the first integral, that we must then have

$$g(y) = \frac{1}{|b|} f\left(\frac{y-a}{b}\right).$$

What is the equivalent for vector variables? We shall consider the general question of transformations of density under transformations of variables in the next chapter.

5. Show that the standard normal distribution has odd order moments equal to zero and even order moments given by

$$\mu_{2j} = \frac{(2j)!}{2^j(j!)} \qquad (j = 0, 1, 2, \ldots).$$

6. Show, by integration of the outer relations of (11), that

$$\frac{1}{2\pi} \int \phi(\theta) \exp(-i\theta x - \tfrac{1}{2}\delta^2\theta^2)\, d\theta = \int f(x - \delta u)h(u)\, du,$$

where $h(u)$ is a standard normal density. By letting δ tend to zero we then see that the inversion formula (12) must hold, at least at continuity points of the density f.

7. The limit property of the normal distribution is related to a closure property. Suppose that $u_n = (\sum_1^n X_j)/\sqrt{n}$ has the same distribution as the individual IID r.v.s X_j. This property holds if the X_j are $N(0, v)$ for some v, and, in fact, only in that case.

   The centred Cauchy distribution shows the same closure property under the transformation $(\sum_1^n X_j)/n$. However, there is not a natural multivariate Cauchy distribution (apart from the case of independent variables) and no analogue of Theorem 7.5.2.

   Show that if the r.v.s X_j are IID with c.f. $\phi(\theta)$ and there exists a sequence of normalizing constants B_n such that $B_n^{-1}\sum_1^n X_j$ also has c.f. $\phi(\theta)$, then $\phi(\theta)$ must necessarily have the form $\exp(-c|\theta|^\gamma)$, where c may take different values for $\theta > 0$ and $\theta < 0$. This is (roughly) the class of *stable laws*; the Cauchy and normal laws correspond to $\gamma = 1, 2$. However, values of γ outside the range $[0, 2]$ are inadmissible, since the function is then not a characteristic function. The normal law then has a unique character, as the only stable law of finite variance.

8. These are many extremal characterizations of the normal distribution. For example, it is the distribution whose density $f(x)$ is 'most uniform' in that it maximizes the 'continuous entropy' $-\int f(x) \log f(x)\, dx$ subject to the scale constraint $\int x^2 f(x)\, dx = 1$.

9. Consider the choice of a scalar function of time $a(t)$ for which $I_1 = \int a(t)^2\, dt$ is prescribed and for which one wishes both $I_2 = \int t^2 a(t)^2\, dt$ and $I_3 = \int (da/dt)^2\, dt$ to be small. It follows from Cauchy's inequality and partial integration that

$$I_2 I_3 \geq \left(\int t \frac{da}{dt} a\, dt\right)^2 = (\tfrac{1}{2}I_1)^2 \tag{30}$$

with equality if da/dt is proportional to $ta(t)$, i.e. if $a(t)$ is of the 'normal' form

$$a(t) = \alpha e^{-\beta t^2/2}$$

for some β and appropriate α.

   The motivating model is that in which $a(t)$ is a signal which modulates a carrier wave, and is to represent a 'pulse' at $t = 0$. Then I_1 is proportional to the total energy in the signal (which is prescribed) and I_2 represents the spread of the signal in time. The term I_3 represents the spread of the signal in frequency, in that it can be written $I_3 \propto \int \omega^2 \tilde{a}(\omega)^2 \, d\omega$, where $\tilde{a}(\omega)$ is the Fourier transform of $a(t)$. A pulse of Gaussian form thus compromises best between spread in the dual variables of time and frequency.

The situation also has quantum-mechanical overtones, the transformations $a(t) \to ta(t)$ and $a(t) \to da(t)/dt$ implying the operators corresponding to position and momentum, and inequality (30) representing an uncertainty principle.

10. Suppose that in Exercise 9 we set $f(t) = a(t)^2$ and assume scales so normalized that $\int a^2 \, dt = \int f \, dt = 1$, so that $f(t)$ can be regarded as a probability density. Then, in terms of f, inequality (30) becomes

$$\left(\int t^2 f(t) \, dt \right) \left(\int \left(\frac{d \log f}{dt} \right)^2 f(t) \, dt \right) \geq 1,$$

again with equality if and only if $f(t)$ is a normal distribution centred on the origin. This is very close to the statisticians' Cramér–Rao inequality, expressing an upper bound on the precision with which the time origin of a pulse can be estimated from observation of the signal.

11. Suppose X standard normal and $a \geq 0$. Then note the useful inequality

$$P(X \geq a) \leq \inf_{\theta \geq 0} Ee^{\theta(X-a)} = \inf_{\theta \geq 0} e^{-\theta a + \theta^2/2} = e^{-a^2/2}.$$

CHAPTER 8

Continuous Random Variables and Their Transformations

1. Distributions with a Density

When we began considering concrete examples in Chapter 4 we restricted ourselves for some time to cases in which the sample space Ω was discrete, so that there were at most countably many realizations. However, we were increasingly led to consider r.v.s X on more general sample spaces, and it became evident in Section 7.5 that we need to be able to calculate the distribution of functions of X in terms of the distribution of X.

For the discrete case the expectation functional is given by the sum (3.15); for the general case it is given by some general type of integral such as (3.17) (although it is central to our approach that the known expectations will not necessarily be given in this form).

However, in this chapter we shall largely restrict ourselves to the special case when any r.v. X of interest takes values in a finite-dimensional Euclidean space ($\mathbb{R}^m$, say) and has a probability density $f(x)$ with respect to Lebesgue measure in this space. That is, expectations are given by

$$E[H(X)] = \int (H(x)f(x)\,dx, \tag{1}$$

where $dx = dx_1\,dx_2\,\ldots\,dx_m$ and the x_j are the components of the vector x. Relation (1) is understood to hold for a class of functions which includes at least the *simple functions* (i.e. functions for which $\mathbb{R}^m$ can be divided into a finite number of m-dimensional intervals, $a_j < x_j \le b_j (j = 1, \ldots, m)$, such that H is constant on each of these intervals).

Theorem 8.1.1. *If A is a set in $\mathbb{R}^m$ consisting of a finite union of intervals, then*

$$P(X \in A) = \int_A f(x)\,dx. \tag{2}$$

Consequently,

$$f(x) = \frac{\partial^m F(x)}{\partial x_1 \partial x_2 \cdots \partial x_m}, \tag{3}$$

where $F(x)$ is the distribution function of X:

$$F(x) = P(X \le x). \tag{4}$$

PROOF. Since the indicator function of A is certainly a simple function equation, (2) follows from (1). Formula (3) (which shows that f is determined if $E(H)$ is known for the class of functions indicated) follows by differentiation of the special case of (2)

$$F(x) = \int_{y \le x} f(y) \, dy$$

with respect to all the elements of x. $\qquad\qquad\qquad\qquad\qquad\qquad\qquad\square$

We recall other consequences of (2): that f is nonnegative almost everywhere and integrates to unity.

By appealing to the expectation axioms we shall see in Section 15.4 that the validity of (1) can be extended to the case when H is a *Borel function* (a limit of simple functions), and that (2) can similarly be extended to the case when A is the set of x satisfying $H(x) \ge 0$ for some Borel function H. This class of functions and sets (which are, respectively, included in the class of *integrable* functions and *measurable* sets) is wide enough for most practical purposes, and we shall assume henceforth (tacitly, unless emphasis is needed) that all r.v.s. and events considered fall in these sets.

Theorem 8.1.2. *If* (1) *holds then the smaller set of r.v.s.* $(X_1, X_2, \ldots, X_r) \, (r \le m)$ *also possesses a probability density, determined by*

$$f(x_1, x_2, \ldots, x_r) = \int \int \cdots \int f(x) \, dx_{r+1} \, dx_{r+2} \ldots dx_m. \tag{5}$$

Note that we use the same notation f for both densities. This is a very common and convenient convention: that sometimes, at least, one uses f to denote 'density of' just as one uses P to denote 'probability of', the relevant variables being indicated within the bracket. If one wishes to indicate a functional form for this density then some more explicit notation is needed. We shall use F in the same generic sense for 'distribution function of'.

PROOF. Relation (5) follows from (1) if we consider an $H(x)$ which is a function only of the first r components of x. $\qquad\qquad\qquad\qquad\qquad\qquad\square$

Expression (5) is sometimes known as the *marginal density* of $X_1, X_2, \ldots,$ X_r, although the term is more usually applied to the single variable densities $f(x_1), f(x_2), \ldots$.

Theorem 8.1.3. *If* $X_1, X_2, \ldots, X_m$ *possess a joint probability density* $f(x)$ *then they are independent if and only if*

$$f(x) = \prod_{j=1}^{m} f(x_j). \tag{6}$$

PROOF. We have the factorization of the indicator function

$$I(X \leq x) = \prod_j I(X_j \leq x_j).$$

Assuming independence and taking expectations we then have

$$F(x) = \prod_j F(x_j), \tag{7}$$

a result which is valid whether or not a density exists. Differentiating (7) we deduce (6). The converse implication (that (6) implies independence) is clear.

□

We have already seen motivated examples of densities: the rectangular (or uniform), exponential and gamma densities listed in Section 7.2, the beta density of Section 5.8 and the normal density of Section 7.5.

EXERCISES AND COMMENTS

1. Suppose X_j ($j = 1, 2, \ldots, m$) are independent r.v.s with respective distribution functions $F_j(x_j)$. Show that $Y = \max(X_1, X_2, \ldots, X_m)$ has distribution function $\prod_{j=1}^{m} F_j(y)$. Hence show that the minimum of m IID exponential r.v.s is also exponentially distributed. (This result has applications when one considers the failure times of complex electronic or mechanical systems.)

2. Suppose that $X_1, X_2, \ldots, X_m$ are independently and rectangularly distributed on $[0, L]$. Show that the rth of these in magnitude has probability density

$$f(x) = \frac{m!}{(r-1)!\,(n-r)!} \frac{x^{r-1}(L-x)^{m-r}}{L^m} \qquad (0 \leq x \leq L).$$

This is just the beta distribution encountered in Section 5.8.

3. A point source of light is placed at the origin of the (x, y) plane and is equally likely to release a photon (following a straight-line path) in the plane) in any direction. Thus, if θ is the angle between the flight path and the y-axis, then θ can be taken as uniformly distributed on $[-\pi, \pi)$.

    Consider the distribution of θ conditional on the fact that the photon strikes the line $y = a$, where a is positive. Show that θ then has a conditional density which is uniform on $(-\frac{1}{2}\pi, \frac{1}{2}\pi)$. (We have taken pains to indicate whether the ends of the relevant θ-interval are closed or open. This is decided, sometimes by convention, and sometimes, as in the second case, by the physics of the situation. It is rarely an important point, since it in any case concerns an event of probability zero.)

4. *The Buffon needle problem.* A needle of length L is dropped onto a floor made of parallel boards of width D. Make the obvious uniformity and independence as-

sumptions concerning the position and orientation of the fallen needle. Show that if $L < D$, then the probability that the needle intersects a crack between two floorboards is $2L/\pi D$. Buffon made an empirical estimate of π from the proportion of times this event occurred in several hundred trials.

5. A particle is released in a cloud chamber. The time until its first collision and the time until it leaves the chamber are independent exponential variables with means $1/\alpha$ and $1/\beta$, respectively. Show that the probability that a collision is observed in the chamber is $\alpha/(\alpha + \beta)$.

2. Functions of Random Variables

One very often wishes to consider r.v.s derived from a given set of r.v.s by a functional transformation. We have already considered functions such as $\sum_1^m X_j$ and $\max(X_1, X_2, \ldots, X_m)$ and shall encounter others. It is natural to ask how probability densities transform under such a change of variable, if, indeed, the new r.v.s possess a density at all. Our guiding principle will be the following. Suppose a vector r.v. Y is derived from a vector r.v. X by a transformation $Y = a(X)$ where the components of a are Borel functions. Suppose also that X has density $f(x)$ and that the expectation integral can be rewritten

$$E[H(Y)] = \int H[a(x)]f(x)\,dx = \int H(y)g(y)\,dy \qquad (8)$$

identically in H. Then $g(y)$ is by definition the probability density of Y. For if $H(y)$ is a simple function of y, then $H[a(x)]$ is a Borel function of x, so that (8) is then valid for a class of functions $H(y)$ including the simple functions.

Theorem 8.2.1. *Suppose that vectors X and Y have dimension m and r, respectively ($r \le m$), and that the transformation $Y = a(X)$ can be complemented by a transformation to a vector $Z = b(X)$ of dimension $(m - r)$ such that the combined transformation $X \to (Y, Z)$ is Borel, one to one, and possesses an inverse and a Jacobian*

$$J(Y, Z) = \frac{\partial(X)}{\partial(Y, Z)}. \qquad (9)$$

Then Y has density

$$g(y) = \int f[x(y, z)]J(y, z)\,dz. \qquad (10)$$

By the Jacobian (9) we mean the absolute value of the determinant whose jkth element is $\partial X_j/\partial Y_k$ ($j, k = 1, 2, \ldots, m$) with Y_k identified with Z_{k-r} for $k > r$. The notation $x(y, z)$ in (10) indicates that the transformation $x \to (y, z)$ is to be inverted to express x in terms of y and z; the Jacobian (9) is correspondingly expressed in terms of Y and Z.

PROOF. This follows from the usual formula for a change of variable under the integral; see, e.g. Apostol; (1957), p. 270. Transforming from x to (y, z) under the integral we have

$$\int H[a(x)]f(x)\, dx = \iint H(y)f[x(y, z)]J(y, z)\, dy\, dz,$$

which is of the form (8) with $g(y)$ given by (10).   □

If the transformation is simply one between scalar variables then (10) reduces to

$$g(y) = f[x(y)] \left| \frac{dx}{dy} \right|, \tag{11}$$

with dx/dy expressed as a function of y.

For example, consider the light-emission problem of Exercise 8.1.3, and ask for the distribution of the x-coordinate $X = a \tan \theta$ of the point of impact of the photon on the line $y = a$ (conditional on the fact that the photon *does* hit the line). We know that the density of θ is $1/\pi$ if $|\theta| < \frac{1}{2}\pi$ and zero otherwise. Thus, by formula (11) (but transforming from θ to X rather than from X to Y) we see that the density of X is

$$f(x) = \frac{1}{\pi}\frac{d\theta}{dx} = \frac{1}{\pi}\frac{d}{dx}\tan^{-1}\left(\frac{x}{a}\right) = \frac{a}{\pi(a^2 + x^2)}. \tag{12}$$

This is the density of the Cauchy distribution we have already encountered in Exercise 7.2.2, although now with a scale parameter a. The distribution there appeared as a handy and cautionary counterexample, illustrating cases of failure of the law of large numbers, etc. Here we see that it has a surprisingly immediate physical origin—the average position of n incident photons really would have the same distribution as does the position of a single photon.

One of the simplest transformations is the linear one, $Y = A + BX$. If this is to be nonsingular then B must be square and nonsingular. In this case (10) will reduce to

$$g(y) = \|B\|^{-1}f[B^{-1}(Y - A)], \tag{13}$$

where $\|B\|$ is the absolute value of the determinant of B. We have already appealed to this when deriving the density of the general normal distribution (7.23) from that of the standard normal distribution (7.22), and the density of the general multivariate normal distribution (7.25) from that of several independent standard normal variables (7.27).

EXERCISES AND COMMENTS

1. Suppose that scalar r.v.s X and Y are related by $Y = a(X)$, where a is a monotonic function. If X and Y have distribution functions $F(x)$ and $G(y)$, respectively, show that $G(y)$ equals $F(x)$ or $1 - F(x-)$ according as h is increasing or decreasing. Show that this result implies (11) when densities exist.

2. Suppose the scalar transformation $x \to y$ is many-to-one, so that there are several x-values yielding the same value of y. Show that, under appropriate regularity conditions, formula (11) can then be generalized to

$$g(y) = \sum_j [f(x)|dx/dy|]_{x=x_j(y)},$$

where the $x_j(y)$ are the values of x corresponding to given y. Hence show that if velocity X is $N(0, 1)$, then kinetic energy $Y = \frac{1}{2}X^2$ has density $e^{-y}/\sqrt{\pi y}$ for $y \geq 0$.

3. Show that if scalar r.v.s X_1 and X_2 have joint density $f(x_1, x_2)$, then their sum $Y = X_1 + X_2$ has density

$$g(y) = \int f(x_1, y - x_1) \, dx_1 \tag{14}$$

However, the distribution of sums is much better treated by the use of c.f.s if the summands are independent. Suppose, for example, that one is calculating the distribution of kinetic energy $Y = \frac{1}{2}\sum_{j=1}^{m} X_j^2$ for an m-dimensional example for which the component velocities X_j are $N(0, v)$ and independent. Show that the c.f. of Y is $(1 - iv\theta)^{-m/2}$, the c.f. of a gamma distribution.

4. Suppose that the position Y of the piston in a car engine is related to the angular position of the crankshaft by $Y = a \cos X$. If X is uniformly distributed over the range $[-\pi, \pi)$, then show that Y has density

$$g(y) = \frac{1}{\pi\sqrt{a^2 - y^2}} \qquad (-a \leq y \leq a).$$

The U-shaped form of this distribution corresponds to the fact that the piston spends most of its time near the ends of its stroke, when the motion is slowest.

5. Suppose that a nonnegative r.v. X has a distribution function $F(x)$, and that Y is the rounding error if X is rounded off to the nearest integer below. Show that Y has the distribution function

$$\sum_{j=0}^{\infty} [F(j + y) - F(j)] \qquad (0 \leq y < 1).$$

6. Show that if X has a differentiable distribution function $F(x)$ and we define the r.v. $Y = F(X)$ then Y is uniformly distributed on $(0, 1)$. This is the *probability integral transformation*, useful in a number of problems.

7. *Personal records.* Consider again the high jumper of Exercise 4.3.6, and his sequence of IID jumps. Suppose that the heights X_j he attains follow a standard exponential distribution. (This is almost a matter of scaling rather than of a fundamental assumption, if the X_j have a density. We see from the previous exercise that there is a change of scale which will take any continuous distribution into any other, and a change of scale will not affect the timing or the ordering of records.) Show that the distribution of height reached when he sets his jth record has the density $f_j(x) = e^{-x}x^j/j!$ of a gamma distribution.

8. Consider the photon-emission problem discussed in the text, but in three dimensions, light being radiated from a point source uniformly in all directions. Consider those photons striking a given plane distant a from the source. Show that if X is the absolute distance of the point of impact of a random photon from the foot

of the perpendicular from the source to the plane, then X has density

$$f(x) = \frac{ax}{(a^2 + x^2)^{3/2}} \qquad (x \geq 0).$$

9. Show that if scalar r.v.s X_1 and X_2 have joint density $f(x_1, x_2)$ then their quotient $Y = X_2/X_1$ has density

$$g(y) = \int f(x_1, x_1 y)|x_1| \, dx_1.$$

Hence show that the quotient of two independent $N(0, 1)$ variables is Cauchy distributed.

10. Suppose that a random vector X has density $f(x)$. Suppose also that we require that X have a distribution which is isotropic (i.e. rotation-invariant), in that $Y = UX$ has the same distribution as X for any orthogonal matrix U. Then Y has density $g(y) = |U|^{-1}f(U^{-1}y) = f(U^{-1}y)$, since $|U| = 1$. Hence demonstrate that the necessary and sufficient condition for isotropy is that $f(x)$ should be a function of $|x|$ alone.

11. Results for the summation operation $S_n = \sum_1^n X_j$ often have an analogue for the maximization operation $T_n = \max(X_1, X_2, \ldots, X_n)$. If the X_j are IID with distribution function $F(x)$ then T_n has distribution function $F(t)^n$, corresponding to the c.f. result for sums. Suppose that sequences of standardizing constants A_n and B_n exist such that $u_n = (T_n - A_n)/B_n$ has a limit distribution function $e^{G(u)}$. Show that there is then a closure relationship

$$nG\left(\frac{u - A_n}{B_n}\right) = G(u).$$

Hence show that, under suitable regularity conditions, $G(u)$ must be of the form $(a + \beta u)^\gamma$; this includes limit forms such as $\delta e^{\varepsilon u}$. These are the *extreme value distributions*, with a theory analogous to that of the stable laws of Exercise 7.5.7. They are useful for the description of extremes in large sets of r.v.s which may not even be independent, just as the central limit theorem can be valid even if the summands show some mutual dependence. One appeals to these distributions when discussing, for example, the distribution of peak floods over a long period of time, or the failure probability in a structure of many parts.

3. Conditional Densities

Theorem 8.3.1. *Suppose that X and Y are vector r.v.s with joint density $f(x, y)$. Then the distribution of X conditioned by the value of Y has density*

$$f(x|y) = \frac{f(x, y)}{\int f(x, y) \, dx} = \frac{f(x, y)}{f(y)}. \tag{15}$$

PROOF. This is the result for which we might hope. What we have to demonstrate is that the conjectured evaluation of a conditional expectation

$$E[H(X)|Y] = \int H(x)f(x|Y) \, dx \tag{16}$$

with $f(x|y)$ given by (15) satisfies the defining characterization

$$E\{E[H(X)|Y]K(Y)\} = E\{H(X)K(Y)\} \qquad (17)$$

(see (5.14)) for at least all simple H, K. The reader will verify directly that, with definitions (15) and (16), both sides of expression (17) are equal to $\iint H(x)K(y)f(x, y)\, dx\, dy$. $\qquad\square$

As an example, suppose that we wish to consider a probability distribution uniform on the surface of a sphere in three dimensions. This is a situation which occurs quite frequently: we might, for example, wish to consider a point source in three dimensions which radiates uniformly in all directions, and so gives a uniform flux through the surface of a sphere centered on the source. However, the distribution, although simple in concept, is rather difficult to express, because there is no set of coordinates on the surface of a sphere which is natural in the sense that it obviously expresses equivalence of all points on the surface.

Almost the simplest way of going about the problem is the following. Consider the sphere as embedded in Cartesian space with coordinates X, Y, Z, so that the surface of interest is

$$X^2 + Y^2 + Z^2 = 1. \qquad (18)$$

Suppose that we consider X, Y, Z to be r.v.s with a density $f(x, y, z)$ which is isotropic in the sense that

$$f(x, y, z) = g(r), \qquad (19)$$

where $r = \sqrt{x^2 + y^2 + z^2}$. That is, the three-dimensional density f depends only upon absolute distance from the origin r, and not on direction; see Exercise 8.2.10. If we consider the density of X, Y, Z conditional on (18), that is, on $R = 1$, then we shall be considering just the density we wish: a density uniform on the surface (18).

It is simplest to transform to polar coordinates R, ϕ, θ, so that

$$X = R \cos \phi, \qquad Y = R \sin \phi \cos \theta, \qquad Z = R \sin \phi \sin \theta$$

$$(R \geq 0, 0 \leq \phi \leq \pi, 0 \leq \theta < 2\pi). \qquad (20)$$

We see from (10) and (19) that the transformed variables have density

$$f(r, \phi, \theta) = g(r)r^2 \, |\sin \phi|.$$

It then follows from (15) that the joint density of ϕ and θ conditional on any prescribed value r of R is

$$f(\phi, \theta | R = r) = \frac{f(r, \phi, \theta)}{\iint f(r, \phi, \theta) \, d\phi \, d\theta} = \frac{1}{4\pi} |\sin \phi|, \qquad (21)$$

the range of the variables being that indicated in (20). This expresses the desired density in terms of the angular coordinates ϕ and θ. As expected, the final density (21) is independent of the particular function g introduced in (19).

In fact, assumption (19) implies that the three polar coordinates are statistically independent.

The idea that the conditioning event $R = 1$ is only meaningful when embedded in a field of events (generated from $R \leq r$ for varying r) comes very much to the fore in this example. In effect, it is by the embedding that one *defines* what is meant by 'uniformity' on the surface of a sphere.

EXERCISES AND COMMENTS

1. Consider two jointly normal random vectors X and Y with c.f. (7.28). Show that the distribution of X conditional on Y is normal with mean and covariance matrix

$$E(X|Y) = \mu_X + V_{XY}V_{YY}^{-1}(Y - \mu_Y), \tag{22}$$

$$\mathrm{cov}(X|Y) = V_{XX} - V_{XY}V_{YY}^{-1}V_{YX}. \tag{23}$$

We recognize in (22) just a vector version of the LLS estimate $\hat{X}$ of X in terms of Y and in (23) a vector version of the mean square estimation error $E[(\hat{X} - X)^2]$ deduced in Section 2.8. This is an indication of the intimate connection between the normal distribution and LLS approximation ideas, which we will explore in Chapter 11.

The form of the density of X conditional on Y follows from appeals to (10) and the form (7.25) of the multivariable normal density. The reader may nevertheless find the need for some rather heavy and unevident matrix reductions. The economical proof of the above assertions is that to be given in Theorem 11.4.1.

CHAPTER 9

Markov Processes in Discrete Time

1. Stochastic Processes and the Markov Property

We now wish to bring in the idea of *dynamics*: that time enters into one's model, and that the model specifies how variables evolve in time. Already it has been natural to consider gaming (Section 4.5), renewal processes (Section 6.1) and population growth (Section 6.4) as situations evolving in time. Almost all real-world phenomena must be considered dynamically.

We shall denote time by t. Most physical models must be framed in *continuous time*, for which t may take any real value. We shall take this formulation in Chapter 10, but it is simpler to begin with models in *discrete time*, when time advances in unit steps, and t may be assumed to take integral values ..., $-2, -1, 0, 1, 2, \ldots$. So, for the gaming model of Section 4.5, the unit step marks the passage between rounds of the game; for the branching process of Section 6.4 it marks the passage between consecutive generations. It is also true that many variables (e.g. economic, meteorological) are observed at regular intervals (e.g. daily, quarterly, yearly) although this is more a matter of observation than of true dynamics.

So, the notion of an isolated r.v. X will now be replaced by that of a sequence $\{X_t\}$, where X_t corresponds to the value of the dynamical variable at time t. The variable X_t will often denote the value of some specified physical quantity at time t (e.g. position, velocity, temperature, price, water level) and we shall use X to denote this variable generically. The r.v. X_t is then the value of X at time t. If X is the only variable one considers, so that the model is one which explains the evolution just of the random sequence $\{X_t\}$, then we shall term X the *process variable*. We shall then often speak of the *stochastic process* $\{X_t\}$, meaning the model which specifies the probabilistic evolution of this sequence. A realization ω is a particular value $\{x_t\}$ of the whole sequence.

A deterministic dynamic model would prescribe a relation

$$X_{t+1} = a(X_t, X_{t-1}, \ldots; t) \tag{1}$$

determining X_{t+1} in terms of past values of X. The relation may also be time-dependent, as we have indicated.

If the dynamic equation (1) reduces to the simply recursive form

$$X_{t+1} = a(X_t; t), \tag{2}$$

then the implication is that specification of the value of X at time t constitutes a *dynamically complete description*. It is complete in that at time t one need only specify the current value X_t in order to determine the future course of the process, and can forget the history of earlier values $\{X_\tau; \tau < t\}$. If this situation holds then the process is said to have *state structure*, and X is the *state variable*, in that it indeed sums up the current state of the process. The state-structured case is both formally simpler and conceptually satisfying. Indeed, one generally believes that, for any model which is physically reasonable, one can attain, or at least approach, state structure, by taking a description X which is sufficiently detailed.

If the argument t is absent from the function a in relations (1) or (2) one says that the process is *time-homogeneous*. In this case the variables of course continue to change with time, but the dynamics, the rules by which they change, remain constant. In the time-homogeneous state-structured case the dynamic relation would then reduce to the form

$$X_{t+1} = a(X_t). \tag{3}$$

Models (1)–(3) are deterministic; we should now like to develop the corresponding ideas for a probabilistic model. It is in this latter case that one uses the term 'stochastic process', as distinct from 'deterministic process'.

The analogue of relation (1) is the assignment on the basis of the model of a functional form for the conditional expectation $E[H(X_{t+1})|X_t, X_{t-1}, \ldots]$ for a sufficiently wide class of functions H. If this functional form does not include t as an argument then the process is time-homogeneous. From this expectation one should, in principle, be able to build up the expectation $E(Z_t|X_t, X_{t-1}, \ldots)$ where Z_t is any r.v. (i.e. any function of the realization). Essentially, one repeatedly applies formula (5.17) for the iteration of conditional expectations. This is somewhat the probabilistic analogue of solving recursion (1); see the comments at the end of the section.

The analogue of state-structure is exactly the Markov property. Let Z_t be any r.v. defined on the present-and-future at time t, i.e. a function of $\{X_\tau; \tau \geq t\}$ alone. Then the process $\{X_t\}$ is defined as *Markov* if for all such Z_t and for all t the conditional expectation $E(Z_t|X_t, X_{t-1}, \ldots)$ is a function of X_t (and possibly also of t) alone. In this case, we know from Theorem 5.3.2(iii) that the expectation can be identified with $E(Z_t|X_t)$. Thus, for a Markov process

$$E(Z_t|X_t, X_{t-1}, \ldots) = E(Z_t|X_t) \tag{4}$$

for all r.v.s. Z_t defined on the present-and-future at time t. For a Markov process, X is again referred to as the *state variable*. The space $\mathcal{X}$ in which the X_t may take values is termed the *state space*. This could depend upon t, but we shall assume that it does not, and shall indeed generally restrict ourselves to the time-homogeneous case.

Formally, it should be necessary to demand condition (4) only for Z_t a function $H(X_{t+1})$ of X_{t+1}, the general version then following by repeated appeal to the special version and to formula (5.17) for iterated conditional expectations. However, in order to accord with convention and to avoid an appeal to a possible infinite iteration, we shall formulate the Markov property in its strong form (4).

The Markov property is something which is physically natural to demand (if one's level of description is good enough) and which leads to a powerful theory.

To specify the model we need to be able to specify $E(Z_t | X_t)$ explicitly as a function of X_t and t. In fact, as indicated above, it is enough for most purposes to specify $E[H(X_{t+1})|X_t = x]$ as a function of x and t, for sufficiently general scalar functions of state H. This is a new function of x, derived from H, which we shall denote by $P_t H$. The value of $P_t H$ at x might by some purists be denoted $(P_t H)(x)$. We shall find the notation $P_t H(x)$ much less cumbersome, and perfectly proper if it is understood that operations on the function H are applied *before* the evaluation of the resultant function at a particular argument x. With this understanding we then have the definition of P_t:

$$E[H(X_{t+1})|X_t = x] = P_t H(x). \tag{5}$$

Specification of the model implies specification of the transforming operator P_t, and in fact the reverse is true (see Theorem 9.1.1 below). The operator P_t is known as the *generator* of the process.

We shall henceforth restrict ourselves to the time-homogeneous case, so that the generator P is time-independent. The following theorem indicates both that we now have the basis of a useful formalism and that specification of P determines the stochastic evolution of the process. In fact, specification of P is the probabilistic analogue of the specification of the function a in (3).

Theorem 9.1.1. *The generator P is a linear operator, transforming scalar functions of state to scalar functions of state. It has the property*

$$E[H(X_{t+s})|X_t = x] = P^s H(x) \qquad (s = 0, 1, 2, \ldots). \tag{6}$$

PROOF. The conditional expectation operator P is indeed linear, and plainly transforms as indicated. By P^s we mean its sth iterate.

To demonstrate assertion (6), define the operator $P^{(s)}$ by

$$P^{(s)} H(x) = E[H(X_{t+s})|X_t = x].$$

Then for $r, s = 0, 1, 2, \ldots$ we have

$$E[H(X_{t+r+s})|X_t = x] = E\{E[H(X_{t+r+s})|X_{t+r}, X_t]|X_t = x\}$$

$$= E\{E[H(X_{t+r+s})|X_{t+r}]|X_t = x\}. \qquad (7)$$

The first equality follows from the general formula (5.17) for the iteration of conditioned expectations and the second equality from the Markov property (4).

Now, relation (7) can be given the more attractive operator expression

$$P^{(r+s)}H = P^{(r)}[P^{(s)}H] = P^{(r)}P^{(s)}H,$$

or, even simpler:

$$P^{(r+s)} = P^{(r)}P^{(s)} \qquad (r, s = 0, 1, 2, \ldots). \qquad (8)$$

But (8), with initial condition $P^{(1)} = P$, implies that $P^{(s)} = P^s$, which is just what (6) asserts. □

The property expressed equivalently by (6) or (8) is basic. Stated in words, it says that, for a Markov process, the s-step transition operator is the sth iterate of the one-step transition operator. The assertion is not true generally; the second equality of (7) makes an essential appeal to the Markov property.

Another way of expressing (6) or (8) is to say that the function

$$H_s(x) = E[H(X_{t+s})|X_t = x]$$

obeys the recursion

$$H_s = PH_{s-1}. \qquad (9)$$

Equation (9) is not, as one might perhaps expect, the analogue of the deterministic dynamic equation (3); it differs significantly.

For the deterministic process (3), $H_s(x)$ is the value that $H(X_{t+s})$ would have if $X_t = x$. The generator P is defined by

$$PH(x) = H[a(x)]$$

and relation (9) becomes

$$H_s(x) = H_{s-1}[a(x)]. \qquad (10)$$

Relation (10) is a recursion on the *initial* value of X, rather than, as in (3), on the final value. In the next section we shall see (9) (and its deterministic analogue (10)) as a *backward equation*, and relation (3) (and its stochastic analogue, yet to come) as a *forward equation*. The backward equation relates conditional future expectations as a recursion on initial values; the forward equation relates probability distributions as a recursion on final values.

The 'forward' point of view may seem to be the natural one, corresponding as it does to the view of the dynamic equation (3) as a forward recursion and of the iteration of this recursion as determination of the path and 'solution' of the problem. However, the more one works with these matters, the more one finds oneself impelled to the backward formulation. It is by the specification of conditional expectations over the future (i.e. by the backward approach) that one most easily specifies stochastic dynamics in general. It is by the backward approach that one defines the Markov property, defines the generator P, classifies states (Section 8), and determines cost-minimizing actions (Sections 5.6 and 15.4). The martingale concept (Section 7 and Chapter 14), which ultimately turns out to be fundamental, is intrinsically a 'backward' concept.

The alternatives of an expectation or a probability formulation correspond exactly to the alternatives of a backward or forward treatment. However, as hitherto, we shall work with whichever formulation is convenient for the immediate purpose, excluding neither on doctrinal grounds.

Finally, the characterization (4) of the Markov property can be given a much more attractive form, symmetric in past and future.

Theorem 9.1.2. *The Markov property* (4) *can be alternatively expressed: at any time the past and future of the process are independent conditional on the current value of state. Formally,*

$$E(Y_t Z_t | X_t) = E(Y_t | X_t) E(Z_t | X_t) \qquad (11)$$

for any t, where Y_t is any function of past-and-present $\{X_\tau; \tau \le t\}$ and Z_t is any function of present-and-future $\{X_\tau; \tau \ge t\}$.

PROOF. Let us drop the t-subscripts for simplicity. Then (4) implies (11), because (4) implies that $E(Z|X, Y) = E(Z|X)$ and so that

$$E(YZ|X) = E[E(YZ|X, Y)|X] = E[YE(Z|X, Y)|X] = E[YE(Z|X)|X]$$

$$= E(Y|X)|(Z|X).$$

To deduce the reverse implication, denote $E(Z|X)$ by $\psi(X)$. Then (11) implies that

$$E(YZ) = E[E(YZ|X)] = E[E(Y|X)\psi(X)|X] = E[Y\psi(X)].$$

But this implies, by the defining characterization of a conditional expectation, that $\psi(X)$ can be identified as $E(Z|X_t, X_{t-1}, \ldots)$, and hence that (4) holds. □

Comparing the characterization of the theorem with that of a regeneration point in Section 6.2, we see that the Markov property can also be expressed: *a process is Markov if and only if the occurrence of any value of the process variable constitutes a regeneration point.*

EXERCISES AND COMMENTS

1. Let $\{X_t\}$ be a Markov process and let $\{t_j\}$ be a predetermined and increasing sequence of times. Appeal to Theorem 5.3.2(iii) and the Markov property to

show that if Z_t is a variable defined on the future at time t and $t \geq t_k$, then $E(Z_t | X_{t_j}; j \leq k) = E(Z_t | X_{t_k})$. Hence the *embedded process* $\{X_{t_j}\}$ is also Markov.

2. Suppose that the state space $\mathcal{X}$ of a Markov process is discrete, with elements x. Show then that

$$P(X_t = x_t; t = 1, 2, \ldots, n | X_0 = x_0) = \prod_{t=1}^{n} P(X_t = x_t | X_{t-1} = x_{t-1}).$$

3. One begins to need a reminder of a process which is *not* Markov. Suppose that the r.v.s X_t $(t \geq 0)$ are independent conditional on X_0. The starting value X_0 can thus affect all the subsequent dynamics. Show that in general the process $\{X_t\}$ is not Markov. However, the process with variable $X_t^* = (X_0, X_t)$ is Markov. That is, Markov character is restored if the process variable also tells one which dynamics one is in.

4. Consider a discrete-time renewal process, and define X as the age of the article currently in service. Verify that $\{X_t\}$ is a Markov process with generator P defined by

$$PH(x) = a_x H(x + 1) + b_x H(0),$$

where a_x and b_x are the survival and failure probabilities at age x defined in Exercise 5.1.3.

5. Consider a gaming model prescribing a sequence of independent games with unit stakes and with success probability p for Alex. Let X denote his the current capital. Then, at least for values of capital x for which play continues, the process behaves as a Markov process with generator P defined by

$$PH(x) = pH(x + 1) + qH(x - 1).$$

This should be supplemented by $PH(x) = H(x)$ at those values of x at which the game terminates (e.g. at the ruin of one of the players) if termination is understood as meaning that x freezes at its current value.

6. Note that the branching process of Section 6.4 is a Markov process with generator essentially specified by

$$E(z^{X_{t+1}} | X_t = x) = G(z)^x.$$

Thus, the conditional expectation of $H(X_{t+1})$ is specified only for the functions $H(x) = z^x$, and this is enough.

7. A discrete-time version of the dynamics of a point mass subject to a random force might go follows. One defines a scalar position X and velocity V, and supposes that

$$X_{t+1} = X_t + V_t, \qquad V_{t+1} = V_t + \varepsilon_t,$$

where the ε_t are IID r.v.s, representing random forces accelerating the mass. Note then that the process $\{X_t, V_t\}$ is Markov, as is the process $\{V_t\}$, but that the process $\{X_t\}$ is not.

8. Note that $PH > 0$ if $H \geq 0$. Suppose that H is such that $PH \geq H$. It follows then by iteration that $P^{s+1}H \geq P^s H$ for $s = 0, 1, 2, \ldots$ and so that the sequence $H_s = P^s H$ is nondecreasing. This turns out to be a useful observation.

2. The Case of a Discrete State Space: the Kolmogorov Equations

Suppose that the state space $\mathcal{X}$ is discrete, and that the possible values of the state variable are labelled simply $j = 1, 2, 3, \ldots$. Define the *transition probability*

$$p_{jk} = P(X_{t+1} = k | X_t = j). \tag{12}$$

(We shall assume for simplicity that that this is independent of t, consistent with our general assumption of time-homogeneity.) Then the conditional expectation operator is

$$E[H(X_{t+1}) | X_t = j] = \sum_k p_{jk} H(k),$$

so the generator P has the effect

$$PH(j) = \sum_k p_{jk} H(k). \tag{13}$$

In fact, if we regard the function H as a vector with jth element $H(j)$, then the transformed vector (13) is just PH where $P = (p_{jk})$ is the *transition matrix*. That is, if a function H is seen as a vector whose components are the values of that function at possible arguments, then the generator P, an operator on functions, becomes exactly the transition matrix P, multiplying vectors. We can rephrase Theorem 9.1.1 in these terms as

Theorem 9.2.1. *The conditional expectation $E[H(X_{t+s}) | X_t = j]$ can be identified as the jth element of the vector $P^s H$, where P is the transition matrix of the process.*

Suppose we define the matrix of s-step transition probabilities

$$P^{(s)} = (p_{jk}^{(s)}) = (P(X_{t+s} = k | X_t = j))$$

for $s = 0, 1, 2, \ldots$. (Note that our notation is consistent, as this is a matrix interpretation of the operator $P^{(s)}$ defined in the proof of Theorem 9.1.1.) Since then, by Theorem 9.1.1, $P^{(s)} H = P^s H$ for all H, we have

Theorem 9.2.2. *The s-step transition matrix is related to the one-step transition matrix by*

$$P^{(s)} = P^s \qquad (s = 0, 1, 2, \ldots). \tag{14}$$

Relation (14) is, of course, just a matrix reformulation in the discrete case of the identical operator relation already obtained in Theorem 9.1.1 for the general case.

The particular corollaries of (14)

$$P^{(s)} = PP^{(s-1)}, \qquad p^{(s)} = P^{(s-1)}P,$$

correspond to the recursions for $p_{jk}^{(s)}$

$$p_{jk}^{(s)} = \sum_i p_{ji} p_{ik}^{(s-1)}, \tag{15}$$

$$p_{jk}^{(s)} = \sum_i p_{ji}^{(s-1)} p_{ik}. \tag{16}$$

These are known, respectively, as *Kolmogorov's backward equation* and *Kolmogorov's forward equation*. They amount, as one can see, to recursions on initial state and on final state, respectively. We have already encountered the backward equation in the guise (9). It is the forward equation which obviously corresponds to the deterministic dynamic equation (3).

If in (16) we set $s = t + 1$ and replace $p_{jk}^{(t)}$ by $\pi_k(t)$ then the equation becomes

$$\pi_k(t + 1) = \sum_j \pi_j(t) p_{jk}. \tag{17}$$

One can regard $\pi_k(t)$ as the probability that the process is in state k at time t, given that it started from an arbitrary distribution over states at some earlier time. Then relation (17) determines the distribution over states at time $t + 1$ in terms of that at time t, just as relation (3) determined the value of the variable at time $t + 1$ in terms of that at time t.

Relation (17) can again advantageously be written in vector form. If we define the column vector $\pi(t)$ with kth element $\pi_k(t)$, then (17) becomes

$$\pi(t + 1)^{\mathrm{T}} = \pi(t)^{\mathrm{T}} P, \tag{18}$$

or

$$\pi(t + 1) = P^{\mathrm{T}} \pi(t).$$

Comparing this last relation with (9), we see that it is the adjoint P^{T} of P which governs the forward equation. It is partly this fact which makes the forward formulation less natural on general state spaces.

We should formalize the properties that characterize a transition matrix:

Theorem 9.2.3. *The matrix P is a transition matrix if and only if it is square with nonnegative elements and unit row sums; i.e. $p_{jk} > 0$ and*

$$\sum_k p_{jk} = 1 \tag{19}$$

for $j, k \in \mathcal{X}$.

PROOF. The matrix must be square since we are considering transitions from $\mathcal{X}$ onto itself. The other properties follow from the fact that the rows of P constitute probability distributions over $\mathcal{X}$. Necessity is thus proved, and sufficiency is evident.   □

A matrix with these properties is often termed a *stochastic matrix*.

The powers P^s of the transition matrix obviously play a central role. One might, for example, imagine that the effect of initial conditions would wear off as time progresses, so that $p_{jk}^{(t)}$ becomes independent of both j and t for t large enough. That is,

$$p_{jk}^{(t)} \to \pi_k, \tag{20}$$

where $\{\pi_k\}$ then constitutes an *equilibrium distribution* over states. The convergence (20) may or may not hold; if it does, one says that the process is *ergodic*. Note that if (20) in fact holds then, at least in the case of a finite state space, it and relation (17) will imply that

$$\pi_k = \sum_j \pi_j p_{jk}, \tag{21}$$

the equilibrium form of the Kolmogorov forward equation. A distribution π satisfying (21) is termed a *stationary distribution* or an *invariant measure*. We shall see that such distributions may exist even though (20) does not hold. Note that we can also write (21) in the form

$$\sum_j (\pi_j p_{jk} - \pi_k p_{kj}) = 0, \tag{22}$$

which will later prove significant.

We can also write the relations (19) and (21) in the vector forms

$$P\mathbf{1} = \mathbf{1}, \tag{23}$$

$$\pi^{\mathrm{T}} = \pi^{\mathrm{T}} P. \tag{24}$$

where $\mathbf{1}$ is a vector of units. Relation (23), always valid, implies that P, as a matrix, has an eigenvalue $\lambda = 1$ with corresponding right eigenvector $\mathbf{1}$. Relation (23) would then imply that the equilibrium distribution vector π is the corresponding left eigenvector. The correct full statement of these ideas will begin to emerge in Section 8.

We can obtain considerable insight by considering the two-state case

$$P = \begin{bmatrix} p_{11} & p_{12} \\ p_{21} & p_{22} \end{bmatrix} = \begin{bmatrix} 1-\alpha & \alpha \\ \beta & 1-\beta \end{bmatrix}, \tag{25}$$

say, where α and β are the probabilities of transition. The two states may, for example, correspond to the fact that a molecule is in one of two energy levels, or in one of two regions in space, compartments 1 or 2. To avoid subsequent confusion between the state of a single molecule and the state of a Markov process, we shall speak of the molecule's having two possible positions: being in compartment 1 or compartment 2.

From the point of view of prediction over a time interval of length s, the quantity we need is the matrix power P^s, and we leave it to the reader to verify by induction that

$$P^s = \frac{1}{\alpha+\beta} \begin{bmatrix} \beta & \alpha \\ \beta & \alpha \end{bmatrix} + \frac{(1-\alpha-\beta)^s}{\alpha+\beta} \begin{bmatrix} \alpha & -\alpha \\ -\beta & \beta \end{bmatrix}. \tag{26}$$

The reader who is familiar with the concept of the *spectral representation* of a matrix will recognize (26) as the spectral representation of P^s, derived from that of P.

If $|1 - \alpha - \beta| < 1$ then we see from (26) that

$$P^s \to \frac{1}{\alpha + \beta} \begin{bmatrix} \beta & \alpha \\ \beta & \alpha \end{bmatrix}$$

as $s \to \infty$, so that the process is then *ergodic*. That is, whatever the initial distribution over states, the process tends to an equilibrium distribution with $\pi_1 = \beta/(\alpha + \beta)$ and $\pi_2 = \alpha/(\alpha + \beta)$. The reader can verify that this satisfies the equilibrium equation (21).

We can attain the extreme case $1 - \alpha - \beta = 1$ only if $\alpha = \beta = 0$, when

$$P = \begin{bmatrix} 1 & 0 \\ 0 & 1 \end{bmatrix}. \tag{27}$$

This is the case when no motion is possible, and the molecule is frozen in the compartment in which it started. One says that both states are *absorbing*. In this case, $\lambda = 1$ is a double eigenvalue of P, and

$$\pi = \begin{bmatrix} 1 \\ 0 \end{bmatrix}, \qquad \pi = \begin{bmatrix} 0 \\ 1 \end{bmatrix}, \qquad .$$

are both possible equilibrium distributions (as is any average of them). The right eigenvector is also nonunique; we shall see that this a reflection of the fact that there is more than one class of states in which the process can ultimately become absorbed.

The extreme case $1 - \alpha - \beta = -1$ is attained only if $\alpha = \beta = 1$, when

$$P = \begin{bmatrix} 0 & 1 \\ 1 & 0 \end{bmatrix}, \tag{28}$$

corresponding to deterministic cycling between the two compartments. The process is now certainly not ergodic, because (20) does not hold. On the other hand, equation (21) has the solution $\pi_1 = \pi_2 = \frac{1}{2}$. This makes sense in that the process obviously spends half its time in each compartment over a period of time.

Case (27) and (28) are referred to as *reducible* and *periodic*, respectively, and represent the only two possible sources of nonergodicity for a process with finitely many states.

A modification of the process is to assume that there are N molecules, each moving independently between the two compartments with transition matrix (25). This expanded process is still Markov, as we leave the reader to verify, with 2^N states. If we neglect the identities of molecules, and so simply consider the number X_t of molecules in position 1 at time t, then $\{X_t\}$ is again a Markov process, now with $N + 1$ states. Expressions for the transition probabilities are very ugly for this model; although their evaluation is implied in the much simpler expressions for conditional expectations:

$$\Pi_s(z) = E(z^{X_{t+s}}|X_t = x) = [(1 - \alpha_s)z + \alpha_s]^x [\beta_s z + (1 - \beta_s)]^{N-x}.$$

Here

$$P^s = \begin{bmatrix} 1 - \alpha_s & \alpha_s \\ \beta_s & 1 - \beta_s \end{bmatrix}$$

is determined by (26). In the ergodic case, we find that

$$\Pi_t(z) \to \left[\frac{\alpha + \beta z}{\alpha + \beta} \right]^N$$

at $t \to \infty$, so that the equilibrium distribution of X is binomial.

This model, known as the *Ehrenfest model*, is of considerable interest, since it was used to resolve a celebrated paradox of statistical mechanics; see Section 10.9. It is sometimes less reverently referred to as the 'dog-flea model', for obvious reasons.

EXERCISES AND COMMENTS

1. We have tended to use both s and t to denote the length of a time interval. This is because the fact that we consider only the time-homogeneous case has led to an obscuring of the distinction between the concepts of 'time', 'time elapsed' and 'time-to-go'. If t indicates current time, and $t \geq 0$, then t also measures the time which has elapsed between a start at $t = 0$ and the current moment. If the process is to terminate in some sense at time T, then $s = T - t$ measures 'time-to-go'; the length of time which remains before termination. The point of the distinctions will become clearer in some later applications.

2. We considered a random mating model in Section 5.6, but in plant and animal breeding one will have systematic methods of breeding and selection. Consider the case of *pure inbreeding*, in which an individual is mated with itself, as is possible with plants. If we follow such a breeding line, considering only a single individual in each generation, then this will constitute a Markov process with the three states AA, Aa and aa. Show that

$$P = \begin{bmatrix} 1 & 0 & 0 \\ \frac{1}{4} & \frac{1}{2} & \frac{1}{4} \\ 0 & 0 & 1 \end{bmatrix} \quad \text{and} \quad P^t \to \begin{bmatrix} 1 & 0 & 0 \\ \frac{1}{2} & 0 & \frac{1}{2} \\ 0 & 0 & 1 \end{bmatrix}.$$

That is, a pure (homozygous) line retains its state, while an Aa line ultimately becomes either AA or aa, each with probability $\frac{1}{2}$. The effect of inbreeding is thus to produce a pure line.

3. Consider *inbreeding with selection against a*, so that all aa individuals are discarded, and breeding continues in a given generation until a non-aa individual has been produced. The process is still Markov, but with only the states AA and Aa. Show that

$$P = \begin{bmatrix} 1 & 0 \\ \frac{1}{3} & \frac{2}{3} \end{bmatrix} \quad \text{and} \quad P^t \to \begin{bmatrix} 1 & 0 \\ 1 & 0 \end{bmatrix}.$$

The process thus ends with a pure AA line. For mating schemes which involve crosses it is, in general, not enough to consider just the states of an individual; one

has to consider the actual numbers of the three genotypes in the population as
random variables.

4. Show that the two descriptions of the Ehrenfest model are indeed Markov.

5. Note that the mere fact that the elements of P are bounded uniformly in s implies
that the eigenvalues of P cannot lie outside the unit circle. However, probabilists
regard discussion of the behavior of P^s in terms of spectral representation, etc. as
'nonprobabilistic'.

6. Consider the discrete-time renewal example formulated as a Markov process in
Exercise 9.1.5. Show that the equilibrium distribution of age is

$$\pi_j = \pi_0 a_0 a_1 a_2 \ldots a_{j-1}.$$

Here π_0 is to be determined by normalization. If π_j converges to zero too slowly to
be summable, then this is an indication that articles are so long-lived that age is not
a finite r.v. Note the distinction between π_j and $p_j = a_0 a_1 a_2 \ldots a_{j-1} b_j$. The first gives
the age distribution of the article in service at a randomly chosen time; the second
gives the distribution of lifespan for a randomly chosen article.

7. *The Chapman–Kolmogorov equation.* This is the name given to the relation which
would be written

$$P(X_{t+r+s} = k | X_t = j) = \sum_i P(X_{t+r} = i | X_t = j) P(X_{t+r+s} = k | X_{t+r} = i)$$

in the case of discrete $\mathscr{X}$. In expectation terms the relation is expressed by (8), a
much more compact and transparent expression, and valid as it stands for general
state spaces.

8. Suppose that a Markov process is ergodic and has a finite state space. Show that
X_r and X_s become independent as $|r - s| \to \infty$.

3. Some Examples: Ruin, Survival and Runs

A classic problem is the gambler's ruin problem, already touched upon in
Section 4.5. Gamblers Alex and Bernard play successive independent games
for unit stakes, Alex winning at each stage with probability p, and so losing
with probability q. Suppose that the two gamblers have a combined capital
of a and that Alex has capital X_t at the end of the tth round (both amounts
being multiples of the unit stake). Then, if debt is not allowed, play must end
when X equals either 0 or a, corresponding to ruin of Alex or Bernard,
respectively. For intermediate values of X the model is a Markov process with
transition probabilities $p_{j,j+1} = p$ and $p_{j,j-1} = q$ for $0 < j < a$; see Exercise
9.1.6. These are the only nonzero transition probabilities for j in this range.
We can imagine that X_t does not change once the game has terminated, so
the 'ruin by time t' means 'ruin at or before time t'.

The quantity of greatest interest to Alex is $\rho_j^{(s)} = p_{j0}^{(s)}$; the probability that if
he starts with a capital of j, then he is ruined after a time s. We should view

this ruin probability as a conditional expectation:

$$\rho_j^{(s)} = E[I(X_s = 0)|X_0 = j]. \tag{29}$$

The vector $\rho^{(s)}$ of ruin probabilities then obeys the backward equation

$$\rho_j^{(s)} = p\rho_{j+1}^{(s-1)} + q\rho_{j-1}^{(s-1)} \qquad (0 < j < a),$$

$$\rho_0^{(s)} = \rho_0^{(s-1)}, \qquad \rho_a^{(s)} = \rho_a^{(s-1)},$$

for $s > 0$, with the terminal condition

$$\rho_j^{(0)} = \begin{cases} 1 & (j = 0), \\ 0 & (j > 0. \end{cases} \tag{30}$$

Note an implication of (3) and (4): that $\rho^{(1)} \geq \rho^{(0)}$ and hence that the bounded sequence $\rho^{(s)}$ is nondecreasing in s. It thus has a limit $\rho^{(s)} \uparrow \rho = (\rho_j)$, where we can interpret ρ_j as Alex's *probability of ultimate ruin*, evaluated at a time when his current capital is j.

We have then in the limit to solve the difference equation

$$\rho_j = p\rho_{j+1} + q\rho_{j-1} \qquad (0 < j < a) \tag{31}$$

with end conditions

$$\rho_0 = 1, \qquad \rho_a = 0. \tag{32}$$

The general solution of the difference equation (31) is $\rho_j = c_1 + c_2(q/p)^j$. The end conditions (32) determine the arbitrary constants c and determine the solution as

$$\rho_j = \frac{(q/p)^j - (q/p)^a}{1 - (q/p)^a}, \tag{33}$$

(reducing to $(a - j)/a$ if $p = q$).

Expression (33) gives Alex's ruin probability as a function of his initial capital. One can obtain a better idea of how this function behaves if one considers the extreme case $a = +\infty$, when Alex is opposed by an infinitely rich Bernard. Solution (33) then becomes

$$\rho_j = \begin{cases} (q/p)^j & (p > q), \\ 1 & (p \leq q). \end{cases} \tag{34}$$

That is, if the game is advantageous then his probability of ruin is less than 1, and decreases exponentially fast with increase in initial capital. In the other cases (even the fair one) ruin is certain.

A rather more developed model of this type (see Section 6) is used to calculate ruin probabilities for insurance companies, 'wins' then being the inflow of premiums, 'losses' the outflow of claims and current capital being just that. The company is effectively playing against an infinitely rich opponent, since it never enters an absorbing state of 'victory', no matter how large its capital.

Another variable of interest is the duration τ of the game, if the game continues until ruin of one or the other player. To begin with, suppose that play is stopped after s games, under all circumstances, so that we are considering $\tau \wedge s = \min(\tau, s)$. Consider the expectation

$$\mu_j^{(s)} = E[\tau \wedge s | X_0 = j]. \tag{35}$$

This then obeys the backward equation (see Exercise 1)

$$\mu_j^{(s)} = 1 + p\mu_{j+1}^{(s-1)} + q\mu_{j-1}^{(s-1)} \qquad (0 < j < a), \tag{36}$$
$$\mu_0^{(s)} = \mu_a^{(s)} = 0,$$

for $s > 0$, with terminal condition $\mu_j^{(0)} = 0$.

By the same argument as before (indeed, by definition) $\mu_j^{(s)}$ is nondecreasing in s, and so has a limit value $\mu_j = E(\tau | X_0 = j)$ (possibly infinite) which satisfies

$$\mu_j = 1 + p\mu_{j+1} + q\mu_{j-1} \qquad (0 < j < a) \tag{37}$$

with $\mu_0 = \mu_a = 0$.

If $p \neq q$ then the general solution of (37) is

$$\mu_j = c_1 + c_2(q/p)^j + j/(q - p).$$

Fitting this to the boundary conditions we obtain

$$\mu_j = \frac{a}{p - q}\left[\frac{1 - (q/p)^j}{1 - (q/p)^a} - \frac{j}{a}\right]. \tag{38}$$

In the case of a fair game, $p = q$, this has the confluent form

$$\mu_j = j(a - j). \tag{39}$$

In the case $a \to \infty$ and $p \leq q$, expression (38) has the limit form

$$\mu_j = \frac{j}{q - p}. \tag{40}$$

If $p > q$ then termination is uncertain, so $E(\tau)$ is not defined.

One can continue to vary these themes on a number of standard problems, all having their own interest; see the exercises for a couple of examples.

EXERCISES AND COMMENTS

1. The backward equation from which (36) is derived is $E(\tau \wedge s | X_0 = j) = E[E(\tau \wedge s | X_1) | X_0 = j]$. However, $E[\tau \wedge s | X_1 = k]$ is 1 if k is an absorption state and can be identified with $1 + E[\tau \wedge (s - 1) | X_0 = k]$ otherwise.

2. Note from the ruin example that $\rho = P\rho$ can have solutions other than $\rho \propto 1$ if there is more than one state (or set of states) in which the process can become absorbed. In fact, let $\rho_j(A)$ be the probability of ultimate absorption in an absorbing set A, conditional on $X_0 = j$, and let $\rho(A)$ be the vector with these as components (or the function $\rho_j(A)$ with the j-argument understood). Then show that $\rho(A)$ is the smallest solution of $\rho = P\rho$ for which $\rho \geq I(A)$. (Here by $I(A)$ we mean the indicator function of set A as a function of *state j*.)

3. Consider again the renewal problem of Exercise 9.2.6. Suppose the article in current use has age j, and let τ be the age at which it fails. Show that the condition that $P(\tau < \infty)$ is that $\prod_0^\infty a_k = 0$, and that $E(\tau)$ is then evaluated by $1 + a_0 + a_0 a_1 + a_0 a_1 a_2 + \cdots$. Show also that the equilibrium age distribution deduced in Exercise 9.2.6 can be expressed as $\pi_j = P(\tau > j)/E(\tau)$.

4. *The occurrence of patterns.* Consider a sequence of independent Bernoulli trials with outcomes head or tail (denoted H or T) having respective probabilities p or q. There has always been an interest in the study of the statistics of runs (e.g. a run of heads of a given length). More recently, there has been an interest in the study of the frequency of occurrence of a prescribed pattern, such as HHTH. This has implications for the study of chromosomes, which are sequences of genes, and which can break at a point where a prescribed pattern has been completed.

   One analyses the problem by assigning a state X to the current configuration at every point in the sequence, where the value of state indicates what stage has been reached in the synthesis of the required pattern. The model can then be analysed as a Markov chain. To take the simplest non-trivial example, suppose the required pattern is HH. Then the two stages of completion could be labelled as 0 and 1, state 0 being that in which no progress has been made (i.e. the current sequence ends with a T) and state 1 is that in which the current sequence ends with a single H. For a sequence just beginning one starts in state 0. Let τ denote the number of the trial at which the pattern is first completed, and define $R_j(z) = E(z^\tau | X_0 = j)$. Show that R satisfies the backward equations

   $$R_0 = z(pR_1 + qR_0), \qquad R_1 = z(p + qR_0),$$

   and hence that

   $$R_0(z) = \frac{(pz)^2}{1 - qz - pqz^2}.$$

5. If one wishes to study, not merely the time until first completion of the pattern, but also the frequency of occurrence of the pattern on a continuing basis, then one has to calculate the equilibrium distribution of the Markov chain defined on the process. One has also to decide whether overlapping patterns are counted, e.g. is HHHH counted as two occurrences of HH or three? Let us refer to these as the nonoverlapping and overlapping conventions, respectively. Show that under the two conventions the completion of HH (at a random point in a long sequence) has probability $p^2/(1 + p)$ and p^2, respectively.

6. Complete the calculations of Exercises 4 and 5 for the cases when the prescribed pattern is (i) HTH, and (ii) a run of r heads.

4. Birth and Death Processes: Detailed Balance

There is a class of processes which occurs not infrequently and whose equilibrium behaviour is easily analysed. These are the processes for which the states can indeed be linearly ordered (by integral j, say) and for which passage is only possible between neighbouring states. So, from state j the process can move in a single time step only to states $j + 1$, $j - 1$ or j itself, with respective transition probabilities p_j, q_j and r_j, say. Such processes are called *birth and death processes*, for obvious reasons. It is sometimes convenient to abbreviate this to BD process.

So, the gambling model with unit stakes is a BD process. Models such as the Ehrenfest model and a queueing model would be BD processes if the unit time step were short enough that the possibility of multiple events could be ruled out—this is exactly the case in continuous-time versions. Population growth would itself be a BD process in continuous time if one could rule out the possibility of genuine multiple events (e.g. the birth of twins; mass deaths in accidents).

Let us suppose that $q_0 = 0$, so that the process is confined to the set of states $j \geq 0$ if it starts there. Let us also suppose that $p_j > 0$ and $q_{j+1} > 0$ for $0 \leq j < a$, say, and that these quantities are zero for $j \geq a$ (where a may be infinite). The aim is merely to ensure that all states on an interval $0 \leq j \leq a$ communicate, and that states $j > a$ are irrelevant.

Theorem 9.4.1. *Consider the birth and death process on the set of states $0 \leq j \leq a < \infty$, all states communicating. Then the detailed balance equation*

$$p_j \pi_j = q_{j+1} \pi_{j+1} \qquad (0 \leq j < a) \tag{41}$$

holds for the equilibrium distribution π_j, which then has the evaluation

$$\pi_j = \pi_0 \frac{p_0 p_1 \cdots p_{j-1}}{q_1 q_2 \cdots q_j} \tag{42}$$

with π_0 determined by

$$\pi_0^{-1} = \sum_{j=0}^{\infty} \frac{p_0 p_1 \cdots p_{j-1}}{q_1 q_2 \cdots q_j}. \tag{43}$$

PROOF. The equilibrium equation for π_j is

$$\pi_j = p_{j-1} \pi_{j-1} + r_j \pi_j + q_{j+1} \pi_{j+1} \qquad (0 \leq j \leq a) \tag{44}$$

with $\pi_{-1} = \pi_{a+1} = 0$, so that, in particular,

$$\pi_0 = r_0 \pi_0 + q_1 \pi_1. \tag{45}$$

Since $p_j + q_j + r_j = 1$, equation (45) can be rewritten

$$p_0 \pi_0 = q_1 \pi_1. \tag{46}$$

Equation (41) thus holds at $j = 0$. Now, if relation (41) holds at j, then we see from the balance equation (44) that it also holds at $j + 1$. We thus establish by induction that (41) holds for all relevant j. Assertions (42) and (43) then follow.   □

If we consider the case when a is infinite then, just as in Exercise 9.2.6, the sum (43) must converge if (42) is to define a proper equilibrium distribution. If it diverges then π_j is zero for all finite j, and the state variable is infinite with probability one.

The detailed balance relation (41) is plainly a considerable simplification of the full balance equation (44). One can see why it holds in this case. Let A be the set of states $0, 1, 2, \ldots, j$ and $\bar{A}$ the complementary set $j + 1, j + 2, \ldots$. In equilibrium the 'probability flux' (the expected number of transitions per unit time) from A to $\bar{A}$ must balance that from $\bar{A}$ to A. However, the only way to pass from A to $\bar{A}$ is to make the transition $j \to j + 1$, and correspondingly for the reverse transitions. So, the probability fluxes $j \leftrightarrow j + 1$ must balance, which is what relation (41) states.

The general topic of detailed balance is an important one, which we touch upon briefly in Exercise 2, and consider more thoroughly in the continuous-time context in Section 10.8.

EXERCISES AND COMMENTS

1. Consider the following discrete-time version of a queueing model. The queue may contain j customers ($j = 0, 1, 2, \ldots$). There is a probability α that a single new customer joins the queue in a unit time interval, and a probability β that a single *old* customer (if there are any; i.e. if $j > 0$ at the beginning of the interval) is served and leaves. Multiple arrivals and departures are excluded. Show, under the obvious independence assumptions, that the queue-length has the equilibrium distribution

$$\pi_j = \frac{\pi_0}{1 - \beta} \left[\frac{\alpha(1 - \beta)}{\beta(1 - \alpha)} \right]^j$$

for $j = 1, 2, \ldots$. What is the condition on α and β that this should constitute a proper distribution?

2. Suppose that, for a Markov process with transition probabilities p_{jk}, one can find constants γ_j such that $\gamma_j p_{jk} = \gamma_k p_{kj}$ for all j, k. It then follows that

$$\sum_j (\gamma_j p_{jk} - \gamma_k p_{kj}) = 0,$$

and, as we see from form (22) of the general balance equation, that $\pi_j \propto \gamma_j$ is a possible equilibrium distribution. That is, the simple verification of a detailed balance relation supplies an evaluation of a possible equilibrium distribution.

5. Some Examples We Should Like to Defer

There are a number of models which one can formulate in discrete time, but for which the continuous-time formulation is more natural. It may be physically more natural, but may also be formally more attractive, in that multiple transitions are then excluded. That is, each transition corresponds to a definite physical event, rather than to a superposition of several such events. This is true for the Ehrenfest model, for models of queueing and population growth, and indeed for almost all the models we shall consider in the next chapter. The passage to continuous time may even simplify the model to a birth and death process.

Another model which shows an even greater simplification when one makes the double passage to continuous time and a continuous state space is what is sometimes called the 'heap problem', but which we shall term the 'filing problem'. One takes a book that one wants from a pile, and replaces it at the *top* of the pile when one is finished with it. This process continues. The observation often made is that the process is in some degree self-optimizing, in that the books which are most often referred to will tend to be near the top of the pile. We shall term this the 'filing problem' since we could also regard the problem as one of calling on files in a store. If files, when replaced, are replaced at the front of the store, then we have the same situation.

One can formulate this as a Markov process, with the transition being the random permutation of the files induced by the bringing to the front of a randomly specified file. The state variable is the order of the files, and we should like to evaluate its equilibrium distribution. The discreteness of both time and permutation makes this a difficult problem. We shall see, in Section 10.7, that there is a continuous embedding of the problem which reduces it radically.

We could very well discuss discrete-time models with a nondiscrete state space; the material of Section 1 gives the formal basis. However, again, most such models are more naturally formulated in continuous time.

6. Random Walks, Random Stopping and Ruin

The random walk is a special but important process. The most general example of what one might regard as the classic random walk is a model in which one takes successive steps in some region $\mathscr{C}$ of m-dimensional space, $\mathbb{R}^m$, these steps being IID as long as one remains in $\mathscr{C}$. The region $\mathscr{C}$ is termed the *continuation set*, and the rules may change once one leaves it. The gambler's ruin problem of Section 3 constituted a *simple random walk*: simple in that $m = 1$ and only unit steps were taken (in the continuation set $x = 1, 2, \ldots, a - 1$). However, one may well consider higher dimensions and allow both multiple and continuously-distributed steps. One could relax the IID assumption, but has then lost the distinctive structure of the model and reverted almost to the general Markov process.

In this section we shall deal with the one-dimensional case, and shall suppose only that the step-size must be integral. This case is sufficiently general that it reveals the general solution. It is also of practical interest in that, for example, it leads to a solution (mathematical!) of the so-called *ruin problem* for an insurance company.

We shall continue with the spatial analogy suggested by the term 'random walk' and shall refer to the *position* X_t of the representative point at time t. This is the state variable of a Markov process. The increment in position over t steps is the sum of t IID r.v.s, so one would expect that the p.g.f. or c.f. techniques we have employed for such sums would be useful. This is so, despite the fact that the rules change once X leaves $\mathscr{C}$.

For the one-dimensional lattice example that we consider the value of state variable (position) will be an integer, x. Let us denote the complement of $\mathscr{C}$ in the set of integers by $\mathscr{D}$. We shall assume that the walk stops once it enters $\mathscr{D}$, which is then a *stopping set*. Other rules could be considered (e.g. a random restart, if a company which has become ruined is refloated, or an elevation of insurance premiums) but we shall assume a simple stop, for definiteness. We shall sometimes speak of the event 'stopping' as 'absorption', since the process then indeed becomes absorbed in one of the states of $\mathscr{D}$.

Let τ denote the stopping time; the time at which the walk first enters $\mathscr{D}$. Then there is interest in the determination of the distribution of both the stopping time τ and the stopping coordinate X_τ.

Let

$$\Pi(z) = \sum_k p_k z^k$$

be the p.g.f. of step-size (signed), so that

$$\mu = \Pi'(1) = \sum_k k p_k$$

is the expected displacement at each step. The following simple result turns out to be central.

Theorem 9.6.1. *Suppose that $p_k > 0$ for some positive and some negative k, and that $\Pi(z)$ is finite in some real interval with $z = 1$ as interior point. Then the equation*

$$\Pi(z) = 1 \qquad (47)$$

has exactly two nonnegative real solutions, one of which is $z = 1$. If the smaller solution is $z = \zeta$ then $\zeta = 1$ or $\zeta < 1$ according as $\mu \leq 0$ or $\mu > 0$.

PROOF. Note that the finiteness assumption on Π implies that p_k must decrease exponentially fast as $|k|$ increases, so that μ certainly exists and is finite. The function $\Pi(z)$ is convex on the positive axis; the assumptions that steps of both positive and negative size have positive probability imply that it is strictly so, and indeed approaches $+\infty$ as z approaches 0 or $+\infty$. The course of $\Pi(z)$ must then appear as in Fig. 9.1, and equation (47) has either two solutions in $z \geq 0$ or none. But it certainly has a root $z = 1$, so there is a second, possibly coincident. The value of Π' is negative at the smaller root and positive at the upper (zero if the two roots are coincident) and $\Pi'(1) = \mu$. Hence the last assertion. $\qquad\qquad\square$

Let the p.g.f. of initial position $E(z^{X_0})$ be denoted $\Pi_0(z)$; we suppose $X_0 \in \mathscr{C}$. Define also the generating functions

$$D(w, z) = E(w^\tau z^{X_\tau}) = E\left[\sum_{t=0}^{\infty} w^t I(\tau = t) z^{X_t}\right],$$

$$C(w, z) = \sum_{t=0}^{\infty} w^t E[I(\tau > t) z^{X_t}]. \qquad (48)$$

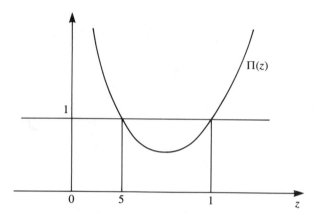

Figure 9.1. A graph of the probability generating function $\Pi(z)$ in the case when the expectation μ of the variable is positive.

That is, $D(w, z)$ is the joint p.g.f. of stopping time and stopping coordinate, and $C(w, z)$ is the generating function (over t) of the p.g.f. of that part of the X_t distribution which is still in the continuation region. The coefficient of $w^t z^x$ in $D(w, z)$ is thus $P(\tau = t, X_\tau = x)$ and will be zero unless $x \in \mathcal{D}^*$, where $\mathcal{D}^*$ is that part of $\mathcal{D}$ which can be reached from $\mathcal{C}$ in a single step. Note also that this distribution need not integrate to unity if stopping is uncertain; so the expectation by which $D(w, z)$ is defined receives no contribution from those realizations which do not stop.

Suppose it can be established (see Exercise 1) that

$$P(\tau > t) \le K\rho^t \tag{49}$$

for some K and ρ, so that the probability that the walk has not stopped by time t is of order ρ^t.

Theorem 9.6.2. *Assume the bound* (49) *for the probability of nonabsorption. The generating functions $C(w, z)$ and $D(w, z)$ certainly converge if $|w| < \rho^{-1}$, if z is such that both $\Pi(z)$ and $\Pi_0(z)$ converge, and if either:*

(a) *$\mathcal{C}$ is bounded and $|z|$ is bounded away from 0 and $+\infty$; or*
(b) *$\mathcal{C}$ is bounded in the direction of negative j, and $|z|$ does not exceed unity and is bounded away from 0.*

In this region of validity the identity

$$[1 - w\Pi(z)]C(w, z) = \Pi_0(z) - D(w, z) \tag{50}$$

holds.

PROOF. The assertions of the regions of validity follow from the facts that the coefficients in both the generating functions C and D constitute probability distributions, and that the probabilities both of being unabsorbed and of becoming absorbed at time t are of order ρ^t. To deduce the central identity (50), note that it follows from the independent nature of the steps that

$$E[I(\tau > t)z^{X_{t+1}}] = \Pi(z)E[I(\tau > t)z^{X_t}],$$

and hence that

$$w\Pi(z)C(w, z) = w\Pi(z)E\left[\sum_{t=0}^{\infty} w^t I(\tau > t)z^{X_t}\right]$$

$$= E\left[\sum_{t=0}^{\infty} w^{t+1} I(\tau > t)z^{X_{t+1}}\right]$$

$$= E\left\{\sum_{t=0}^{\infty} w^{t+1}[I(\tau > t + 1) + I(\tau = t + 1)]z^{X_{t+1}}\right\}$$

$$= C(w, z) - \Pi_0(z) + D(w, z),$$

whence (50) follows.   □

The Wald–Miller identity (50) is central; it relates the two undetermined functions C and D. Complete determination follows from

Theorem 9.6.3. *For the values of w and z prescribed in Theorem 9.6.2 and also satisfying*

$$1 - w\Pi(z) = 0, \tag{51}$$

the relation

$$D(w, z) = \Pi_0(z) \tag{52}$$

or

$$E(w^\tau z^{X_\tau}) = E(z^{X_0}) \tag{53}$$

holds. In particular, the Wald identity

$$E(z^{X_\tau}\Pi(z)^{-\tau}) = E(z^{X_0}) \tag{54}$$

holds for z satisfying $|\Pi(z)| > \rho^{-1}$ and the conditions of Theorem 9.6.2.

PROOF. The left-hand member of (50) is zero if w, z are such that (51) holds and all the generating functions are convergent. Then the right-hand member must also be zero, which is what relations (52) and (53) state. In particular, using relation (51) to determine w in terms of z, we derive the Wald identity (54).   □

One regards relations (51) and (52) as determining $D(w, z)$, which is the quantity of principal interest, determining as it does the distribution of stopping time and coordinate. Relation (50) then determines $C(w, z)$, if desired. Note that (54) is indeed the identity already encountered as the Wald identity in Exercise 6.1.9.

We shall consider a series of applications of these assertions in the exercises. However, they lead to one other significant result.

Theorem 9.6.4 (Bounds on the Ruin Probability). *Suppose that the step-size p.g.f.* $\Pi(z)$ *satisfies the conditions of Theorem 9.6.1 and that the continuation region is* $x > 0$. *Characterize the event of entering the stopping set* $x \leq 0$ *as 'ruin'. Then*

$$\zeta^x \leq P(\text{ruin}|X_0 = x) \leq \zeta^{x-c}, \qquad (55)$$

where ζ *is the smaller positive root of* (47) *and* c *is the maximal possible overshoot into the stopping set.*

That is, $x = -c$ is the least value that can be reached in a single step from $x = 1$, so that the maximal step length which is possible in the negative direction is $c + 1$.

PROOF. Since $0 < \zeta \leq 1$ and $-c \leq X_\tau \leq 0$, then

$$D(\zeta, 1) \leq D(1, 1) \leq \zeta^{-c}D(\zeta, 1).$$

But $P(\text{ruin}) = D(1, 1)$ and $z = \zeta$, $w = 1$, satisfy the conditions of Theorem 9.6.3. Thus (52) implies that $D(\zeta, 1) = \Pi_0(\zeta)$, which with this last relation implies that

$$\Pi_0(\zeta) \leq P(\text{ruin}) \leq \zeta^{-c}\Pi_0(\zeta). \qquad (56)$$

This reduces to (55) in the particular case $\Pi_0(z) = z^x$.   □

The importance of this conclusion is in the insurance context, when X represents the current capital of the company and the steps represent the changes in X due to the inflow of premiums and the outflow of claims. Passage to $X \leq 0$ then indeed represents ruin, and bounds (55) give $P(\text{ruin}) \sim \zeta^X$ as an assessment of the ruin probability. (Bounds (55) are quoted for a prescribed initial X, but it is right to regard the probability as changing as one observes changes in the r.v. X, i.e. in the current capital.) The dependence of the company's prospects on the balance between claims and premiums is reflected in the value of ζ, which decreases from a value of unity as the expected net income rate μ increases from zero.

Of course, matters are much more complicated in the real world: increments are not independent, premiums would be loaded if times became bad, and the companies have mutual reinsurance arrangements. Nevertheless, Theorem 9.6.4 provides a clean solution to the essential problem, and some form of this conclusion will continue to hold for more elaborate versions.

EXERCISES AND COMMENTS

1. *Bounds on survival probability.* Suppose that the continuation set is the interval (a, b), that $X_0 = 0$, that $0 \leq b < \infty$ and $\mu > 0$. Let S_t be the *free* sum of t IID steps.

Then $P(\tau > t) \leq P(S_t \leq b) \leq \Pi(z)^t/z^b$ for any z in $[0, 1]$. Since $\mu = \Pi'(1) > 0$ we can find a z for which $\Pi(z) < 1$. In fact, we can achieve (49) with $\rho = \inf_z \Pi(z)$, the infimum being over nonnegative z.

2. The gambler's ruin problem of Section 3 is just the case when the continuation set is $(1, 2, 3, \ldots, a - 1)$ and $\Pi(z) = pz + qz^{-1}$. We can write

$$D(w, z) = D_0(w) + D_a(w)z^a,$$

where the coefficient of w^t in $D_x(w)$ is the probability that, conditional on initial conditions, the game terminates at time t and at coordinate (Alex's capital) x. If the initial coordinate is $X_0 = j$, then relation (52) implies the two equations

$$D_0(w) + D_a(w)z_k(w)^a = z_k(w)^j \qquad (k = 1, 2), \tag{57}$$

where the z_k are the two roots

$$z_k(w) = \frac{1 \pm \sqrt{1 - 4pqw^2}}{2pw} \tag{58}$$

of (51). Equations (57) determine D_0 and D_a; e.g.

$$D_0(w) = \frac{z_1^a z_2^j - z_1^j z_2^a}{z_1^a - z_2^a}. \tag{59}$$

As $w \uparrow 1$ then the z_k tend to 1 and q/p and the evaluation of $D_0(1)$ yielded by (59) is just the expression (33) for Alex's ruin probability. The quotient $D_0(w)/D_0(1)$ is the p.g.f. of the time to his ruin, conditional on the event of ruin.

3. Continuing Exercise 2, consider the case $a = \infty$ and $p \leq q$, which is one of certain ruin for Alex. Solution (59) then becomes $D_0(w) = z_2(w)^j$, where z_2 corresponds to the minus option in (58). Since $D_0(w)$ is the p.g.f. of the time τ to ruin (i.e. to make a net loss of j), then $z_2(w)$ must be the p.g.f. of the time taken to make a net loss of unity. Show that $E(\tau) = j/(q - p)$ and $\text{var}(\tau) = jpq/(q - p)^3$.

4. Consider the simple random walk without boundary restrictions, starting from the origin and in the symmetric case $p = q$. Let τ be the time of first return to the origin. Show from Exercise 3 that $E(w^\tau) = 1 - \sqrt{1 - w^2}$ for $|w| \leq 1$.

5. Consider again the ruin problem of Theorem 9.6.4, with $X_0 = x$. A formal no-overshoot approximation yields $\zeta(w)^x$ as the approximate p.g.f. (not necessarily proper) of time to ruin, where $\zeta(w)$ is the smaller positive root of (51) for $0 \leq w \leq 1$. This generalizes the assertion of the theorem.

6. For the general random walk (multi-dimensional, arbitrary step distribution) the Wald–Miller identity (50) would become

$$[1 - w\phi(\theta)]C(w, \theta) = \phi_0(\theta) - D(w, \theta),$$

where $\phi(\theta)$ and $\phi_0(\theta)$ are the c.f.s of step-size and initial coordinate and

$$D(w, \theta) = E(w^\tau e^{i\theta^T X_\tau}), \qquad C(w, \theta) = \sum_{t=0}^{\infty} w^t E[I(\tau > t)e^{i\theta^T X_t}].$$

7. Auguries of Martingales

Suppose that $\{X_t\}$ is a Markov process, and suppose that there is another process $\{M_t\}$ which has the property that

$$E(M_{t+1}|X_t) = M_t, \tag{60}$$

at least over some range of t of interest. Then $\{M_t\}$ is a Markov version of what will later be termed a *martingale* with respect to $\{X_t\}$. The concept seems to have no particular motivation at the moment, but we can see that it is at least a recasting of a familiar idea.

The conditional expectation in (60) must be a function of X_t, and possibly also of t; let us denote it by $\psi_t(X_t)$. Then (60) states that we know a sequence of functions $\psi_t(x)$ with the property

$$E[\psi_{t+1}(X_{t+1})|X_t] = \psi_t(X_t), \tag{61}$$

or

$$P\psi_{t+1} = \psi_t.$$

In the particular case when $\psi_t(x)$ is independent of t (61) becomes simply

$$P\psi = \psi. \tag{62}$$

We know that (62) holds in the trivial case $\psi(x) = 1$, and have seen in Section 3 that it also holds in the case when $\psi(x)$ is the probability of ultimate absorption in some prescribed set of states conditional on a start from state x. If we can find functions ψ such that (62) holds (i.e. 'find a martingale') then this could be very useful. The relation $P^t\psi = \psi$ $(t = 0, 1, 2, \ldots)$ then implies that we have the evaluation of an expectation

$$E[\psi(X_t)|X_0] = \psi(X_0).$$

The result might even continue to hold if t is replaced by a random stopping time τ. That is,

$$E[\psi(X_\tau)|X_0] = \psi(X_0). \tag{63}$$

The *optional stopping theorem* (see Sections 14.4 and 14.5) gives conditions under which the stronger conclusion (63) holds. We have in fact already found martingales (see Exercise 1) and shall later find new and useful ones.

EXERCISES AND COMMENTS

1. Consider the random walk of the last section without stopping restrictions. Then $\psi(x) = z^x$ is a martingale if (47) holds, and $\psi_t(x) = w^t z^x$ is a martingale if (51) holds. If we now stop the process at the time τ when it first leaves $\mathscr{C}$ and apply the t-dependent version of (63) to the second martingale we deduce just

$$E[\Pi(z)^{-\tau} z^{X_\tau}|X_0] = z^{X_0},$$

i.e. Wald's identity.

8. Recurrence and Equilibrium

In the last few sections we have considered processes which ended in an absorbing state. However, the other behaviour of interest is that in which the process continues to some kind of equilibrium behaviour. We shall now give some of the formal theory for such processes, assuming a discrete state space. The treatment is kept brief; those interested in applications may resort to it for reference rather than reading.

Let us say that state j communicates with state k if passage from j to k is possible. That is, if $p_{jk}^{(s)} > 0$ for some s. A *closed class* of states $\mathscr{C}$ is a set of states, all of which communicate with each other and with no state outside the class. The set $\mathscr{X}$ of all states can obviously be partitioned into closed classes $\mathscr{C}_1$, $\mathscr{C}_2$, $\mathscr{C}_3$, ... and a set of states $\mathscr{T}$ from which passage into the closed classes is possible. A process whose states form a single closed class is *irreducible*.

Let τ_{jk} be the *first passage time* from state j to state k. This is a r.v. defined as the smallest $t > 0$ for which $X_t = k$ if $X_0 = j$. If $k = j$ then this is just the *recurrence time* for state j. State j is said to be *recurrent* (or *persistent*) if recurrence is certain, i.e. if $P(\tau_{jj} < \infty) = 1$. A state which is not recurrent is *transient*. If a state is recurrent, it is said to be *positive recurrent* if also $E(\tau_{jj}) < \infty$, *null recurrent* otherwise.

Define the generating functions

$$U_{jk}(w) = E\left(\sum_{t=0}^{\infty} I(X_t = k)w^t \mid X_0 = j\right),$$

$$F_{jk}(w) = E(w^{\tau_{jk}}),$$

which certainly exist for $|w| < 1$ and $|w| \le 1$, respectively. Define also the matrices $U = (U_{jk})$, $F = (F_{jk})$ with these as elements.

Theorem 9.8.1. *The generating functions $U(w)$ and $F(w)$ obey the relations*

$$U(w) = (I - wP)^{-1}, \tag{64}$$

$$U_{jj}(w) = \delta_{jk} + F_{jk}(w)U_{kk}(w). \tag{65}$$

PROOF. Since $U_{jk}(w) = \sum_{t=0}^{\infty} w^t p_{jk}^{(t)}$ then $U(w) = \sum_{t=0}^{\infty} (wP)^t = (I - wP)^{-1}$. To establish equation (65), suppose that $X_0 = j$ and define τ as the first passage time (recurrence time) to state k. Then

$$\sum_{t=0}^{\infty} I(X_t = k)w^t = \delta_{jk} + w^{\tau} \sum_{t=0}^{\infty} I(X_{t+\tau} = k)w^t,$$

where the δ_{jk} counts the initial occupation of the state k if $j = k$. Note that τ is independent of the final sum, which begins with an occupation of state k. Taking expectations conditional on $X_0 = j$ we deduce (65). □

Theorem 9.8.2. *State j is recurrent if and only if* $\lim_{w \uparrow 1} U_{jj}(w) = \infty$.

PROOF. For recurrence we require that $\lim_{w \uparrow 1} F_{jj}(w) = 1$. But we see from (65) that

$$F_{jj}(w) = \frac{U_{jj}(w) - 1}{U_{jj}(w)},$$

and this can have the limit value unity if and only if U_{jj} diverges as indicated. □

The result agrees with intuition; we would interpret $U_{jj}(1)$ as the expected number of visits to state j in $t \geq 0$, conditional on $X_0 = j$.

Theorem 9.8.3. *Suppose state j recurrent. Then it is positive recurrent if and only if*

$$\pi_j = \lim_{w \uparrow 1} (1 - w) U_{jj}(w) \tag{66}$$

is positive, and one can then make the identification

$$E(\tau_{jj}) = 1/\pi_j. \tag{67}$$

In denoting the limit (66) by π_j we are anticipating the interpretation of this quantity as the equilibrium probability that state j is occupied. Certainly limit (66) is the Abel mean (over time) of the probability of occupation.

PROOF. We can rewrite identity (65) in the case $k = j$ as

$$\frac{1 - F_{jj}(w)}{1 - w} = \frac{1}{(1 - w) U_{jj}(w)}. \tag{68}$$

and can write the left-hand member as $\sum_{t=0}^{\infty} P(\tau_{jj} = t)(1 + w + w^2 + \cdots + w^{t-1})$. As $w \uparrow 1$ this quantity converges monotonically to $\sum_{t=0}^{\infty} t P(\tau_{jj} = t) = E(\tau_{jj})$, so the right-hand member of (68) must similarly converge. We thus deduce relation (67), and $E(\tau_{jj})$ is finite or infinite according as π_j is positive or zero. □

It is now simplest to specialize to the irreducible case. This scarcely amounts to any loss of generality, since the general case can be built up from it.

Theorem 9.8.4. *Suppose the process irreducible. Then all the $U_{jk}(w)$ behave in the same way as functions of w in the limit $w \uparrow 1$, in that one can find finite nonzero constants c_{jkhi} such that*

$$U_{jk}(w) \geq c_{jkhi} U_{hi}(w) \tag{69}$$

for all j, k, h, i and for w in a nonempty interval $[\rho, 1)$.

PROOF. Consider w in the real interval indicated. Then from the relations $U(w) - wPU(w) = U(w) - wU(w)P = I$ we deduce that $U(w) \geq \rho PU(w) = \rho U(w)P$, and hence that

$$U(w) \geq \rho^{r+s} P^r U(w) P^s.$$

That is,

$$U_{jk}(w) \geq \rho^{r+s} p_{jh}^{(r)} U_{hi}(w) p_{ik}^{(s)}.$$

But, by the communication hypothesis, we can find r and s such that both transition probabilities are positive. We thus have the bound. (69). $\qquad \square$

Theorem 9.8.5 ('Solidarity'). *Suppose the process irreducible. Then all states fall into the same class (i.e. recurrent or transient, and, if recurrent, positive recurrent or null recurrent).*

PROOF. This follows from the comparison results of Theorem 9.8.4 and the state characterizations of Theorems 9.8.2 and 9.8.3 in terms of the U_{jj}. $\qquad \square$

Theorem 9.8.6. *Suppose the process irreducible and recurrent. Then $F_{jk}(1) = 1$ for all j, k, and*

$$\lim_{w \uparrow 1} \frac{U_{jk}(w)}{U_{kk}(w)} = 1. \tag{70}$$

That is, ultimate passage from j to k is certain, for any j, k.

PROOF. Relations (65) can be written

$$F(w) - wPF(w) = -wPD(w)^{-1},$$

where $D(w)$ is a diagonal matrix with jth entry $U_{jj}(w)$. Since $D(w)^{-1} \downarrow 0$ as $w \uparrow 1$ we thus deduce that $F(1) = PF(1)$, and so that $F(1) = P^s F(1)$ for any $s = 0, 1, 2, \ldots$. That is, for any k we have

$$1 = f_{kk} = \sum_j p_{kj}^{(s)} f_{jk}, \tag{71}$$

where $f_{jk} = F_{jk}(1)$ is the probability of ultimate passage to k, conditional on a start from j. But, since $f_{jk} \leq 1$, relation (71) can hold only if $f_{jk} = 1$ for any j such that $p_{kj}^{(s)} > 0$. Since this latter inequality will hold for some s for given j, k, we thus deduce that indeed $f_{jk} = 1$ for all j, k. Letting w tend to unity in the relation

$$F_{jk}(w) = \frac{U_{jk}(w)}{U_{kk}(w)}$$

for distinct j, k, we then deduce (70). $\qquad \square$

We now come to the result which most explicitly asserts convergence to an equilibrium.

Theorem 9.8.7. *Suppose the process irreducible and positive recurrent. Then:*

(i) *The limit*

$$\pi_k = \lim_{w \uparrow 1} (1 - w) U_{jk}(w)$$

exists, is independent of j, and interpretable both as $[E(\tau_{kk})]^{-1}$ and as the Abel mean of the probability that state k is occupied.

(ii) *The vector π constitutes the unique stationary distribution for the process.*

PROOF. The first assertion follows from Theorem 9.8.3 and (70). For the second, it follows from (64) that

$$(1 - w)[U(w) - wU(w)P] = (1 - w)I,$$

so that

$$(1 - w)U_{kk}(w) - w \sum_j (1 - w)U_{kj}(w)p_{jk} = 1 - w.$$

Letting w tend to 1 we deduce that

$$\pi_k - \sum_j \pi_j p_{jk} = 0,$$

so that π is indeed a stationary distribution. It is also normalized; by letting w tend to 1 in the relation

$$(1 - w)U(w)\mathbf{1} = \mathbf{1}$$

one deduces that $\pi^{\mathrm{T}}\mathbf{1} = 1$.

Let a be any other stationary distribution, so that $a^{\mathrm{T}} = a^{\mathrm{T}}P$. Then $a^{\mathrm{T}} = a^{\mathrm{T}}P^s$, and so $a^{\mathrm{T}} = a^{\mathrm{T}}(1 - w)U(w)$. In the limit $w \uparrow 1$ this yields $a_k = \sum_j a_j \pi_k = \pi_k$. Thus π is the unique stationary distribution. $\square$

Irreducibility and positive recurrence thus imply the existence of a unique stationary distribution, with π_j identifiable with $[E(\tau_{jj})]^{-1}$. However, we can prove only that

$$(1 - w) \sum_{t=0}^{\infty} p_{jk}^{(t)} w^t \to \pi_k \tag{72}$$

as $w \uparrow 1$, and need stronger hypotheses to attain the stronger conclusion that

$$p_{jk}^{(t)} \to \pi_k.$$

at $t \to \infty$.

Finally, the following theorem is useful.

Theorem 9.8.8. *Suppose the process irreducible. Then it is positive recurrent if either:*

(a) *the state space is finite, or*

(b) *a stationary distribution exists.*

The second assertion is particularly useful. If the process is irreducible and one can find a stationary distribution, then one can conclude, by the previous theorem, that it is unique, and that the process converges to it at least in the sense (72).

PROOF. (a) If the process is not positive recurrent, then $\pi_j = 0$ for all j, which is incompatible with $\sum_j \pi_j = 1$.

(b) Let a be the stationary distribution. Then $a_j > 0$ for some j, and the relation $a^{\mathsf{T}} = a^{\mathsf{T}} P^s$ implies that $a_j > 0$ for all j. The relation $a^{\mathsf{T}} = a^{\mathsf{T}}(1 - w)U(w)$ then implies that $\lim_{w \uparrow 1} (1 - w)U_{jk}(w)$ is positive for some j for every k, and hence by (65), that $\lim_{w \uparrow 1} (1 - w)U_{kk}(w)$ is positive for all k. That is, all states are positive recurrent. □

EXERCISES AND COMMENTS

1. Suppose that state 1 is recurrent. Then show that the set of all states which can be reached from state 1 form a closed recurrent class.

2. Note, for Theorem 9.8.8 (b), that it is not enough to find a nonnegative solution π of the balance equation. This solution must also be summable.

9. Recurrence and Dimension

The simplest example of an irreducible process which can show all possible characters is the simple random walk on $x \geq 0$ with an impermeable barrier at the origin (i.e. the transition $x \to x - 1$ of probability q becomes a transition $0 \to 0$ if $x = 0$). One could regard the process as Alex's game against an infinitely rich Bernard who is willing to forego his win if this would take Alex into debt. It is a birth and death process with equilibrium distribution

$$\pi_x = \pi_0 (p/q)^x \qquad (x = 0, 1, 2, \ldots). \tag{73}$$

Distribution (73) is a proper distribution, and all states are positive recurrent, if $p < q$. If $p = q$ then all states are still recurrent (as we will show below), but in fact null-recurrent, as indicated by the fact that the distribution is uniform on the half-axis, and $\pi_x = 0$ for all x. If $p > q$ then all states are transient (the probability mass escapes to $+\infty$).

If we consider the simple random walk on the *whole* axis then $p_{0x}^{(t)} = P(X_t = x | X_0 = 0)$ is the coefficient of z^x in $(pz + qz^{-1})^t$. In the notation of the last section we thus have, in particular,

$$U_{00}(w) = \text{absolute term in } z \text{ in } [1 - w(pz + qz^{-1})]^{-1}$$

$$= \sum_{r=0}^{\infty} \binom{2r}{r} (pq)^r w^{2r} = (1 - 4pqw^2)^{-1/2}. \tag{74}$$

(Here, by 'the absolute term in z', we mean the coefficient of z^0 in an expansion on the unit circle in powers of z.) The expected number of times the initial state $x = 0$ is occupied is then

$$U_{00}(1) = \frac{1}{\sqrt{1 - 4pq}} = \frac{1}{|p - q|}.$$

This is finite if $p \neq q$; the state $x = 0$ (and all states) are then transient. If $p = q$ then $U_{00}(1)$ is infinite, so recurrence is certain. However, in this case $\lim_{w \uparrow 1} (1 - w)U_{00}(w) = 0$, as is easily verified from (74), so that the state is null-recurrent, and its expected recurrence time is infinite.

The extension of this analysis to that of the free symmetric random walk in several dimensions is interesting, because it gives us our first indication of the qualitative effects of dimension. As for the one-dimensional case above we have

$$U_{00}(w) = \text{absolute term in } z \text{ in } [1 - w\Pi(z)]^{-1},$$

where $\Pi(z)$ is the multivariable p.g.f. of the step in the random walk. For the symmetric simple random walk in d dimensions this takes the form

$$\Pi(z) = \frac{1}{2d} \sum_{j=1}^{d} (z_j + z_j^{-1}). \tag{75}$$

An argument which needs some expansion but which gives the correct order of magnitude of $p_{00}^{(t)}$ for large t is the following. If the walk has returned to the origin after t steps, then on the jth coordinate axis the walk must have taken an equal number of steps, say r_j, in each direction ($j = 1, 2, \ldots, d$). Suppose that $t = 2dr$ for large integral r. Then, by the central limit theorem, the r_j will differ from r only by a term of probable order $r^{1/2}$, and the probability of a return to the origin will be approximately what it would have been if the r_j equalled r for all j, i.e.

$$\left[\binom{2r}{r} 2^{-2r} \right]^{d} = O(r^{-d/2})$$

for large r. This is summable only for $d > 2$. That is, recurrence remains certain in two dimensions, but is uncertain in three or more.

If ρ_d is the probability of return to the origin for the simple symmetric random walk in d dimensions, then we know now that $\rho_1 = \rho_2 = 1$. In fact, ρ_3 is close to 0.3, and it can be shown that

$$\rho_d = \frac{1}{2d} + O(d^{-2})$$

for large d.

In Section 10.12 we shall consider what is in some senses a more natural version of the problem: that in which both time and space are continuous.

Markov Processes in Continuous Time

1. The Markov Property in Continuous Time

In continuous time we shall use the notation $X(t)$ rather than X_t for functions of time.

The definition of the Markov property given in Section 9.1 still holds for the continuous-time version of the problem, with the obvious rephrasing. That is, suppose that at time t the random variables $Y(t)$ and $Z(t)$ are arbitrary functions of, respectively, past-and-present $\{X(\tau); \tau \le t\}$ and present-and-future $\{X(\tau); \tau \ge t\}$. Then we can express the Markov property as

$$E[Z(t)|X(\tau); \tau \le t] = E[Z(t)|X(t)],$$

or equivalently (see Theorem 9.2.1), as the independence of past and future conditional on the present:

$$E[Y(t)Z(t)|X(t)] = E[Y(t)|X(t)]E[Z(t)|X(t)]. \tag{1}$$

As in Section 9.1 we can define the conditional expectation operator $P(s)$ over a step s into the future by

$$P(s)H(x) = E[H(X(t + s)|X(t) = x] \qquad (s \ge 0). \tag{2}$$

(Again, we shall consider only the time-homogeneous case, for which the expectation (2) depends only on time-difference s.)

The argument of Theorem 9.11 will again establish the validity of the Chapman–Kolmogorov equation, expressed in the form

$$P(r + s) = P(r)P(s) \qquad (r, s \ge 0). \tag{3}$$

The definition (2) implies that $P(0) = I$, as would indeed be necessary for consistency of (3). Suppose we can make the stronger assertion: that one can attach a meaning to the operator

$$\Lambda = \lim_{s \downarrow 0} \frac{P(s) - I}{s}. \tag{4}$$

(That is, that $P(s)H = H + s\Lambda H + o(s)$ for small positive s, uniformly in H and its argument for all H of interest.) Then Λ is termed the *infinitesimal generator* of the process. Relations (3) and (4) imply formally that

$$\frac{\partial P(s)}{\partial s} = \Lambda P(s) = P(s)\Lambda$$

with solution

$$P(s) = e^{s\Lambda}. \tag{5}$$

We would interpret the solution (5) as

$$e^{s\Lambda} = \sum_{j=0}^{\infty} \frac{(s\Lambda)^j}{j!}.$$

Relation (5) is the analogue of the relation

$$P(s) = P(1)^s \qquad (s = 0, 1, 2, \ldots)$$

of the discrete case, which we see from (3) to be equally valid in continuous time.

2. The Case of a Discrete State Space

If state is discrete and labelled by j then the operator $P(s)$ has the effect

$$P(s)H(j) = \sum_k p_{jk}(s)H(k), \tag{6}$$

where $p_{jk}(s) = P[X(t + s) = k | X(t) = j]$. As in the discrete-time case, the operator $P(s)$ acting on functions can as well be re-interpreted as a matrix acting on vectors. The infinitesimal generator Λ can then also be interpreted as a matrix, with jkth element identifiable, by (4) and (6), with

$$\lambda_{jk} = \lim_{s \downarrow 0} \frac{p_{jk}(s) - \delta_{jk}}{s}. \tag{7}$$

Plainly $\lambda_{jk} \geq 0$ for $j \neq k$, and the relation $\sum_k p_{jk}(s) = 1$ implies that

$$\sum_k \lambda_{jk} = 0. \tag{8}$$

For $j \neq k$ one terms λ_{jk} the *probability intensity* of the transition $j \to k$. The definition (7) implies that $p_{jk}(s) = \lambda_{jk}s + o(s)$ for $j \neq k$. This is the consequence of assuming the existence of an infinitesimal generator: if the process is in state j then the probability of a transition to another state k over a short time interval s is, to first order, proportional to s. One can regard λ_{jk} as a rate for the discontinuous transition $j \to k$; see Exercise 1.

The value of λ_{jj} is negative, and one sometimes writes

$$\lambda_j = -\lambda_{jj} = \sum_{k \neq j} \lambda_{jk},$$

so that λ_j is interpretable as the intensity of transition out of state j by any means.

In the discrete-state case we shall indeed understand the infinitesimal generator Λ as a matrix (the *transition intensity matrix*) operating on vectors, although this is purely a matter of language. Its action is

$$\Lambda H(j) = \sum_k \lambda_{jk} H(k) = \sum_k \lambda_{jk}[H(k) - H(j)]. \tag{9}$$

The second form follows from (8), and has the advantage that one sums only over actual transitions, and needs no exceptional understanding of the special case $k = j$.

As in Section 9.2, we have the Kolmogorov forward and backward equations for the transition probabilities:

$$\frac{\partial P(s)}{\partial s} = P(s)\Lambda, \qquad \frac{\partial P(s)}{\partial s} = \Lambda P(s),$$

respectively. Written in full, they read

$$\frac{\partial p_{jk}}{\partial s} = \sum_i (p_{jl}\lambda_{ik} - p_{jk}\lambda_{ki}), \tag{10}$$

$$\frac{\partial p_{jk}}{\partial s} = \sum_i \lambda_{ji}(p_{ik} - p_{jk}). \tag{11}$$

If we define

$$\pi_k(t) = \sum_j \pi_j(0) p_{jk}(t),$$

the distribution over states at time t for an arbitrary distribution at time 0, then by (10) this will obey the equation

$$\frac{\partial \pi_k}{\partial t} = \sum_j (\pi_j \lambda_{jk} - \pi_k \lambda_{kj}), \tag{12}$$

which is the more familiar form of the Kolmogorov forward equation.

We have already dwelt, in Section 9.5, on a particular advantage of the continuous-time formulation: that transitions correspond to *single* events. For example, consider the Ehrenfest model of Section 9.2, in which N particles migrate independently and by the same rules between two compartments, 1 and 2. In discrete time, the very independence of the migrations means that the transition in state over unit time is the net effect of a random number of migrations in each direction. In the continuous time version a state transition will consist of the single migration of a given particle. The pattern of transition is then much simplified, and traces the actual course of events in detail.

If, for such a model, one *wishes* to allow a multiple event in that, for example, one wishes to allow pairs of particles to make the migration simultaneously, then one must introduce a transition, with its own intensity, which represents such a pair-migration.

Suppose one wishes to consider *alternative* transitions, in that one wishes to consider transitions out of state j into a set of states A, say. Let the transition and its intensity be written $j \to A$ and λ_{jA}. Since $P[X(t + s) \in A | X(t) = j] = \sum_{k \in A} P_{jk}(s)$ then

$$\lambda_{jA} = \sum_{k \in A} \lambda_{jk}. \tag{13}$$

For example, in the Ehrenfest model the state variable X is a listing of the locations of each of the N identified particles. In the model as we have imagined it, the possible transitions are just the migrations of single particles; suppose these all have the same intensity: λ. Suppose there are n particles in compartment 1. There are then n possible transitions which would take some particle to compartment 2; i.e. which would induce the transition $n \to n + 1$. Since these all have intensity λ, it follows from (13) that the transition $n \to n - 1$ has intensity $n\lambda$. Correspondingly, the transition $n \to n + 1$ has intensity $(N - n)\lambda$. Since these are functions of n alone, it seems clear that the derived process $\{n(t)\}$ is also Markov, with these transitions and intensities. The conclusion is correct, although the argument needs to be made more explicit; see Exercise 3.

EXERCISES AND COMMENTS

1. Consider N independent replicas of the Markov process with intensity matrix Λ; let $n_j(t)$ be the number which are in state j at time t, and $m_j(t)$ the expected value of this quantity, for arbitrary initial conditions. Show that the m_j also obey the forward equations (12): $\dot{m}_k = \sum_j (m_j \lambda_{jk} - m_k \lambda_{kj})$. In this sense λ_{jk} is a rate: $m_j \lambda_{jk}$ is the expected number of transitions $j \to k$ per unit time.

2. The independence of the particles in the Ehrenfest model is assured by the assumptions that (i) coupled migrations are not permitted, and (ii) the intensity of migration for a given particle does not depend upon the positions of the other particles. The identity of the particles is reflected in the fact that this latter intensity is also independent of the particle. We need not have assumed the same intensity of migration in the two directions.

3. Suppose that $\{X(t)\}$ is a Markov process and $\{Y(t)\}$ a process derived from it by the transformation $Y(t) = g[X(t)]$. Then the Y-process will in general not be Markov. Let $\mathcal{X}(y)$ be the set of states x such that $g(x) = y$, and suppose that for any pair y, y' of possible Y-values and any x in $\mathcal{X}(y)$ the intensity of the transition $x \to \mathcal{X}(y')$ is a function only of y and y'. Show that the Y-process is then Markov.

4. Note that the matrix Λ has eigenvalue 0, with corresponding right and left eigen-vectors **1** and π, where π is a stationary distribution (if such exists).

5. Determine $P(s)$ explicitly in the case

$$\Lambda = \begin{bmatrix} -\lambda & \lambda \\ \mu & -\mu \end{bmatrix}$$

(see the analogous discrete-time case (9.25)).

3. The Poisson Process

We have already considered the Poisson process in Chapter 4: in its spatial version as a process representing an independent and uniform distribution of particles over infinite space at a prescribed density, and, in its temporal form (see Section 4.8), as an analogous distribution of events over time. Let us now see the temporal version as a Markov process. This adds nothing in the way of results, but we shall see in Section 10 that the process has a particular character as a Markov process.

As stated in Section 4.8, the Poisson process is a model for the occurrence of events at a constant rate λ (with 'events' understood in the colloquial and nontechnical sense). These events might be births, accidents or registrations on a Geiger counter.

Let $X(t)$ be the number of events up to time t; this can then take values $j = 0, 1, 2, \ldots$. We specify the model by supposing that the only possible transition out of state j is to $j + 1$, with constant intensity λ. The Kolmogorov forward and backward equations for $p_{jk}(t)$ then become

$$\dot{p}_{jk} = \lambda(p_{j,k-1} - p_{jk}), \tag{14}$$

$$\dot{p}_{jk} = \lambda(p_{j+1,k} - p_{jk}), \tag{15}$$

respectively, where $\dot{p} = dp/dt$. One can verify directly that

$$p_{jk}(t) = e^{-\lambda t} \frac{(\lambda t)^{k-j}}{(k-j)!} \qquad (k \geq j) \tag{16}$$

solves either of these equations, with, of course, $p_{jk}(t) = 0$ for $k < j$. That is, the number of events in the time interval $(0, t]$ is Poisson distributed with parameter λt.

The easy way to actually deduce solution (16) is to introduce the p.g.f. $\Pi_j(z, t) = \sum_{k=0}^{\infty} p_{jk}(t)z^k$, in terms of which equations (14) and (15) become

$$\dot{\Pi}_j = \lambda(z - 1)\Pi_j,$$

$$\dot{\Pi}_j = \lambda(\Pi_{j+1} - \Pi_j),$$

respectively, with initial condition $\Pi_j(z, 0) = z^j$. The first equation, the forward version, then obviously has the solution

$$\Pi_j(z, t) = z^j e^{\lambda t(z-1)}. \tag{17}$$

The backward version yields the solution less immediately, but actually is the more helpful for the processes considered next.

We know, of course, from Section 4.8, that one has the very much more powerful version of (17):

$$E\left\{\exp\left(i \int \theta(t)\, dX(t)\right)\right\} = \exp\left(\lambda \int (e^{i\theta(t)} - 1)\, dt\right). \tag{18}$$

This will prove useful when we come to consider the whole course of the process: the 'Poisson stream'.

4. Birth and Death Processes

As in Section 9.4, we understand a birth and death process to be a Markov process with integer-valued state variable such that transitions of type $j \to j + 1$ and $j \to j - 1$ are the only ones possible. Let these have intensities λ_j and μ_j, respectively. Then, just as in Theorem 9.4.1, we deduce that the process shows detailed balance in that $\lambda_j \pi_j = \mu_{j+1} \pi_{j+1}$, and, if irreducible, has equilibrium distribution

$$\pi_j = \pi_0 \frac{\lambda_0 \lambda_1 \dots \lambda_{j-1}}{\mu_1 \mu_2 \dots \mu_j}. \tag{19}$$

A special case of interest is the *simple* birth, death and immigration process, for which

$$\lambda_j = v + j\lambda, \qquad \mu_j = j\mu. \tag{20}$$

The motivation for the choice of intensities (20) is that, if we really do interpret j as the size of a population, then *individuals* in the population give birth or die with respective intensities λ and μ, and immigration also occurs with intensity v (births and immigrations being single). This then implies the rates (20) for the population, as (13) demonstrates. The equilibrium distribution for process (20) would be, by (19),

$$\pi_j \propto \frac{1}{j!} \prod_{k=0}^{j-1} \left(\frac{\lambda k + v}{\mu}\right).$$

If $v > 0$ then this is summable only if $\lambda < \mu$; the net reproduction rate $\lambda - \mu$ must be negative if there is to be an equilibrium population distribution in the presence of immigration. If $\lambda = 0$ then the distribution is Poisson. If there is no immigration, so that $v = 0$, then it would seem that $\pi_j = 0$ for $j > 0$. In fact, as we shall see, the situation is more complicated. The state $j = 0$ is always an absorbing state—once the population dies out it cannot recover. However, if $\lambda > \mu$ then one can also say that $j = +\infty$ is an absorbing state.

If the population can avoid early extinction then it will become so large that it is subsequently safe from extinction.

One calculates transient behaviour for process (20) by the same methods as for the Poisson process of the last section. Define again the p.g.f. $\Pi_j(z, t) = E(z^{X(t)} | X(0) = j)$. Then we leave the reader to verify that this obeys the forward and backward equations

$$\frac{\partial \Pi_j}{\partial t} = (\lambda z - \mu)(z - 1)\frac{\partial \Pi_j}{\partial z} + v(z - 1)\Pi_j, \tag{21}$$

$$\frac{\partial \Pi_j}{\partial t} = (\lambda j + v)(\Pi_{j+1} - \Pi_j) + \mu j(\Pi_{j-1} - \Pi_j). \tag{22}$$

The partial differential equation (21) can be solved, but the easier course is to resort to (22). Since $X(t)$ is the sum of contributions at time t from immigration and from each of the initial ancestors, and these contributions are independent, then Π_j must have the form

$$\Pi_j(z, t) = A(z, t)B(z, t)^j.$$

Here $A(z, t)$ is the p.g.f. of the component derived from immigration and $B(z, t)$ that from a given initial ancestor. Substituting this expression into (22) we find that A and B satisfy the equations

$$\dot{A} = v(B - 1)A, \qquad \dot{B} = (\lambda B - \mu)(B - 1).$$

These must be solved with initial conditions $A = 1$ and $B = z$ at time 0. One finds the solutions

$$A(z, t) = \left[\frac{\lambda(1 - z)e^{(\lambda - \mu)t} - (\mu - \lambda z)}{\lambda - \mu}\right]^{-v/\lambda},$$

$$B(z, t) = \left[\frac{\mu(1 - z)e^{(\lambda - \mu)t} - (\mu - \lambda z)}{\lambda(1 - z)e^{(\lambda - \mu)t} - (\mu - \lambda z)}\right].$$

If there is no immigration then the probability of extinction by time t is

$$P[X(t) = 0 | X(0) = j] = B(0, t)^j = \left[\frac{\mu e^{(\lambda - \mu)t} - \mu}{\lambda e^{(1 - \mu)t} - \mu}\right]^j.$$

For large t this tends to 1 if $\lambda \leq \mu$ and to $(\mu/\lambda)^j$ if $\lambda > \mu$, so the conclusions are much as for the branching process of Section 6.4: ultimate extinction is uncertain if and only if the net reproduction rate is strictly positive. Indeed, the simple birth and death process is a continuous-time version of a branching process, with a progeny distribution over a positive time which, as we see from the formula for $B(z, t)$ above, is modified geometric.

EXERCISES AND COMMENTS

1. The results for the probability of extinction for the simple birth and death process recall those for the gambler's run problem (with an infinitely rich adversary) of

Section 9.3. In fact, let ρ_j denote the extinction probability for the birth and death process conditional on a current population size of j. Show that this obeys the equation $j[\lambda\rho_{j+1} + \mu\rho_{j-1} - (\lambda + \mu)\rho_j] = 0$. This reduces to the corresponding equation for the gambler's ruin problem if we define $p = \lambda/(\lambda + \mu)$. The probability of a given sequence of transitions is the same for the two processes; the only effect of the factor j in the population case is to increase the speed of the transitions when the population is large. That is, the two processes differ only by a random change of time scale.

2. One obtains a plausible model of radioactive decay by assuming that a molecule can be in states $j = 0, 1, 2, \ldots$, and can suffer only the transitions $j \to j - 1$ with intensity μ_j ($j > 0$). The state $j = 0$ is absorbing and final. This is then a *pure death process*. Show that, if one defines the Laplace transform

$$\bar{p}_{jk}(\alpha) = \int_0^\infty e^{-\alpha t} p_{jk}(t)\, dt,$$

then

$$\bar{p}_{jk}(\alpha) = \frac{\mu_j \mu_{j-1} \cdots \mu_{k+1}}{(\mu_j + \alpha)(\mu_{j-1} + \alpha) \ldots (\mu_k + \alpha)} \qquad (k \le j).$$

This is closely related to the more probabilistic assertion, that the c.f. of the passage time from state j to state k is $\prod_{h=k+1}^{j} (1 - i\theta/\mu_h)^{-1}$.

3. Consider the normalized r.v. $Y(t) = X(t)/E[X(t)]$ for the simple birth and death process with $X(0) = 1$. Show from the formula derived for $B(z, t)$ above that, if $\lambda > \mu$, then in the limit of large t this has the c.f. $\rho + (1 - \rho)[1 - i\theta/(1 - \rho)]^{-1}$, where $\rho = \mu/\lambda$. Interpret the formula.

4. Consider a model of a queue in which the two possible transitions are arrival of a new customer (intensity λ) and the completion of service for an existing customer (intensity μ). Show that the equilibrium distribution of queue size is geometric starting from zero: $\pi_j = (1 - \gamma)\gamma^j$ ($j = 0, 1, 2, \ldots$) where $\gamma = \lambda/\mu$ is the *traffic intensity*.

5. The slightest modification of a birth and death process in general destroys the detailed balance property. Consider an irreducible model with states $j = 0, 1, 2, \ldots$, $d - 1$ and intensities λ_j and μ_j for the transitions $j \to j + 1$ and $j \to j - 1$, but with the state d identified with 0. That is, transitions are single displacements along a *ring* of states. Show that in equilibrium the net probability flux $\lambda_j \pi_j - \mu_{j+1} \pi_{j+1}$ between states j and $j + 1$ is constant in j, but is zero if and only if $\prod_{j=0}^{d-1} (\lambda_j/\mu_j) = 1$. What is the equilibrium distribution if the μ_j are zero?

6. The birth and death process is a stochastic analogue of a deterministic process $\dot{x} = a(x)$, where $a(x) = \lambda_x - \mu_x$. The deterministic process would have possible equilibria at x-values determined by $a(x) = 0$, these being stable or unstable according to whether $a(x)$ is locally decreasing or increasing in x. Show from equation (19) that π_x correspondingly has a local maximum or minimum, respectively.

7. A telephone exchange has a channels; suppose that n of these are engaged. Suppose that incoming calls arrive in a Poisson steam of intensity ν, and are accepted if there is a free channel and are otherwise lost. Suppose also that any busy channel has a constant intensity μ of becoming free. What is the equilibrium distribution of

the number of busy channels? What is the probability, in equilibrium, that the exchange is saturated? This probability, as a function of v and a, is known as the *Erlang function*. It is important in telecommunication contexts, even for the approximate analysis of more complicated networks.

8. Consider the example of Exercise 5, but suppose that a call, once accepted by a channel, goes through several stages of service, labelled $j = 1, 2, \ldots, m$. The transition rate out of stage j is μ_j; this takes the channel to stage j if $j < m$ and to stage 0 (channel free) if $j = m$. Let $\pi(\mathbf{n})$ be the equilibrium distribution of $(n_1, n_2, \ldots, n_m)$; the numbers of busy channels in different stages of service. Show that

$$\pi(\mathbf{n}) \propto \prod_{j=1}^{m} (v/\mu_j)^{n_j}/n_j! \qquad \left(n = \sum_{j=1}^{m} n_j \leq a \right).$$

The equilibrium distribution of the number of busy channels n is then $\pi(n) \propto (v\kappa)^n/n! \; (n = 0, 1, 2, \ldots, a)$, where $\kappa = \sum_j (1/\mu_j)$. This is exactly what the distribution is for the case of Exercise 5, except that κ has replaced $1/\mu$. Both quantities represent the *expected service time* for a customer. So, at least for the class of service time distributions which can be represented by the stage model, the distribution of the number of busy channels seems to depend upon service time distribution only through its mean. The result is in fact valid generally, and is an example of *insensitivity*, a frequently observed phenomenon in these contexts.

5. Processes on Nondiscrete State Spaces

We shall deal with the case of a general state space $\mathscr{X}$ only formally. The line to be followed is clear, even if there are points of rigour to be covered.

Let us for the moment revert to the case of discrete time, and so consider a time-homogeneous Markov process with transition operator P. Suppose that the state variable X has a probability density $f(x, t)$ at time t relative to a fixed measure M (see Section 3.3). Prescription of initial conditions at time 0 then means the prescription of $f(x, 0)$. The central result is the following:

Theorem 10.5.1. *Define the* adjoint P^T *of* P *as the operator for which*

$$\int f(x)[PH(x)]M(dx) = \int H(x)[P^T f(x)]M(dx) \tag{23}$$

holds identically in H and f. Then the density $f(x, t)$ obeys the Kolmogorov forward equation

$$f(x, t + 1) = P^T f(x, t). \tag{24}$$

This is the general version of equation (9.17)

PROOF. We have, by the definition of f,

$$E[H(X_t)] = \int H(x)f(x, t)M(dx).$$

Thus

$$\int H(x)f(x, t + 1)M(dx) = E[H(X_{t+1})] = E\{E[H(X_{t+1})|X_t]\} = E[PH(X_t)]$$

$$= \int f(x, t)[PH(x)]M(dx)$$

$$= \int H(x)[P^T f(x, t)]M(dx). \tag{25}$$

But (25) is to hold identically in $H(x)$, which it can do only if f obeys the recursion (24).   □

As a formal corollary one expects that the Kolmogorov forward equation in continuous time would be

$$\frac{\partial f}{\partial t} = \Lambda^T f, \tag{26}$$

where the adjoint Λ^T of the infinitesimal generator Λ is defined as in (23). This is the generalization of (12).

For an example, consider a continuous-time renewal process, with the age of the article in service taken as the state variable X. Suppose that $\mu(x)$ is the hazard rate at age x. That is, this is the probability intensity for failure of the article, and so for the transition $x \to 0$. If no failure takes place then x simply increases at unit rate. The infinitesimal generator then has the action

$$\Lambda H(x) = \lim_{s \downarrow 0} (1/s)[\mu(x)sH(0) + (1 - \mu(x)s)H(x + s) - H(x)]$$

$$= \mu(x)[H(0) - H(x)] + H'(x),$$

where the prime indicates a differentiation with respect to x. If $f(x)$ is a probability density with respect to Lebesgue measure than an integration by parts yields

$$\int_0^\infty f(x)\Lambda H(x)\, dx = -\int_0^\infty H(x)[\mu(x)f(x) + f'(x)]\, dx$$

$$+ H(\infty)f(\infty) - H(0)f(0) + H(0)\int_0^\infty \mu(x)f(x)\, dx,$$

which implies that

$$\Lambda^T f(x) = -\mu(x)f(x) - f'(x) \qquad (x > 0). \tag{27}$$

The definition of Λ^T at $x = 0$ needs special discussion, but equations (26) and (27) yield

$$\dot{f} = -(\mu f + f') \qquad (x > 0),$$

which is all we need—the value of $f(0, t)$ follows from normalization of the density. In particular, this last equation yields the equilibrium distribution of age

$$f(x) = f(0) \exp\left[-\int_0^x \mu(u) \, du \right] \tag{28}$$

which $f(0)$ determined by normalization.

EXERCISES AND COMMENTS

1. Let us denote expression (28) by $f_A(x)$, to designate it as the probability density of age. We know the probability density of lifespan for an article to be

$$f_L(x) = \mu(x) \exp\left[-\int_0^x \mu(u) \, du \right]$$

(Exercise 5.4.1). If we define $G(x) = \int_x^\infty f_L(u) \, du$ as the probability that lifetime exceeds x and m as the expected lifetime then expression (28) can be written $f_A(x) = G(x)/m$.

2. Note that $f_A \equiv f_L$ only if $\mu(x)$ is constant, when both distributions are exponential.

3. However, consider the density $f_R(y)$ of *residual lifetime* Y. This is the lifetime remaining for the article in service, when the process has reached equilibrium. The density of Y conditional on age x is $f_L(x + y)/G(x)$, so that

$$f_R(y) = \int_0^\infty \frac{G(x)}{m} \frac{f_L(x + y)}{G(x)} \, dx = \frac{G(y)}{m} = f_A(y).$$

We have then the seemingly remarkable conclusion that age and residual lifetime have the same distribution. However, one sees that the conclusion in inevitable if one realizes that a time reversal will not change the nature of the process, but will change the roles of the two variables.

4. *The Pareto distribution of wealth.* Let $\int_A n(x, t) \, dx$ be the expected number of people who have wealth x in the set of values A at time t. Then $n(x, t)$ can be regarded as a kind of unnormalized density; unnormalized because the total number of people $n(t) = \int n(x, t) \, dx$ may be changing in time.

   Suppose that an individual's wealth grows at rate β by compound interest, but that the individual may die with probability intensity γ. If we assume that he has m heirs, all of whom benefit equally, then the estate of size x becomes m estates of size x/m. If $H(x)$ were the 'value' of an estate of size x then the rate of increase of expected value derived from a single estate of size x is

$$\Lambda H(x) = \beta x H'(x) + \gamma[mH(x/m) - H(x)]$$

and $n(x, t)$ obeys

$$\frac{\partial n(x, t)}{\partial t} = \Lambda^T n(x, t) = -\beta \frac{\partial}{\partial x}[xn(x, t)] - \gamma n(x, t) + \gamma m^2 n(mx, t). \tag{29}$$

Note that the generator does not satisfy the condition $\Lambda 1 = 0$; in fact, $\Lambda 1 = \gamma(m - 1)$. This is because the population is not constant in size, but is growing at rate $\gamma(m - 1)$. Try a steady-state solution

$$n(x, t) \propto x^{-\alpha-1} e^{\gamma(m-1)t}$$

of (29). One finds that this indeed satisfies (29) if α satisfies

$$\alpha\beta/\gamma = m(1 - m^{-\alpha}). \tag{30}$$

The x-dependence $x^{-\alpha-1}$ represents the *Pareto distribution* of wealth, a distribution confirmed by data from many sources before this model (due to Wold) was proposed. The distribution cannot hold for all nonnegative x, not being integrable, but a realistic modification of the model produces the same distribution for x exceeding a positive threshold value (see Wold and Whittle, 1957). The Pareto parameter α is observed to lie in the range 1.4–2.0 (for wealth; rather higher for income). Equation (30) relates this parameter to interest rate, mortality rate and family size; it is consistent with observed values. Note that emergence of the Pareto power law is a consequence simply of the fact that the process is multiplicative: all changes in x are proportional to x.

5. Suppose that, at death, an individual has m heirs with probability p_m, and that a fraction $1 - q$ of his estate is removed by death duties. Show that equation (30) becomes

$$\alpha\beta/\gamma = \sum_m m p_m [1 - (q/m)^\alpha].$$

6. The Filing Problem

Consider again the filing problem proposed in Section 9.5. It is greatly simplified if we make both time and the state variable continuous. Suppose that there are N files, which can be filed in any position in 'file space' $x \geq 0$. The state variable is then $X = (X_1, X_2, \ldots, X_N)$, a listing of the positions of the N files.

We shall suppose that the following transitions can take place:

(i) In the absence of any other event, each file moves at unit rate in file space: $\dot{X}_j = 1$ ($j = 1, 2, \ldots, N$).
(ii) For each j, file j can be called with probability intensity λ_j; it is then immediately brought to the front of file space: $X_j \to 0$.

The steady movement of unrequested files can be regarded simply as a stratagem for keeping files in order, while making space at the front of file-space for translation of a requested file.

The equilibrium distribution of the process is simple: X_j is the 'age' variable of a renewal process with constant hazard rate λ_j, and the N variables are independent. The X_j are thus independent exponential r.v.s with respective expectations $1/\lambda_j$; see equation (28). On the other hand, the *ordinal* positions of the files are not independent; one can obtain the simplification of independence only by going to a continuum description.

We can derive the distribution of ordinal positions. The probability that file j is first in file space is

$$p_j = \int_0^\infty \lambda_j e^{-\lambda_j u} \exp\left(-u \sum_{i \neq j} \lambda_i\right) du = \frac{\lambda_j}{\sum_i \lambda_i},$$

the second exponential in the integral being the value of $P(X_i > u; i \neq j)$. Correspondingly, the probability that files j and k occupy first and second positions, respectively, is

$$\int_0^\infty du_1 \int_{u_1}^\infty du_2 \, \lambda_j \lambda_k \exp\left(-\lambda_j u_1 - \lambda_k u_2\right) \exp\left(-u_2 \sum_{i \neq j, k} \lambda_i\right) = \frac{\lambda_j \lambda_k}{(\sum_i \lambda_i)(\sum_{i \neq j} \lambda_i)}.$$

This expression indicates the lack of independence already, in that it is not equal to $p_j p_k$. In general, the probability of observing the files in a given order is

$$\prod_{i=1}^{N-1} \left(\mu_i / \sum_{h \geq i} \mu_h\right),$$

where μ_i is the value of λ for the file in the ith position. This formula recalls sampling without replacement: given the identities of the files in the first $i - 1$ places and that file j is not among them, the probability that file j is in the ith place is the ratio of λ_j to the sum of λ-values for files as yet unplaced.

7. Some Continuous-Time Martingales

As a Section 9.7, if we can find a function $\psi(x)$ such that $\Lambda \psi = 0$, then $\psi[X(t)]$ is a martingale, and

$$E\{\psi[X(t)] | X(0)\} = \psi[X(0)] \tag{31}$$

for any $t \geq 0$. Under conditions to be clarified by the Optional Stopping Theorem in Section 14.4, relation (31) generalizes to

$$E\{\psi[X(\tau)] | X(0)\} = \psi[X(0)], \tag{32}$$

where τ is a random stopping time.

We shall give some examples to demonstrate the usefulness of these ideas. There are, of course, continuous-time analogues of the random walk examples of Section 9.7. Consider the following variant on these; the calculation of the p.g.f. of the total number of descendants of the initial stock for a population model.

Theorem 10.7.1. *Consider a simple birth and death process; let $X(t)$ denote the size of the population at time t and $Y(t)$ the total number of descendants by time t of the $X(0)$ initial ancestors. Let τ denote the time when the population becomes extinct, and define Y as $Y(\tau)$ if the population indeed becomes extinct,*

+∞ otherwise. Then

$$E[z^Y | X(0)] = w(z)^{X(0)},\qquad(33)$$

where

$$w(z) = \frac{\lambda + \mu - \sqrt{(\lambda + \mu)^2 - 4\lambda\mu z}}{2\lambda z},\qquad(34)$$

and realizations for which $Y = \infty$ make no contribution to the expectation (33).

PROOF. The generator of the (X, Y)-process has action

$$\Lambda H(x, y) = \lambda x[H(x + 1, y + 1) - H(x, y)] + \mu x[H(x - 1, y) - H(x, y)].$$

The stopping time τ is defined as the smallest value of t for which $X(t) = 0$. If we can find a martingale $\psi(X(t), Y(t))$ then equation (32) will become, in this special case,

$$E\{\psi[0, Y)]|X(0)\} = \psi[X(0), 0]\qquad(35)$$

The equation $\Lambda\psi = 0$ is solved by

$$\psi(x, y) = w^x z^y\qquad(36)$$

if w, z satisfy

$$\lambda w^2 z - (\lambda + \mu)w + \mu = 0.\qquad(37)$$

Taking the smaller root (34) of the equation (37) for w, and substituting (36) into (35), we deduce exactly the assertion (33). The reason for taking the smaller root is to meet the conditions of the Optional Stopping Theorem. However, note also that the larger root of (37) does not have an expansion in nonnegative powers of z, and so could not represent a p.g.f.

Note also that

$$w(1)^j = \left[\frac{\lambda + \mu - |\lambda - \mu|}{2\lambda}\right]^j = \begin{cases}(\mu/\lambda)^j & (\lambda > \mu),\\ 1 & (\lambda \le \mu);\end{cases}$$

exactly the extinction probability determined earlier. The p.g.f (33) does not represent the whole probability mass, but only that for finite Y. The Optional Stopping Theorem (32) holds only if realizations for which stopping does not occur make no contribution to the expectation in the left-hand number. □

For a rather different example, consider a model of an epidemic, in which X and Y respectively represent the numbers of susceptibles and infecteds. Let us suppose the generator

$$\Lambda H(x, y) = axy[H(x - 1, y + 1) - H(x, y)] + by[H(x, y - 1) - H(x, y)].$$

This represents a situation in which new infections take place at rate axy and infecteds are removed from the population, by one means or another, at rate by. Let τ denote the time when the epidemic has run to completion; this is the smallest value of t for which $Y(t) = 0$. One is interested in the total size of the epidemic; the number of people who became infected during its course. Alternatively, one could consider the number of survivors $S = X(\tau)$. One can verify that the functions

$$\psi_k(x, y) = \binom{x}{k}\left[\frac{b}{ak + b}\right]^{x+y} \qquad (k = 0, 1, 2, \ldots)$$

(understood as zero if $x < k$) satisfy $\Lambda\psi = 0$, and hence that

$$E\left[\binom{S}{k}\alpha_k^S \,\Big|\, X(0) = x, Y(0) = y\right] = \binom{x}{k}\alpha_k^{x+y} \qquad (k = 0, 1, 2, \ldots, x), \quad (38)$$

where $\alpha_k = b/(ak + b)$. Relations (38) are of some interest in themselves. In principle, they determine the distribution of S, and represent in any case a considerable reduction of the original two-variable problem.

8. Stationarity and Reversibility

In almost any mathematical study the notion of invariance sooner or later begins to show its significance. Let X be a r.v. taking values in arbitrary space—it could be, for example, a scalar, a vector, or even a process. Suppose it can suffer a transformation U which converts it into UX. Then we shall say that X is *statistically invariant under U* if X and UX have the same distributional characteristics, in that

$$E[H(UX)\} = E[H(X)]$$

for any H for which the right-hand side has meaning. (Note an implication: that UX takes values in the same set as does X.)

For a stochastic processes $\{X(t)\}$ there are two transformations which are of interest. One is that of *time-translation*, which converts $X(t)$ into $X(t + s)$, if the translation is by an amount s. The other is that of *time-reversal*, which converts $X(t)$ into $X(a + (u - t)) = X(2u - t)$ if the reversal is about the pivot $t = u$. These motivate the definitions:

(i) A stochastic process is *stationary* if it is statistically invariant under any time translation.
(ii) A stochastic process is *reversible* if it is statistically invariant under any time reversal.

The important and interest of the two concepts will emerge. Let us first collect three fairly immediate conclusions.

Theorem 10.8.1.

(i) *A process is stationary if and only if it is time homogeneous and in equilibrium.*
(ii) *A reversible process is stationary.*

(iii) *The Markov property is preserved under both time translation and time reversal.*

PROOF. Assertion (i) is immediate. Assertion (ii) holds because any time translation can be realized by two successive time reversals about different pivots. Assertion (iii) holds because the characterization (1) of the Markov property is invariant under both operations.   □

We are particularly interested in the implications of these properties for Markov processes. For definiteness, we shall take the case of discrete state and continuous time. Then the principal result is the following.

Theorem 10.8.2. *A Markov process is reversible if and only if it shows detailed balance*

$$\pi_j \lambda_{jk} = \pi_k \lambda_{kj} \tag{39}$$

for all states j, k under its stationary distribution π.

PROOF. The stationary distribution may be nonunique, but the assumption is that (39) holds under the stationary distribution that prevails.

Reversibility implies that $P(X(t) = j, \; X(t + s) = k) = P(X(t) = k, X(t + s) = j)$. This relation plus stationary implies that $\pi_j p_{jk}(s) = \pi_k p_{kj}(s)$, for $s \geq 0$. Letting s tend to zero we deduce the necessity of (39).

Condition (39) would be without content unless π_j were positive for some j: we can restrict ourselves to the set of states j for which this is true. The reversed process is Markov, also with equilibrium distribution π, and its transition probability is $p_{jk}^*(s) = P[X(t) = k | X(t + s) = j] = \pi_k p_{kj}(s)/\pi_j$. The transition intensity for the reversed process is thus $\lambda_{jk}^* = \pi_k \lambda_{kj}/\pi_j$. But, by (39), this is equal to λ_k. Since the stochastic structure of a Markov process is determined by its transition intensities and the specification of which stationary distribution prevails (if there is more than one), it then follows that the direct and reversed processes are statistically identical   □

We have already seen the advantage of reversibility in Section 9.4: if we can verify the detailed balance relation (39) for any $\{\pi_j\}$ which is nonnegative and summable, then $\{\pi_j\}$ constitutes a stationary distribution (after possible normalization).

As an example, consider a generalization of the telephone exchange model of Exercise 10.4.6. The model is one of *circuit-switched networks*. Suppose that r indexes possible *routes* through a telephone network; a call requests a particular route, and we suppose for simplicity that no other route can be substituted. Calls requesting route r arrive with probability intensity v_r; a call already established on route r terminates with probability intensity μ_r ($r = 1, 2, 3, \ldots$). However, there are physical capacity constraints. Let n_r be the number of calls currently established on route r, and let $\mathbf{n} = \{n_r\}$ be the vector of these numbers. Then $\mathbf{n}$ is restricted to a feasible set $\mathcal{N}$; an arriving call which would take **n** out of $\mathcal{N}$ is refused and is lost. The reader can verify that this process has equilibrium distribution

$$\pi(\mathbf{n}) \propto \prod_r [(v_r/\mu_r)^{n_r}/n_r!] \qquad (\mathbf{n} \in \mathcal{N}) \tag{40}$$

and shows detailed balance. That is, the n_r are distributed as independent Poisson variables, except that they are restricted to the feasible set.

The deterministic characterization of reversibility would be that, if a particular function $x(t)$ satisfies the dynamic equations, then so does $x(-t)$. (There are caveats, however; see Exercise 2.) There is thus no mention of equilibrium; reversibility is a property of the general solution rather than of a particular solution. The stochastic equivalent of this demand (at least in the Markov case) would seem to be a requirement of symmetry of the transition probability, $p_{jk}(s) = p_{kj}(s)$, and so of the intensities:

$$\lambda_{jk} = \lambda_{kj}. \tag{41}$$

We shall term condition (41) the condition of *micro-reversibility*. It is a more stringent demand than the condition (39) of reversibility. It does not invoke equilibrium concepts, but, interestingly, does imply that $\pi_j = $ constant is a possible stationary distribution.

EXERCISES AND COMMENTS

1. Suppose that the feasible set $\mathcal{N}$ for the circuit-switched network is specified by

$$\sum_k a_{kr} n_r \leq M_k \qquad (k = 1, 2, 3, \ldots).$$

These constraints represent the situation that only M_k circuits are available on link k of the network, and that a call on route r demands a_{kr} of these. Suppose that $v_k/\mu_r = V\gamma_r$ and $M_k = V\rho_k$, for all r and k. Then V is a scale parameter for the network, in that an increase in V increases all demands and capacities in proportion. Show by the methods of Section 6.3 that the most probable value of $c_r = n_r/V$ in the limit of large V is $\gamma_r \prod_k w_k^{a_{kr}}$, where the parameters w_k are to be adjusted to meet the constraints, and take values in $[0, 1]$. This result in interesting, in that it implies that for a large-scale network a circuit is available on link k with probability w_k, and that these availabilities are effectively independent statistically.

2. A few comments on the notion of reversibility for a deterministic system might be helpful. Consider the state-structured time-homogeneous system with dynamic equation $\dot{x} = a(x)$. Stationarity of the solution $x(t)$ implies that $x(t)$ is constant; the system has reached equilibrium. Reversibility of the solution, so that $x(2u - t) = x(t)$ for any u and t, also implies that the solution is constant.

   The apparently weaker condition of micro-reversibility implies reversibility of the *system* rather than of the *solution*: that if $x(t)$ solves the equation then so does $x(-t)$ (or indeed $x(2u - t)$). But this condition requires that if x satisfies $\dot{x} = a(x)$ then it also satisfies $-\dot{x} = a(x)$, which implies that $\dot{x} = 0$. The only deterministic system which shows micro-reversibility is then again the trivial one for which $x(t)$ is constant. The corresponding notion (41) of statistical micro-reversibility in nevertheless nontrivial.

The situation is saved by the fact that, for Newton's equations, for example, the state variable x partitions into components (p, q) and the property observed is that, if $[p(t), q(t)]$ is a solution, then so also is $[-p(-t), q(-t)]$. More generally, a reversibility condition with nontrivial consequences requires that a time-reversal *plus a conjugation operation* converts a solution into a solution (or preserves the dynamic equations). In the statistical context this translates into the concept of *dynamic reversibility*. However, there are reasons why it is possible to avoid this concept for a time (essentially because one can consider position variables in isolation from momentum variables). These ideas are discussed further in Chapter 4 of Whittle (1986).

9. The Ehrenfest Model

The simplest Ehrenfest model specifies two compartments, 1, and 2, between which N molecules move independently and symmetrically. In the continuous-time version this means that the only possible transitions are the migrations of single molecules, and that these all have the same intensity λ.

The molecules are regarded as identifiable, and a complete description X lists the positions of all N of them. The process $\{X(t)\}$ is a Markov process with 2^N states. Let us denote the intensity of the transition $x \to x'$ by $\lambda(x, x')$. The process is micro-reversible in that

$$\lambda(x, x') = \lambda(x', x) \tag{42}$$

for all x, x', both sides having values λ or 0. All 2^N state values have equilibrium probability 2^{-N}. This is the unique equilibrium distribution, since the process is irreducible with finite state space.

A less complete description is that in which one no longer identifies individual molecules, but merely specifies the number of molecules in compartment 1: n, say. Then, as we saw in Section 1, the process $\{n(t)\}$ is also Markov, with transition intensities

$$\lambda(n, n + 1) = \lambda(N - n), \qquad \lambda(n, n - 1) = \lambda n, \tag{43}$$

and $N + 1$ states.

Let us refer to the x- and n-descriptions as the *micro-* and *macro-descriptions*, respectively. Define the *degeneracy* of a macro-state as the number of micro-states to which it corresponds. Thus macro-state n has degeneracy $\binom{N}{n}$ and equilibrium probability

$$\pi(n) = \binom{N}{n} \pi(x) = \binom{N}{n} 2^{-N}.$$

The point of the Ehrenfest model is that it illuminates a paradox exposed by Zermelo. The paradox is that statistical mechanics is supposedly derivable from classical mechanics, but the laws of classical mechanics are time-reversible (see Exercise 10.8.2) whereas statistical mechanics manifests irreversibility ('entropy always increases'—see Exercise 3).

In the Ehrenfest model we see the resolution. The model is reversible in the strong sense that it obeys the micro-reversibility condition (42). It exhibits a type of irreversibility at the macro-level in that the transition rates (2) will always tend to bring n towards the central value $N/2$. More specifically, suppose that x and x' are two configurations with corresponding state values n and n' in the macro-description. Suppose that n is small and n' near $N/2$, so that in the first configuration most molecules are in compartment 2 and in the second they are fairly equally divided. Then, for any given $t > 0$,

$$P[X(t) = x'|X(0) = x] = P[X(t) = x|X(0) = x'], \tag{44}$$

but

$$P[n(t) = n'|n(0) = n] > P[n(t) = n|n(0) = n']. \tag{45}$$

Relation (44) follows from micro-reversibility; the margin of inequality in (45) will be very great if N is large.

Relations (44) and (45) make the point. It is all a question of level of description. The process is micro-reversible at the micro-level, but not at the macro-level. Movement in either direction between prescribed micro-states is equally probable, but the tendency will be to move to macro-states of large degeneracy.

The resolution often proposed of the Zermelo paradox is that the n-process is irreversible in the conditional sense (45) but is indeed statistically reversible in that

$$P[n(0) = n, n(t) = n'] = P[n(0) = n', n(t) = n] \tag{46}$$

in equilibrium. (This can be verified, but follows from the fact that $n(t)$ is a function of $X(t)$, and the X-process is statistically reversible.) However, this brings in the concept of equilibrium, which was not invoked when classical mechanics was described as reversible. The stochastic analogue of the statement that the deterministic dynamic equations are invariant under time-reversal would seem to be (42), not (46).

So, the resolution of the Zermelo paradox that we would rather adopt is to say that the Ehrenfest model illustrates that micro-reversibility may hold at the micro-level, but not at the macro-level, and that this is indeed the behaviour we would generally expect.

EXERCISES AND COMMENTS

1. Note that it is because the X-process is invariant under permutation of molecules that the n-process is Markov.

2. The n-process has generator Λ with action $\Lambda H(n) = \lambda(N - n)H(n + 1) + \lambda n H(n - 1)$. Let us write

$$E(\dot{H}|n) = \lim_{s \downarrow 0} \left[\frac{E\{H[n(t + s)] - H[n(t)]|n(t) = n\}}{s} \right]$$

for the rate of change of the conditional expectation of $H[n(t)]$. Then $E(\dot{H}|n) = \Lambda H(n)$, by the definition of Λ. Note then that $E(\dot{n}|n) = \Lambda n = \lambda(N - 2n)$. This relation indicates the tendency for n to move towards the central values $N/2$.

3. Define the *entropy* S as the logarithm of the degeneracy of the current macro-state. For the Ehrenfest model this is then

$$S(n) = \log \binom{N}{n} = \text{constant} - n_1 \log n_1 - n_2 \log n_2 + O(1),$$

where we have written the number of molecules in compartment j as n_j, for symmetry, and the remainder term is $O(1)$ for large N. Verify then that

$$E(\dot{S}|n) = \Lambda S(n) = \lambda(n_1 - n_2)\log(n_1/n_2) - \lambda N^2/(2n_1 n_2) + o(1). \tag{47}$$

The first term in the right-hand member of (47), of order N, is zero only for $n_1 = n_2$ and is otherwise positive. This corresponds to the classical assertion of increasing entropy. The second term, $-\lambda N^2/(2n_1 n_2)$, represents an expected decrease of entropy due to fluctuation. This term is of order 1 and so of relative order N^{-1}, and becomes insignificant as N becomes large.

4. *The heat bath.* Suppose that a system contains j quanta of energy; that it can gain one with intensity v and lose one (if $j > 0$) with intensity μ. Then the system is just like a queue and, in equilibrium, contains j energy quanta with probability $\pi_j \propto (v/\mu)^j$, which we shall write rather as $\pi_j \propto e^{-\beta j}$. This is just the Gibbs distribution of Section 6.3 in the case when energy levels are multiples of a basic quantum, the distribution now being derived from a dynamical model. Note that the fact that the quanta are regarded as indistinguishable is reflected in the fact that they leave the system by behaving as a queue rather than leaving independently!

Suppose that, at a given energy level, the system has other states between which it moves micro-reversibly, so that $\lambda(x, x') = \lambda(x', x)$ if the two states have the same energy and $v\lambda(x, x') = \mu\lambda(x', x)$ if x' is one energy level higher than x. Show that the system has equilibrium distribution $\pi(x) \propto e^{-\beta\varepsilon(x)}$, where $\varepsilon(x)$ is the number of energy quanta associated with state x.

10. Processes of Independent Increments

The idea of a sequence of IID r.v.s is a central one for probability theory and models. This is a discrete-time notion if we regard the sequence as being ordered in time. The corresponding continuous-time notion presents interesting challenges to intuition and analysis. It is equally valuable as a standard elementary process: the engineers' 'shot noise' and 'white noise' are versions of it. However, it is so extreme in character that it requires special mathematical handling, and to some extent can only be approximated physically.

In discrete time we found it natural to consider the sums

$$S_n = \sum_{j=1}^{n} X_j,$$

where the X_j are IID r.v.s. The formal analogue in continuous time would be an integral

$$S(t) = \int_0^t X(s) \, ds,$$

where what is meant by the 'IID character' of the X-process has yet to be clarified. However, such IID character would have two consequences:

(i) $\{S(t)\}$ would be a *process of independent increments*, in that for $0 = t_0 < t_1 < t_2 < \cdots$ the increments $S(t_j) - S(t_{j-1})(j = 1, 2, 3, \ldots)$ would be independently distributed, and

(ii) $\{S(t)\}$ would be time-homogeneous (or, simply, *homogeneous*) in that the distribution of an increment would be unaffected by a time translation.

So, we can approach the study of $\{X(t)\}$ by examining $\{S(t)\}$ as a homogeneous process of independent increments (which we shall often abbreviate to HPII).

Theorem 10.10.1. *Suppose $\{S(t)\}$ a homogeneous process of independent increments. Then:*

(i) *If* $\mathrm{var}[S(t)] < \infty$ *then*

$$\mathrm{var}[S(t)] = \sigma^2 t \tag{48}$$

for some constant σ^2, *and*

$$\mathrm{var}\left(\int \theta(t) \, dS(t)\right) = \sigma^2 \int \theta(t)^2 \, dt. \tag{49}$$

(ii) *Similarly*

$$E(e^{i\theta S(t)}) = e^{t\psi(\theta)} \tag{50}$$

for some function $\psi(\theta)$, *and*

$$E(e^{i \int \theta(t) \, dS(t)}) = e^{\int \psi[\theta(t)] \, dt}. \tag{51}$$

PROOF. Let us write $\mathrm{var}[S(t)]$ as $v(t)$. It follows then from the HPII assumption that

$$v(t_1 + t_2) = \mathrm{var}\{S(t_1) + S(t_1 + t_2) - S(t_1)]\} = v(t_1) + v(t_2) \tag{52}$$

for any nonnegative t_1, t_2, whence (48) follows. It follows likewise that

$$\mathrm{var}\left(\sum_j \theta_j [S(t_j) - S(t_{j-1})]\right) = \sigma^2 \sum_j \theta_j^2 (t_j - t_{j-1}),$$

and (49) then follows by a limiting argument.

The proof of assertion (ii) follows similarly, an additive relation such as (52) being replaced by a multiplicative one. □

Condition (51) is plainly also sufficient. In fact, the 'IID' character of $\{X(t)\}$, the HPII character of $\{S(t)\}$ and the validity of (51) for some ψ are all equivalent. The only question left unresolved is: Within what class of functions may ψ lie?

Relation (48) reveals the difficulty in the direct characterization of the X-process. We have, for $\delta t > 0$,

$$\text{var}\left((\delta t)^{-1} \int_t^{t+dt} X(s)\,ds\right) = (\delta t)^{-2}\,\text{var}[S(t+\delta t) - S(t)] = \sigma^2/\delta t. \quad (53)$$

Formally, one would expect the left-hand member of relation (53) to converge to $\text{var}[X(t)]$ as $\delta t \downarrow 0$. The right-hand member certainly converges to $+\infty$. So, it seems that an 'IID process' whose integral has positive variance must itself necessarily have infinite variance—an indication that its character has something exceptional.

One might then query whether such processes exist at all. However, homogeneous processes with independent increments certainly do. Let $S(t)$ be the number of events in $(0. t]$ for a Poisson process of intensity λ. We have seen already in (18) that for this process relation (51) holds with

$$\psi(\theta) = \lambda(e^{i\theta} - 1). \quad (54)$$

The time-homogeneous Poission process is then indeed a HPII. The sample paths (realizations) of $S(t)$ will consist of constant sections (when there is no event) and unit steps (when there is).

The formal derivative $X(t)$ would consist of a sequence of Dirace δ-functions. Physically interpreted, it is 'shot noise', a random sequences of unit impulses. This example sets us on familiar ground, in that the Poisson process is perfectly proper and well understood. We can see how it is now that the formal derivative must have infinite variance. The process $\{X(t)\}$ can be regarded as a proper mathematical entity, however. Rude as it may be in its raw state, linear functionals $\int \theta(t)X(t)\,dt = \int \theta(t)\,dS(t)$ are certainly proper for appropriately regular $\theta(\cdot)$, and expression (51) can be regarded as the c.f. of such a linear functional.

If the impulses in the Poisson process had magnitude a then $\{S(t)\}$ would still have been HPII, but with $\psi(\theta) = \lambda(e^{i\theta a} - 1)$. If one had superimposed independent Poisson streams of intensities λ_r and scale a_r then the resulting compound Poisson process would still have been a HPII, with

$$\psi(\theta) = \sum_r \lambda_r(e^{i\theta a_r} - 1). \quad (55)$$

The same idea is expressed somewhat differently in Exercise 1. We shall see in Exercise 2 that, by going to a limit for such processes in which the impulses become ever weaker and more frequent, we can achieve

$$\psi(\theta) = -\tfrac{1}{2}\theta^2. \quad (56)$$

That is, the S-process in this limit is a Gaussian process, and we have a process version of the central limit theorem.

In fact, these are the only possible homogeneous processes of independent increments which also have finite variance: the compound Poisson processes and their Gaussian limit.

The Gaussian limit has a special character: it is the only HPII for which $S(t)$ is a continuous function of t. The S-process in this case referred to as *Brownian motion*, and often denoted $\{B(t)\}$. It formal derivative $\{X(t)\}$ is 'Gaussian white noise' and we shall often write it as $\{\varepsilon(t)\}$.

Theorem 10.10.2. *The homogeneous process of independent increments specified by* (51) *is a Markov process whose generator has the action*

$$\Lambda(e^{i\theta S}) = \psi(\theta)e^{i\theta S}. \tag{57}$$

In particular, standard Brownian motion has generator

$$\Lambda = \frac{1}{2}\left[\frac{\partial}{\partial B}\right]^2. \tag{58}$$

The proof of (57) is direct, and (58) then follows formally. The usual direct formal derivation of (58) is to start from the assumptions that, if $\delta B = B(t + \delta t) - B(t)$, then $E[(\delta B)^j]$ is zero for $j = 1$, δt for $j = 2$ and $o(\delta t)$ for $j > 2$. It then follows from a formal Taylor expansion in powers of δB that, conditional on $B(t) = B$,

$$E[H(B + \delta B) - H(B)] = \tfrac{1}{2}H''(B)\delta t + o(\delta t),$$

whence (58) follows.

EXERCISES AND COMMENTS

1. Suppose that $S(t) = \sum_j I(t_j \le t)\xi_j$, where the t_j are the times at which events occur in a Poisson process of rate λ and the ξ_j are r.v.s with c.f. $\phi(\theta)$, independent of the t_j and of each other. Show that $S(t)$ is a HPII with $\psi(\theta) = \lambda[\phi(\theta) - 1]$.

2. Suppose that the ξ_j take values $\pm 1/\sqrt{\lambda}$ each with probability $\tfrac{1}{2}$. Show that $\psi(\theta) \to -\tfrac{1}{2}\theta^2$ as λ becomes large.

3. Note that $\psi(\theta) = -|\theta|$ is also a possibility, corresponding to Cauchy-distributed increments in $S(t)$. However, this process does not have finite variance and is not continuous (see Exercise 5).

4. We can give a plausibility argument that Brownian motion is continuous by dividing the increment over a time interval of length t into N equal parts: $\Delta_{jN} = S[jt/N] - S[(j-1)t/N]$. These are independent normal with zero mean and variance $1/N$, so that

$$P(|\Delta_{jN}| \ge D) \le e^{-ND^2/(2t)}$$

if $D \ge 0$ (see Exercise 7.5.10) and

$$P(|\Delta_{jN}| \le D; j = 1, 2, 3, \ldots) \ge (1 - e^{-ND^2/(2t)})^N.$$

This bound tends to unity as $N \to \infty$ for any fixed positive D. That is, the probability of a jump of D anywhere in the interval approaches zero as we consider a denser sequence of time points (although admittedly a particular sequences).

5. Conversely, we can see that the Cauchy process of Exercise 3 is *not* continuous. The Δ_{jN} are independent (for varying j, fixed N), with probability density $f(\Delta) \propto [\Delta^2 + (t/N)^2]^{-1}$. Suppose $D > 0$. Then we can find a positive constant K such that $P(|\Delta_{jN}| \geq D) \geq Kt/ND$ for large enough N. Hence

$$P(|\Delta_{jN}| \leq D; j = 1, 2, \ldots, N) \leq (1 - Kt/ND)^N \to e^{-Kt/D}.$$

This upper bound is less than unity, and tends to zero as t increases. The probability of a discontinuity of any given size is thus positive, and approaches unity if the time interval studied is long enough.

11. Brownian Motion: Diffusion Processes

Brownian motion holds a special place among processes of independent increments, as the only one with continuous sample paths. The name associates the process with the observations of the botanist R. Brown around 1827 on the erratic motion of pollen particles suspended in water, the motion being due to impulses received by impact with individual molecules of the medium.

We see from (51) and (56) that $\{B(t)\}$ is Gaussian. More specifically, $B(t)$ has c.f. $\exp(-t\theta^2/2)$ and so is normally distributed with zero mean and variance t. The infinitesimal generator of the process (58) is self-adjoint with respect to Lebesgue measure. The density function $f(B, t)$ of $B(t)$ thus obeys the Kolmogorov forward equation

$$\frac{\partial f}{\partial t} = \Lambda^{\mathrm{T}} f = \Lambda f = \frac{1}{2} \frac{\partial^2 f}{\partial B^2}, \tag{59}$$

which indeed has the $N(0, t)$ density as its solution for $f(B, 0) = \delta(B)$. Equation (59) is the equation which holds for the diffusion of heat in a homogeneous conductor. In our case it describes the diffusion of probability mass in B-space. For this reason, Brownian motion is the simplest example of what is termed a *diffusion process*.

Standard Brownian motion could be regarded as the solution of the differential equation $\dot{B} = \varepsilon$, where $\{\varepsilon(t)\}$ is standard Gaussian white noise. One can derive more general diffusion processes by allowing white noise to drive more general differential equations. Consider, for example, the system

$$\dot{X} = a(X) + b(X)\varepsilon, \tag{60}$$

where X could be a vector, although we shall suppose it scalar for the moment. Some would write (60) more circumspectly in the incremental form

$$\delta X = a(X)\delta t + b(X)\delta B, \tag{61}$$

but it is really better to accept the white noise process as an idealization which is legitimate. Relations (60) or (61) might be regarded as a way of writing the assumptions

$$E\{[\delta X(t)]^j | X(t) = x\} = \begin{cases} a(x)\delta t & (j = 1), \\ b(x)^2 \delta t & (j = 2), \\ o(\delta t) & (j > 2). \end{cases} \tag{62}$$

Theorem 10.11.1. *The process* $\{X(t)\}$ *solving the stochastic differential equation* (60) *has infinitesimal generator*

$$\Lambda = a(x)\frac{\partial}{\partial x} + \tfrac{1}{2} b(x)^2 \left[\frac{\partial}{\partial x}\right]^2. \tag{63}$$

The probability density $f(x, t)$ *of* $X(t)$ *with respect to Lebesgue measure thus satisfies*

$$\frac{\partial f}{\partial t} = \Lambda^T f = -\frac{\partial}{\partial x}[a(x)f] + \frac{1}{2}\frac{\partial^2}{\partial x^2}[b(x)^2 f]. \tag{64}$$

The evaluation (63) can be deduced from assumptions (62), just as we deduced the form for the generator of Brownian motion at the end of the last section. Alternatively, relation (61) would imply that

$$e^{-i\theta x}\Lambda e^{i\theta x} = i\theta a(x) + \tfrac{1}{2}[i\theta b(x)]^2.$$

Equation (64) represents a nonhomogeneous diffusion equation with *drift rate* $a(x)$ and *diffusion rate* $b(x)^2$.

If one calls the process a 'diffusion process' then one rather emphasizes the equation (64) obeyed by its density function at the expense of emphasis on the stochastic differential equation (60) obeyed by the variable itself. However, under any name, the process is an important one. The stochastic differential equation (60) is often regarded as the natural stochastic generalization of a deterministic model $\dot{x} = a(x)$, and as such in much employed in technological contexts. It is also often derived as a limit form of some other stochastic version, just as the normal distribution is itself a limit form.

Consider, for example, the Ehrenfest model of Section 9 in the macro-description. Its infinitesimal generator is given in Exercise 10.9.2. Consider now the scaled variable $X = (n - \tfrac{1}{2}N)/\sqrt{N}$. Setting $1/\sqrt{N} = k$ to simplify notation, we see that the generator of the X-process has action

$$\Lambda H(x) = \lambda\left(\frac{N}{2} + \frac{x}{k}\right)(H(x - k) - H(x)) + \lambda\left(\frac{N}{2} - \frac{x}{k}\right)(H(x + k) - H(x))$$

$$= -2\lambda x \frac{\partial H}{\partial x} + \frac{\lambda}{2}\frac{\partial^2 H}{\partial x^2} + O(N^{-1/2}).$$

So, for large N and with this rescaling, the process is approximated by a diffusion process. In this case there is really no advantage, since the original process is more easily treated. However, the diffusion approximation is widely used for genetic models of selection and mutation, for example.

A diffusion process changes interestingly when we transform the variable.

Theorem 10.11.2. *Consider the process* $\{X(t)\}$ *generated by the stochastic differential equation* (60). *Suppose that one makes the variable transformation* $X \rightarrow Y = y(X)$, *where the function* $y(x)$ *is invertible and twice differentiable. Then* $\{Y(t)\}$ *obeys the stochastic differential equation*

$$\dot{Y} = a^*(Y) + b^*(Y)\varepsilon, \tag{65}$$

where

$$a^* = ay_x + \tfrac{1}{2}b^2 y_{xx}, \qquad b^* = by_x. \tag{66}$$

The x and y arguments are understood in (66) as appropriate. We have used the convenient subscript notation for differentials, so that $y_x = dy(x)/dx$, etc. The white noise process ε in (65) is not in general identical to that in (60); we have just used the notation ε to indicate that it *is* white noise.

PROOF. The assertion follows immediately if we calculate the change induced in the infinitesimal generator by the variable transformation. Suppose we set $H(x) = G(y)$. The induced generator Λ^* is then determined by

$$\Lambda^*G = \Lambda H = aH_x + \tfrac{1}{2}b^2 H_{xx} = aG_y y_x + \tfrac{1}{2}b^2 [G_{yy}(y_x)^2 + G_y y_{xx}].$$

But this is the generator of a diffusion process with the drift and diffusion rates asserted in (66).   □

The intuitively surprising aspect of the result is that the new drift term a^* contains a contribution $\tfrac{1}{2}b^2 y_{xx}$ from the old diffusion term. Had the driving input ε in (60) been a conventional deterministic function of time then this simply would not have occurred. The term is, however, a direct consequence of a change of variable in a second-order differential equation. One may get an intuitive feeling for the effect by examining the graph in Fig. 10.1. Here we have assumed y to be a concave function of x, so that $y_{xx} < 0$. If the model had been deterministic then the change δX in X in a time increment δt would have corresponded to the increment δY in Y; see Fig. 10.1(i). However, for the stochastic case, the value of δX in fact varies around the value $E(\delta X)$ (a conditioning on $X(t) = x$ is understood). Because of the concavity of $y(x)$ we have $E[y(x + \delta X)] < y[x + E(\delta X)]$, see Fig. 10.1(ii), and the difference is just $\tfrac{1}{2}b^2 y_{xx}\delta t + o(\delta t)$.

We have in fact already seen an example of this in Exercise 10.9.3, when we found that the deterministic increase in entropy was modified by a term indicating a decrease due to fluctuation. Near the equilibrium point $n = N/2$ these effects become comparable. In the diffusion approximation to this situation we have seen that $X = (n - \tfrac{1}{2}N)/\sqrt{N}$ obeys the equation

$$\dot{X} = -2\lambda X + \sqrt{\lambda}\varepsilon.$$

In terms of the re-scaled variables the re-scaled entropy can be taken as

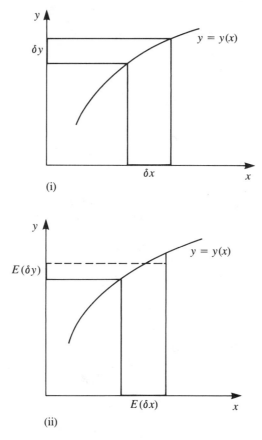

Figure 10.1. The quantities δx and δy are the increments in a short time interval δt of functions of time x and y related by $y = y(x)$. The graphs illustrate the relations between the expected values of these quantities in the case when (i) x is deterministic and (ii) x is random.

$Y = -\frac{1}{2}X^2$. Relations (65) and (66) imply that Y effectively follows the stochastic differential equation

$$\dot{Y} = -4\lambda Y - \tfrac{1}{2}\lambda + \sqrt{2\lambda|Y|}\varepsilon.$$

Since Y is negative, the first term $-4\lambda Y$ represents the deterministic increase in entropy. The second term $-\frac{1}{2}\lambda$ derives from the diffusion term, and represents the decrease in entropy due to fluctuations.

EXERCISES AND COMMENTS

1. Consider the stochastic differential equation $\dot{X} + \alpha X = \varepsilon$, where $\alpha > 0$. Show that X has an equilibrium distribution which is normal. Show that the covariance between $X(t)$ and $X(t-s)$ in equilibrium is $e^{-\alpha|s|}/(2\alpha)$.

202 10. Markov Processes in Continuous Time

2. In equilibrium, the diffusion equation (64) can be integrated once to $-af + \frac{1}{2}(b^2f)x = $ constant. The constant represents the probability flux in the negative direction along the x-axis. If the boundary conditions are such that this flux is zero, then the equation can be integrated once more to yield the equilibrium density (if legitimate)

$$f(x) \propto b(x)^{-2} \exp\left(\int^x 2a(u)b(u)^{-2} \, du \right).$$

3. Suppose the components X_j of a vector X obey the stochastic differential equations

$$\dot{X}_j = -\frac{\partial}{\partial X_j} V(X) + b\varepsilon_j,$$

where b is a constant and the ε_j are independent standard white noise processes. Show that, at least formally, the process has an equilibrium probability density $f(x) \propto \exp[-2V(x)/b^2]$.

12. First Passage and Recurrence for Brownian Motion

Brownian motion in a Euclidean space of d dimensions is the natural analogue in continuous time and space of the symmetric random walk of Section 9.9. Indeed, it is the only analogue which has continuous sample paths, and would be the limit version of a random walk in which the steps become ever shorter and more frequent.

Let us use $x = (x_1, x_2, \ldots, x_d)$ to denote the coordinate vector of the point which performs the motion. If we make the walk isotropic (i.e. statistically invariant to rotations of space) then, just as shown for Theorem 10.10.2, the generator of the proceess will have the action

$$\Lambda H(x) = \tfrac{1}{2}\sigma^2 \nabla^2 H(x) = \tfrac{1}{2}\sigma^2 \sum_{j=1}^d \frac{\partial^2 H(x)}{\partial x_j^2}. \tag{67}$$

We have introduced a scaling factor σ^2 for a slight increase in generality. It could be normalized out, but its presence helps to keep one aware of spatial and temporal scales.

If we consider a function of $r = |x| = \sqrt{\sum x_j^2}$ alone, then Λ will have the action

$$\Lambda H(r) = \tfrac{1}{2}\sigma^2 \left(\frac{d-1}{r}\frac{\partial H}{\partial r} + \frac{\partial^2 H}{\partial r^2} \right), \tag{68}$$

as can be verified. We may then conclude

Theorem 10.12.1. *The r.v.* $R(t) = |X(t)|$ *follows a diffusion process with drift and diffusion coefficients* $\tfrac{1}{2}\sigma^2(d-1)/r$ *and* σ^2.

That is, for $d > 1$ there is a drift in the direction of increasing r. Even though all directions of motion of the current X-variable are equally likely, there are more directions which increase the displacement from any arbitrary origin than those which decrease it. This is another manifestation of the effect of a change of variable.

Passage or recurrence to a point value of x mean little. The specimen set we shall generally consider is the ball $\mathcal{B}(a)$: the set $r \leq a$. One can of course consider the limit $a \downarrow 0$, but this makes very little difference to conclusions, because of the invariance of the character of Brownian motion to scale changes.

There are difficulties with the notion of recurrence in the continuous space/time case. Even in the case when ultimate clear escape from $\mathcal{B}(a)$ is certain, for the time that X is in the close neighbourhood of $\mathcal{B}(a)$ it will make many (indeed, infinitely many) returns to $\mathcal{B}(a)$ before it makes a clear escape. For this reason, we have to use quite direct arguments, rather than the general theory of Sections 9.8 and 9.9.

Theorem 10.12.2. *Consider the Brownian motion $\{X(t)\}$ starting from an initial coordinate $X(0) = x$ distant r from the origin. Then:*

(i) *If x is the interior to $\mathcal{B}(a)$ then passage to the surface of $\mathcal{B}(a)$ is certain and takes expected time*
$$T(x) = (a^2 - r^2)/(d\sigma^2). \tag{69}$$

(ii) *If x is exterior to $\mathcal{B}(a)$ then passage to the surface of $\mathcal{B}(a)$ has probability*
$$A(x) = \begin{cases} 1 & (d \leq 2), \\ (a/r)^{d-2} & (d > 2). \end{cases} \tag{70}$$

PROOF. Whether x is interior or exterior, the passage probability will obey the backward equation
$$\nabla^2 A = 0 \qquad (|x| \neq a), \tag{71}$$

with boundary condition $A = 1$ on $|x| = a$. Plainly $A(x)$ is a function of r alone, $u(r)$ say, so obeying
$$u'' + \frac{d-1}{r} u' = 0,$$

with solution
$$u(r) = \begin{cases} c_1 + c_2 r^{2-d} & (d \neq 2), \\ c_1 + c_2 \log r & (d = 2), \end{cases} \tag{72}$$

where the constants c are to be determined from boundary conditions.

In the case $r < a$ we must have $c_2 = 0$ if (71) is to hold at $x = 0$; the only solution satisfying the surface boundary condition is then $A(x) = u(r) = 1$. Passage to the surface of $\mathcal{B}(a)$ from an interior point is thus certain. The expected time taken, $T(x)$, will satisfy the backward equation
$$1 + \tfrac{1}{2}\sigma^2 \nabla^2 T = 0 \qquad (|x| < a), \tag{73}$$

with boundary condition $T(x) = 0$ on $|x| = 1$. We find that expression (69) satisfies these conditions. Equation (73) would still be satisfied if we added an expression of form (72) to (69). However, c_2 is zero for the same reason as before, and the surface condition implies that $c_1 = 0$.

To determine $A(x)$ in the exterior case we impose the condition $A(x) = 1$ on $|x| = a$ and $A(x) = 0$ on a boundary $|x| = b$ which we then allow to recede to infinity. In this way we obtain the solution (70). ☐

So, the results of Section 9.9 are confirmed, in that passage to $\mathscr{B}(a)$ is certain if $d \leq 2$, uncertain if $d > 2$. Recurrence to $\mathscr{B}(a)$ is then also plainly certain in the case $d \leq 2$. It is also certain in the case $d \geq 2$ in the sense that expression (70) will tend to unity as $r \downarrow a$. That is, one can make passage to $\mathscr{B}(a)$ as near certain as one likes by starting from a point outside $\mathscr{B}(a)$, but sufficiently close to it. However, further recurrence to $\mathscr{B}(a)$ once one has got away from its neighbourhood is uncertain, as formula (70) makes clear. The following assertion is perhaps clarifying.

Theorem 10.12.3. *Consider a Brownian motion starting from the origin:* $X(0) = 0$. *Let $\beta(t)$ be the probability that the representative point is ever in $\mathscr{B}(a)$ after time t. Then $\beta(t) = 1$ for $d < 2$, and*

$$\beta(t) \leq \left[\frac{a^2}{2\sigma^2 t} \right]^{(d/2)-1} = O(t^{1-d/2}). \tag{74}$$

for $d > 2$.

PROOF. The assertion is plain for the case $d \leq 2$. For the other, note that $R(t) = |X(t)|$ has probability density

$$f(r, t) \propto r^{d-1} \exp[-r^2/(2\sigma^2 t)].$$

We have then, for $d > 2$,

$$\beta(t) \leq E\{\min[1, (a/R)^{d-2}]\} \leq E[(a/R)^{d-2}].$$

Evaluating this last expression by appeal to the expression for the probability density $f(r, t)$ of R above, we deduce the evaluation asserted in (74). ☐

That is, however, many recurrences there may be after the first emergence from $\mathscr{B}(a)$, the probability of further recurrence declines to zero as $t^{1-d/2}$.

EXERCISES AND COMMENTS

1. Consider the Brownian motion of the text, starting from the origin. Show, by making the variable transformation $t \to u = r^2/(2\sigma^2 t)$, that the expected time spent during $t \geq 0$ in $\mathscr{B}(a)$ is

$$U(0) = \int_0^\infty dt \int_0^\infty dr\, f(r, t) = \begin{cases} +\infty & (d \leq 2), \\ (a/\sigma)^2/(d-2) & (d > 2). \end{cases}$$

If we identified $U(0)/T(0)$ as the 'expected number of recurrences' and made the further identification $1 + U(0)/T(0) = (1 - \rho_d)^{-1}$, where ρ_d is an effective 'recurrence probability' (as would be valid in renewal contexts) then we would obtain the evaluation ρ_d as 1 or $2/d$ according as $d \leq 2$ or $d > 2$. The argument is appealing but fallacious: even an honest recurrence (i.e. one that implies passage from a clearly exterior point to a clearly interior one) does not imply a restart from $X = 0$.

2. Even in the case of certain ultimate escape the number of recurrences is infinite. For a process starting from within $\mathscr{B}(a)$, the number of passages into $\mathscr{B}(a)$ from points on the shell $|x| = b \ (> a)$ is geometrically distributed with repetition probability $(a/b)^{d-2}$. The probability of at least N such recurrences is then $(a/b)^{N(d-2)}$. For any N, this can be made arbitrarily close to unity by taking b close enough to a.

CHAPTER 11

Second-Order Theory

1. Back to L_2

We introduced least square approximation in Section 2.8 as an immediate and useful technique which required virtually no probabilistic baggage. We saw in Section 5.3 that it had, nevertheless, extensive probabilistic implications. The time has come to take up both of these themes in earnest.

The r.v.s X we deal with will mainly be real vectors which possess second moments: $E(|X|^2) \leq \infty$. (However, the vector case includes the scalar case, and there may be occasion to consider complex vectors.) One could work entirely in terms of product-moment matrices: $U_{XY} = E(XY^T)$. However, it makes sense for various reasons to assume that all r.v.s have been reduced to zero mean (by replacing X by $X - E(X)$, if necessary). In this case, the product-moment matrix is replaced by the covariance matrix, variously written

$$\text{cov}(X, Y) = V_{XY} = E(XY^T).$$

In fact, the matrix V_{XY} would usually be termed the *cross-covariance matrix* between the random vectors X and Y, and the term *covariance matrix* reserved for $V_{XX} = \text{cov}(X, X)$, which we shall also write simply as $\text{cov}(X)$.

An immediate consequence of the definition of a covariance is then that

$$\text{cov}\left(\sum_j A_j X_j, \sum_k B_k Y_k \right) = \sum_j \sum_k A_j \, \text{cov}(X_j, Y_k) B_k^T,$$

where the A_j and B_k are constant matrix coefficients of appropriate dimensions.

Theorem 11.1.1. *A matrix V is a covariance matrix if and only if it is symmetric and nonnegative definite.*

PROOF. We have seen the direct part before. If V is a covariance matrix then $V = E(XX^T)$ for some X, whence symmetry follows. Also, if c is a real vector then

$$c^T V c = E(c^T X X^T c) = E[(c^T X)^2] \geq 0,$$

whence nonnegative definiteness follows.

For the converse, note first that I is a possible covariance matrix; the covariance matrix of a random vector ε whose components are uncorrelated and of unit variance. Suppose a matrix is symmetric and nonnegative definite. It can then be wrtten as $V = MM^T$ for some real matrix M, not necessarily square. But this means that we could make the identification $V = \text{cov}(M\varepsilon)$. $\square$

We shall use the notation $Q > 0$ and $Q \geq 0$ to indicate that the matrix Q is positive definite and nonnegative definite, respectively. If used, they will imply that Q is square and symmetric.

One says that the random vectors X and Y are *uncorrelated* if $V_{XY} = 0$. One indicates this lack of correlation by using the orthogonality sign, $X \perp Y$, and even also speaks of X and Y as being mutually orthogonal. This is consistent mathematically: $\text{cov}(X, Y)$ defines a matrix inner product and $\text{cov}(X)$ a matrix norm.

One readily verifies that independence plus existence of second moments implies orthogonality. The converse is not in general true, as we have seen from Exercise 4.3.4. It is true in the case of normal r.v.s, however: if X and Y are jointly normally distributed then they are are independent if and only if they are uncorrelated (Theorem 7.5.3).

Any representation of a random vector

$$X = M\eta, \tag{1}$$

in terms of uncorrelated r.v.s η_j (so that $\text{cov}(\eta)$ is diagonal), is termed an *orthogonal representation*. There are many such representations (see Exercise 5), some of which have a particular significance and a particular usefulness. Among these we can note the *innovations* representation of Section 3 and the *principal components* or *spectral* representation of Exercise 11.4.2.

One can always find on orthogonal representation by the method of *Gram–Schmidt orthogonalization*. Suppose that X has components X_1, X_2, ..., X_n; now determine linear forms

$$\eta_1 = X_1, \quad \eta_2 = X_2 + c_{21}X_1, \quad ..., \quad \eta_j = X_j + \sum_{k=1}^{j-1} c_{jk}X_k, \quad ..., \tag{2}$$

recursively, by requiring that η_j be orthogonal to $X_1, X_2, ..., X_{j-1}$. Inversion of these equations to express X in terms of η then supplies an orthogonal representation.

Theorem 11.1.2. *The Gram–Schmidt procedure produces an orthogonal representation for which M is lower triangular.*

PROOF. Since $\eta_j \perp X_1, X_2, \ldots, X_{j-1}$ then $\eta_j \perp \eta_1, \eta_2, \ldots, \eta_{j-1}$, and the η_j are mutually orthogonal. Relations (2) can be solved recursively for $X_1, X_2, \ldots$. This inversion yields X_j in terms of $\eta_1, \eta_2, \ldots, \eta_j$, so the effective M is indeed lower triangular.   □

This seems a rather formal procedure, but, if the order of the components in the vector is significant, then the orthogonalization can be significant; see Section 3.

EXERCISES AND COMMENTS

1. Consider a random vector X with elements $X_j = \eta + \varepsilon_j$ ($j = 1, 2, \ldots, n$), where η and all the ε_j are mutually uncorrelated, η has variance β and all the ε_j have common variance α. Show that $V = \text{cov}(X) = \alpha I + \beta E$, where E is the square matrix with units as elements. Show that V has an $(n - 1)$-fold eigenvalue α and a simple eigenvalue $\alpha + n\beta$. Show that V has inverse

$$V^{-1} = \alpha^{-1}I - \frac{\beta E}{\alpha(\alpha + n\beta)}.$$

2. Show, for the example of Exercise 4.3.4, that $\text{cov}(X^2, Y^2) = -\frac{1}{8}$.

3. Note the identity $E(X^{\mathsf{T}}GX) = \text{tr}(GV_{XX})$, useful when discussing quadratic functions of a random vector.

4. Suppose that $\text{cov}(\varepsilon) = I$. Note that, if U is an orthogonal matrix, then $\text{cov}(U\varepsilon) = I$ also. That is, an orthogonal transformation takes orthonormal variables into orthonormal variables.

5. Consider the orthogonal representation (1). One can always achieve $\text{cov}(\eta) > 0$ by omitting η_j of zero variance from the representation. One can then achieve $\text{con}(\eta) = I$ by a rescaling of the η_j. Note that $X = MU^{-1}\varepsilon$ is also an orthogonal representation, where U is any orthogonal matrix and $\varepsilon = U\eta$.

6. The extension of second-order theory to the case of complex vector X is immediate if the operation ^{T} of transposition is replaced by the operation † of transposition *and* complex conjugation. Symmetric matrices then become Hermitian matrices, orthogonal matrices become unitary matrices, etc.

7. Suppose that the scalar r.v.s constituted by the components of a vector X are *exchangeable* in that their joint distribution is invariant under any permutation of the variables. Show that the covariance matrix of X must have the form deduced in Exercise 1, and that one could therefore represent the components as the X_j were represented there.

2. Linear Least Square Approximation

The material of this section has been covered to some extent already in Sections 2.8 and 5.3. However, we repeat it, for the sake of completeness, and to make sure that it is stated for the vector case.

The situation is that one wishes to estimate an unobserved r.v. X from an observed r.v. Y and considers linear estimates

$$\hat{X} = AY. \tag{3}$$

Since all r.v.s are measured from their mean values, one gains nothing by including a constant term in (3) (see Section 2.8), and a mean square measure such as $E[(X - \hat{X})(X - \hat{X})^\mathsf{T}]$ becomes simply $\operatorname{cov}(X - \hat{X})$.

If one could demonstrate that

$$\operatorname{cov}(X - \hat{X}) \le \operatorname{cov}(X - \overline{X}), \tag{4}$$

where $\overline{X}$ is any other linear estimate, then this would show that $\hat{X}$ was an optimal estimate in quite a strong sense, because it would imply that

$$E[(X - \hat{X})^\mathsf{T}G(X - \hat{X})] \le E[(X - \overline{X})^\mathsf{T}G(X - \overline{X})] \tag{5}$$

for any nonnegative definite G. Such estimates exist, and they are the *linear least square (LLS)* estimates.

Theorem 11.2.1.

(i) *The necessary and sufficient condition for the linear estimate $\hat{X} = AY$ to be optimal in the sense (5) for a prescribed positive definite G is that*

$$X - \hat{X} \perp Y, \tag{6}$$

or, equivalently,

$$AV_{YY} = V_{XY}. \tag{7}$$

(ii) *Any $\hat{X}$ satisfying (6) is optimal in the sense (4), in that, for any linear estimate $\overline{X}$,*

$$\operatorname{cov}(X - \overline{X}) = \operatorname{cov}(X - \hat{X}) + \operatorname{cov}(\overline{X} - \hat{X}) \ge \operatorname{cov}(X - \hat{X}), \tag{8}$$

and all such $\hat{X}$ are mean square equivalent.

(iii) *The covariance matrix of the estimation error has the evaluation*

$$\operatorname{cov}(X - \hat{X}) = V_{XX} - V_{XY}V_{YY}^{-1}V_{XY}. \tag{9}$$

PROOF. Minimization of $E[(X - AY)^\mathsf{T}G(X - AY)]$ with respect to A yields relation (7) for A, which is equivalent to (6) for $\hat{X}$. These relations thus have solutions, and are necessary conditions for optimality in the sense (5).

Since $\overline{X} - \hat{X}$ is a linear function of Y, then (6) implies that $(X - \overline{X}) \perp (\overline{X} - \hat{X})$, whence (8) follows. Any $\hat{X}$ satisfying (6) is thus optimal in both the senses (4) and (5). If $\hat{X}_{(1)}$ and $\hat{X}_{(2)}$ are two solutions of (6) then $(X - \hat{X}_{(i)}) \perp (\hat{X}_{(1)} - \hat{X}_{(2)})$ for $i = 1, 2$, whence we deduce that $\operatorname{cov}(\hat{X}_{(1)} - \hat{X}_{(2)}) = 0$.

Finally, if A satisfies (7) then $\operatorname{cov}(X - AY) = \operatorname{cov}(X - AY, X) = V_{XX} - AV_{YX}$, which yields (9) by a further appeal to (7), at least if we understand the inverse of V_{YY} correctly in the case when this matrix is singular. $\square$

If V_{YY} is singular then there is a linear relationship between the elements of Y (in the mean square sense) and the simplest course would seem to be to subtract elements until one had a nondegenerate set. However, with the correct formulation, this is not necessary; see the exercises.

EXERCISES AND COMMENTS

1. Suppose that V_{YY} is singular. Let c be any element of the null-space of this matrix, so that $V_{YY}c = 0$. Then $c^T V_{YY} c = 0$ and $c^T Y \overset{m.s.}{=} 0$. Note also that $|c^T V_{YX} d|^2 = |\text{cov}(c^T Y, d^T X)|^2 \le \text{var}(c^T Y)\, \text{var}(d^T X) = 0$ for any d, so that $c^T V_{YX} = 0$.

2. Consider again the case of singular V_{YY}. We know that equation (7) has solutions for A; denote any one of them by $V_{XY} V_{YY}^{-1}$. Then the general solution is $A = V_{XY} V_{YY}^{-1} + C^T$, where C is a matrix whose columns lie in the null space of V_{YY}. Thus $C^T V_{YX} = 0$, and we can write $V_{XX} - A V_{YX}$ as expression (9), no matter what solution of (7) we take for A.

3. Note that the LLS estimate of $\sum_j c_j X_j$ is just $\sum_j c_j \hat{X}_j$, where the c_j are constant. In particular, the components of $\hat{X}$ are the individual LLS estimates of the components. The generalization achieved by passage to the vector case is then largely illusory, except that one does have the stronger characterization (4) of optimality.

4. Suppose $X(t)$ is a random function of time. Note, as a consequence of the assertion in the last exercise, that the LLS estimate of $dX(t)/dt$ is $d\hat{X}(t)/dt$, at least if the set of observations Y does not change with time.

3. Projection: Innovation

The orthogonality relation (6) is just that which occurs in all L_2 approximation problems. In Fig. 11.1 we have represented X by a point and the manifold $\mathcal{Y}$ of all linear functions of Y (of the dimension of X) by a plane. We seek the

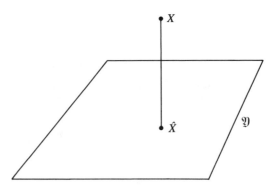

Figure 11.1. The projection $\hat{X}$ of a random vector X onto the manifold $\mathcal{Y}$ generated by all linear functions of Y (of the dimension of X).

point $\hat{X}$ in $\mathfrak{Y}$ which is nearest to X. This point is the foot of the perpendicular form X on to $\mathfrak{Y}$, and relation (6) is just the condition of perpendicularity: that the vector $\hat{X} - X$ be orthogonal to the plane $\mathfrak{Y}$. Otherwise expressed, $\hat{X}$ is the *projection* of X on to $\mathfrak{Y}$.

If we wish to express the dependence of $\hat{X}$ upon Y then it is convenient to write it as $\mathscr{E}(X|Y)$. The notation is appropriate because, as we shall see in the next section, $\mathscr{E}(X|Y)$ is in fact identifiable with the conditional expectation $E(X|Y)$ in the case when X and Y are jointly normal. We read $\mathscr{E}(X|Y)$ as 'the linear least square estimate of X in terms of Y' or 'the projection of X on to the manifold with Y as basis'.

A projection has well-recognized properties.

Theorem 11.3.1. *$\mathscr{E}(X|Y)$ has the properties of a projection; e.g.*

$$\mathscr{E}[\mathscr{E}(X|Y)|Y] = \mathscr{E}(X|Y),$$

and if X, Y_1 and Y_2 are random vectors such that $Y_1 \perp Y_2$, then

$$\mathscr{E}(X|Y_1, Y_2) = \mathscr{E}(X|Y_1) + \mathscr{E}(X|Y_2). \tag{10}$$

We leave verification, which is direct, to the reader.

Often one has a sequence of vector observations $Y_1, Y_2, Y_3, \ldots$, where Y_t is interpretable as the observation obtained at time t. Then quantities which are of interest are the *innovations*

$$\zeta_t = Y_t - \mathscr{E}(Y_t|Y_1, Y_2, \ldots, Y_{t-1}). \tag{11}$$

The innovation ζ_t is the difference between the new observation Y_t and the LLS prediction one had made of it from previous observations. In this sense it is the part of Y_t which is not predictable from previous observations, and represents the genuinely new information gained at time t.

Theorem 11.3.2. *The innovations are mutually orthogonal, and are just the sequence of random vectors determined by Gram–Schmidt orthogonalization of the sequence of observations $\{Y_t\}$.*

This is immediate; formation of the form (11) is equivalent to formation of the forms (2). One should note, however, that the Y_t and ζ_t are in general vectors, whereas one normally thinks of the Gram–Schmidt procedure as being applied to a sequence of scalars. The ordering of the variables Y_t is natural, as being an ordering in time and the order in which the values of the variables is revealed. The concept of an innovation has proved very valuable in recent years, and has given the rather dated Gram–Schmidt procedure a new lease of life.

The inversion of the linear transformation (11) to

$$Y_t = \sum_{s=1}^{t} c_{ts}\zeta_s$$

provides the *innovations representation* of the original observation series. Rather more generally, if one considers the LLS estimate $\hat{X}_t$ of a fixed r.v. X based upon observations up to time t, then, by virtue of the mutual orthogonality of the innovations, this can be written

$$\hat{X}_t = \mathscr{E}(X|Y_1, Y_2, \ldots, Y_t) = \mathscr{E}(X|\zeta_1, \zeta_2, \ldots, \zeta_t) = \sum_{s=1}^{t} \mathscr{E}(X|\zeta_s) = \sum_{s=1}^{t} H_s \zeta_s, \quad (12)$$

where

$$H_s = \text{cov}(X, \zeta_s) \, \text{cov}(\zeta_s)^{-1}. \quad (13)$$

The great advantage of representation (12) is that the coefficients H depend only upon s and not upon t. So, when one moves forward one time-step, the previous estimate of X is updated simply by the recursion

$$\hat{X}_t = \hat{X}_{t-1} + H_t \zeta_t. \quad (14)$$

More general versions of recursive updatings such as (14) are of considerable technological importance for the following and control of dynamic systems; see Exercise 1. The sequence of estimates $\{\hat{X}_t\}$ of a fixed r.v. X turns out to be an interesting object theoretically; it shows strong convergence properties, and virtually constitutes the martingale concept (see Sections 7 and 8).

EXERCISES AND COMMENTS

1. *The Kalman filter.* Consider a dynamic system, in fact, a Markov process with a vector state variable X which obeys the equation

$$X_t = AX_{t-1} + \varepsilon_t, \quad (15)$$

where $\{\varepsilon_t\}$ is vector white noise. (That is, $\varepsilon_s \perp \varepsilon_t$ for $s \neq t$.) This is often known as the *plant equation*, because it represents the dynamics of the 'plant'; the system with which one is concerned. Suppose one cannot observe the state variable, but at time t can observe only Y_t, where this is generated by the equation

$$Y_t = CX_{t-1} + \eta_t,$$

and $\{\varepsilon_t, \eta_t\}$ jointly constitute vector white noise. Let $W_t = (Y_t, Y_{t-1}, Y_{t-2}, \ldots)$ denote the observations available at time t, and define $\hat{X}_t = \mathscr{E}(X_t|W_t)$; the LLS estimate of the state of the system at time t based on the observations available at time t.

Confirm the following train of reasoning. If ζ_t is the innovation in the Y-process then

$$\hat{X}_t = \mathscr{E}(X_t|W_t) = \mathscr{E}(X_t|W_{t-1}, \zeta_t) = \mathscr{E}(X_t|W_{t-1}) + \mathscr{E}(X_t|\zeta_t).$$

But

$$\mathscr{E}(X_t|W_{t-1}) = \mathscr{E}(AX_{t-1} + \varepsilon_t|W_{t-1}) = A\hat{X}_{t-1}$$

and $\mathscr{E}(X_t|\zeta_t) = H_t \zeta_t$ for some matrix H_t. Furthermore,

$$\zeta_t = Y_t - \mathscr{E}(Y_t|W_{t-1}) = Y_t - \mathscr{E}(CX_{t-1} + \eta_t|W_{t-1}) = Y_t - C\hat{X}_{t-1}.$$

Collecting all these relations, we find that the estimate $\hat{X}_t$ obeys the updating relation

$$\hat{X}_t = A\hat{X}_{t-1} + H_t(Y_t - C\hat{X}_{t-1}). \tag{16}$$

Relation (16) is the *Kalman filter*. It has the attractive form that the estimates also obey the plant equation, but a plant equation driven by the innovations ζ_t rather than by the 'plant noise' ε_t.

4. The Gauss–Markov Theorem

Second-order ideas are particularly natural in the Gaussian context, and LLS estimates have then a stronger characterization.

Suppose again that X and Y are random vectors, that Y has been observed and X has not. Denote their joint and conditional probability densities by $f(x, y)$ and $f(y|x)$. Then the *maximum likelihood estimator* (MLE) of X is the value of X maximizing $f(Y|X)$ and the *Bayes estimator* (BE) of X is the value of X maximizing $f(X, Y)$. The ML approach simply regards x as a parameter of the conditional distribution, whereas the Bayes approach takes account of the distribution of X as a r.v.

If X and Y are jointly normally distributed then $f(X, Y) \propto \exp(-\tfrac{1}{2}\mathbb{D})$, where

$$\mathbb{D} = \begin{bmatrix} X \\ Y \end{bmatrix}^{\mathrm{T}} \begin{bmatrix} V_{XX} V_{XY} \\ V_{YX} V_{YY} \end{bmatrix}^{-1} \begin{bmatrix} X \\ Y \end{bmatrix} = \begin{bmatrix} X \\ Y \end{bmatrix}^{\mathrm{T}} \begin{bmatrix} J_{XX} J_{XY} \\ J_{YX} J_{YY} \end{bmatrix} \begin{bmatrix} X \\ Y \end{bmatrix} \tag{17}$$

and J is the *information matrix*, the inverse of the joint covariance matrix of X and Y.

The Bayes estimate $\overline{X}$ then minimizes $\mathbb{D}$, and so is determined by

$$J_{XX}\overline{X} + J_{XY}Y = 0. \tag{18}$$

If we regard $\mathbb{D}$ as a quadratic form in X and complete the square then we have

$$\mathbb{D} = (X - \overline{X})^{\mathrm{T}} J_{XX}(X - \overline{X}) + K(Y), \tag{19}$$

where $K(Y)$ is a function of Y alone.

Theorem 11.4.1 (The Extended Gauss–Markov Theorem). *If X and Y are jointly normally distributed then the LLS estimate $\hat{X}$ of X can be identified with the Bayes estimate and also with the conditional expectation $E(X|Y)$. The covariance matrix $\mathrm{cov}(X - \hat{X})$ of the estimation error can be identified with the conditional covariance matrix $\mathrm{cov}(X|Y)$.*

PROOF. Certainly the Bayes estimate $\overline{X}$ is a linear function of Y. But the partition (19) implies that $X - \overline{X} \perp Y$. and we know from Theorem 11.2.1 that this implies that $\overline{X}$ is an LLS estimate, and so is at least mean square equivalent to $\hat{X}$.

The independence of $X - \overline{X}$ and Y implies that the distribution of $X - \overline{X}$

is the same whether conditioned by Y or not. This implies that $E(X|Y) = \overline{X} = \hat{X}$ and $\text{cov}(X|Y) = \text{cov}(X - \overline{X}) = \text{cov}(X - \hat{X})$. $\qquad\square$

It is interesting that the two estimates $\hat{X}$ and $\overline{X}$, now proved identical, should be calculated in such different ways. We have

$$\hat{X} = V_{XY}V_{YY}^{-1}Y = -J_{XX}^{-1}J_{XY}Y \tag{20}$$

and

$$\text{cov}(X - \hat{X}) = V_{XX} - V_{XY}V_{YY}^{-1}V_{XY} = J_{XX}^{-1}, \tag{21}$$

where we have written the LLS evaluation first and the Bayes evaluation second, and have assumed nonsingularity throughout, for simplicity. Relations (20) and (21) imply some matrix identities which the reader might like to verify from the relation $J = V^{-1}$. However, the difference in the two modes of calculation really has some substance; the Bayes calculations can prove much the more natural and useful. To set up a convincing example would demand more of a detour than we can afford; we give an example which just escapes triviality in Exercise 1.

EXERCISES AND COMMENTS

1. Consider the example of Exercise 2.8.5, for which

$$\mathbb{D} = (X - \mu)^2/v + \sum_{j=1}^{n} (Y_j - X)^2/v_j.$$

   The LLS estimate of X deduced in Exercise 2.8.5, whose calculation is nontrivial, is immediately deduced as the value of X minimizing $\mathbb{D}$.

2. *Principal components.* This and the following exercises concern the topic of principal components, a purely second-order concept, which can be developed without reference to the normal distribution. However, the treatment at certain points can be shortened if one makes the full distributional assumption of joint normality. The topic concerns the intrinsic statistics of a random vector X rather then the relation between random vectors X and Y.
   Suppose that X has covariance matrix V. Consider the determination of the linear function $c^T X$ which has maximal variance, subject to the normalization of coefficients $c^T c = 1$. Show that c must be an eigenvector of V and that the variance of the form is the corresponding eigenvalue.

3. Suppose that V has eigenvalues $\lambda_1 \geq \lambda_2 \geq \cdots \geq \lambda_n$, with corresponding normalized (and orthogonalized, if necessary) eigenvectors $c_1, c_2, \ldots, c_n$. Then the forms $\eta_j = c_j^T X$ are the *principal components* of X. Show that $\text{cov}(\eta_j, \eta_k) = \delta_{jk}\lambda_j$. Show also that $X = \sum_{j=1}^{n} \eta_j c_j$; this is a particular orthogonal representation of X, the *spectral representation.*

4. Show that η_j is the normalized linear function of X which has maximal variance, given that it is orthogonal to η_k $(k < j)$. Alternatively, if X is normally distributed, then η_j is the form that has maximal variance, conditional on prescribed values η_k $(k < j)$.

5. The spectral representation has the following particular significance. The LLS estimate of X in terms of $\eta_1, \eta_2, \ldots, \eta_r$ is $X^{(r)} = \sum_{j=1}^{r} \eta_j c_j$, a truncated form of the spectral representation. This is an r-dimensional approximation to X (in the sense that it approximates X linearly in terms of r scalar r.v.s). It is the best r-dimensional approximation in a sense to be clarified in the next exercise. This is the type of representation used in factor analysis, when one attempts to explain scores in n tests in terms of r ($<n$) factors.

6. Suppose that one forms the normalized linear function of X which has maximal variance, conditional on the values of r prescribed linear functions of X. Let v be this variance. Then v measures the size of that part of X which is not explained by the observations on the r linear forms. Show that $v \geq \lambda_{r+1}$, with equality when the linear forms are the first r principal components.

5. The Convergence of Linear Least Square Estimates

Suppose that one has an infinite sequence of observations $\{Y_n; n \geq 1\}$ and that

$$\hat{X}_n = \mathcal{E}(X \mid Y_1, Y_2, \ldots, Y_n), \tag{22}$$

is the LLS estimate of X in terms of the first n observations. (One would perhaps prefer to use t rather than n, to indicate a sequence ordered in time. However, the time element is not always present, and the use of n in this context is hallowed in the literature.) The estimate $\hat{X}_n$ is a r.v., and one imagines that it would converge in some sense to a limit r.v. $\hat{X}_\infty$ as n increases, that limit being the LLS estimate of X deduced from the whole observation sequence. Such convergence indeed holds, and provides one of the most immediate, important and natural limit laws.

We can restrict ourselves to the case of scalar X, since the LLS estimate of a vector is just the vector of LLS estimates. If u and v are scalar r.v.s then $E[(u-v)^2]$ is the natural measure of squared distance between them, and

$$\|u - v\| = \sqrt{E[(u-v)^2]} \tag{23}$$

is the natural measure of distance. Indeed, $\|u\|$ is just the L_2 norm, familiar in so many contexts. Note that, by Cauchy's inequality,

$$E[(u-v)^2] = E(u^2) + E(v^2) + 2E(uv) \leq E(u^2) + E(v^2) + 2\sqrt{E(u^2)E(v^2)}$$

or

$$\|u - v\| \leq \|u\| + \|v\|, \tag{24}$$

so that the norm obeys the triangular inequality.

Returning to the estimation problem, let us suppose that $\|X\| < \infty$ (i.e. $E(X^2) < \infty$) and define

$$D_n = \|X - \hat{X}_n\|.$$

Lemma 11.5.1. D_n decreases to a finite limit D as $n \to \infty$, and $E(\hat{X}_n^2)$ increases to $E(X^2) - D^2 \leq E(X^2)$.

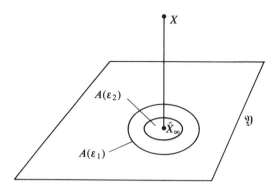

Figure 11.2. The convergence of $\hat{X}_n$ with increasing n into ever-smaller neighbour-hoods of the limiting estimate $\hat{X}_\infty$.

PROOF. Certainly $0 \le D_n \le \|X\|$, the second inequality following because $\hat{X}_n$ cannot estimate X worse than does the estimate 0. Thus D_n is finite and nonnegative. Furthermore, it is nonincreasing in n (since estimation can only improve with increasing n) and so has a limit. The second assertion follows from the fact that $(X - \hat{X}_n) \perp \hat{X}_n$, and so $\|X\|^2 = \|X - \hat{X}_n + \hat{X}_n\|^2 = D_n^2 + \|\hat{X}_n\|^2$. □

Now, let $\mathfrak{Y}$ denote the manifold with $\{Y_n\}$ as basis, i.e. the set of scalar r.v.s generated linearly from the whole observation sequence. Let $A(\varepsilon)$ be the set of elements Z of $\mathfrak{Y}$ for which

$$\|X - Z\| \le D + \varepsilon$$

for prescribed positive ε.

Existence of a limit $\hat{X}_\infty$ to which $\hat{X}_n$ converges must follow from the extremal characterization of $\hat{X}_n$. We present the essence of the argument in Fig. 11.2. There *is* an element $Z = \hat{X}_\infty$ in $\mathfrak{Y}$ which minimizes $\|Z - X\|$; the foot of the perpendicular in the figure. The set $A(\varepsilon)$ is an ε-neighbourhood of it, and $\hat{X}_n$ enters this neighbourhood, never to leave it, as n increases.

Lemma 11.5.2.

(i) *The set $A(\varepsilon)$ is nonempty for $\varepsilon > 0$.*

(ii) *The set $A(\varepsilon)$ decreases monotonically with decreasing ε.*

(iii) *For prescribed positive ε all elements of the sequence $\{\hat{X}_n\}$ belong to $A(\varepsilon)$ for n larger than some finite $n(\varepsilon)$.*

(iv) *For any two elements Z_1 and Z_2 of $A(\varepsilon)$, $\|Z_1 - Z_2\| \le 2\varepsilon$.*

PROOF. All $\hat{X}_n$ belong to $\mathfrak{Y}$, and we know that $\|X - \hat{X}_n\|$ decreases monotonically to D with increasing n. Assertions (i) and (iii) thus follow. Assertion (ii) follows from the definition of $A(\varepsilon)$. Finally, assertion (iv) follows from the triangular inequality

$$\|Z_1 - Z_2\| \le \|X - Z_1\| + \|X - Z_2\| \le 2\varepsilon. \qquad \square$$

We thus deduce

Theorem 11.5.3. *The sequence $\{\hat{X}_n\}$ has a limit $\hat{X}_\infty$, to which it converges in mean square. Moreover,*

$$E(X_\infty^2) = \lim_{n \to \infty} E(\hat{X}_n^2) \le E(X^2).$$

PROOF. We refer to the assertions of Lemma 11.5.2. The sets $A(\varepsilon)$, being monotone (assertion (ii)), converge with decreasing ε to a limit $A(0)$ which is nonempty (assertion (i)). Define $\hat{X}_\infty$ as any member of this set. Then $\hat{X}_\infty$ is essentially unique in that it is mean square equivalent to any other member of $A(0)$ (assertion iv)). Also, $\hat{X}_n$ converges to it in mean square, by assertions (iii) and (iv). The first assertion of the theorem is thus established.

The triangular inequality (24) implies that $\|\hat{X}_n\| \pm \|\hat{X}_n - \hat{X}_\infty\|$ constitute bounds on $\|\hat{X}_\infty\|$. Since $\|\hat{X}_n - \hat{X}_\infty\|$ decreases to zero with increasing n we then have $\|\hat{X}_\infty\| = \lim_{n \to \infty} \|\hat{X}_n\| \le \|X\|$. $\qquad \square$

Theorem 11.5.3 is the first intimation of a result which will increasingly emerge as fundamental.

6. Direct and Mutual Mean Square Convergence

It was very natural to consider the sequence of LLS estimates $\{\hat{X}_n\}$ as we did in the last section. However, the proof of Theorem 11.5.3 suggests a considerably more general assertion.

A sequence $\{X_n\}$ of scalar r.v.s converges in mean square to X if

$$E[(X_n - X)^2] \to 0$$

as $n \to \infty$. The sequence is said to converge *mutually* in mean square if

$$E[(X_m - X_n)^2] \to 0$$

as m and n tend to infinity independently. Note that the second assertion makes no mention of a limit r.v., and it is indeed not clear that there is one.

Convergence obviously implies mutual convergence, since

$$\|X_m - X_n\| \le \|X_m - X\| + \|X_n - X\|.$$

The converse also holds; a conclusion both substantial and useful. It is useful because it enables one to assert convergence to a limit even if one cannot specify the limit r.v. (see Exercise 1).

Theorem 11.6.1. *If a sequence $\{X_n\}$ converges mutually in mean square then it has a mean square limit X_∞.*

PROOF. Define $B(\varepsilon)$ as the set of r.v.s Z for which $\|Z - X_n\| \leq \varepsilon$ for all n greater than some value $n(\varepsilon)$. Then we will establish that $n(\varepsilon)$ is finite for prescribed positive ε and that $B(\varepsilon)$ has all the properties established for $A(\varepsilon)$ in Lemma 11.5.2. We shall refer to the assertions of that lemma.
  Define

$$\delta_n = \sup_{j,k \geq n} \|X_j - X_k\|.$$

Because of mutual convergence δ_n converges to zero with increasing n; mono-tonically, by its definition. Let $n(\varepsilon)$ be the smallest value of n for which $\delta(n) \leq \varepsilon$.
  Then $B(\varepsilon)$ contains X_n for all $n \geq n(\varepsilon)$. This is the analogue of assertion (iii), and implies the analogue of assertion (i): that $B(\varepsilon)$ is nonempty. The analogue of assertion (ii) (that $B(\varepsilon)$ is monotone nonincreasing in ε) is clear, and the analogue of assertion (iv) follows as before.
  We have thus established that all the assertions of Lemma 11.5.2 hold for $B(\varepsilon)$; the conclusion of the theorem then follows just as did that of Theorem 11.5.3.   □

  If follows, also as in Theorem 11.5.3, that if one has a uniform bound on $E(X_n^2)$ then $E(X_n^2)$ has a limit which we can identify with $E(X_\infty^2)$.

EXERCISES AND COMMENTS

1. Suppose the scalar r.v.s η_j mutually uncorrelated. By considering the partial sums $X_n = \sum_{j=1}^n \eta_j$ show that the infinite sum $\sum_{j=1}^\infty \eta_j$ exists as a mean square limit if and only if $\sum_{j=1}^\infty E(\eta_j^2) < \infty$.

7. Conditional Expectations as Least Square Estimates: Martingale Convergence

Suppose the condition $E(X^2) \leq \infty$ holds for the scalar r.v. X. Then we have already seen in Section 5.3 that, just as $\mathscr{E}(X|Y)$ is the least square estimate of X in terms of *linear* functions of Y, so the conditional expectation $E(X|Y)$ is the least square estimate of X in terms of *arbitrary* functions of Y. Moreover, all conclusions for the first case must have an immediate analogue for the second, as $E(X|Y)$ can formally be regarded as the least square estimate of X linear in a sufficiently large class of 'basis functions' of Y (e.g. the indicator functions of sets).

Suppose that, in analogue to (22), we consider the sequence

$$X_n = E(X \mid Y_1, Y_2, \ldots, Y_n) \tag{25}$$

generated by conditioning X on (or estimating X from) an increasing set of observations Y. The characterizing condition for the conditional expectation and its least square property is the extended orthogonality condition

$$X - X_n \perp H(Y_1, Y_2, \ldots, Y_n).$$

valid for arbitrary scalar H.

Theorem 11.7.1. *Suppose that $E(X^2) \leq \infty$. Then $\{X_n\}$ has a mean square limit X_∞.*

The proof goes through exactly as for Theorem 11.5.3. The character of X_n as a least square estimate gives it all the properties to which appeal was made in that theorem.

It is not necessary that X should be square integrable for the conditional expectation (25) to be well defined, but the condition does make possible the powerful and simplifying appeal to second-order ideas.

It follows from the definition (25) that

$$E(X_{n+1} \mid Y_1, Y_2, \ldots, Y_n) = X_n. \tag{26}$$

A sequence $\{X_n\}$ having the property (26) is termed a *martingale* with respect to $\{Y_n\}$. The property may not seem a particularly significant or natural one, but it is both of these. The fact that the conditional expectation (25) generates a martingale might indicate some of the significance, and we have in fact already found martingales useful in Sections 9.7 and 10.7. We shall draw all this material together in Chapter 14, and show that, at least under mild conditions, condition (26) implies that X_n can indeed be represented as in (25), and that $\{X_n\}$ has a limit to which it converges almost certainly as well as in mean square. For the moment, we content ourselves with the following.

Theorem 11.7.2. *Suppose that $\{X_n\}$ is a martingale and that $E(X_n^2)$ is bounded, uniformly in n. Then $\{X_n\}$ has a mean square limit.*

PROOF. The martingale property (26) will equally imply that

$$X_n = E(X_m \mid Y_1, Y_2, \ldots, Y_n) \qquad (m > n),$$

and hence that

$$E[(X_m - X_n)^2] = E(X_m^2 - 2X_m X_n + X_n^2) = E(X_m^2 - X_n^2) \qquad (m > n). \tag{27}$$

We thus see that $E(X_n^2)$ is increasing in n. Since it is bounded above, it must then converge to a finite limit. It follows then from (27) that $E[(X_m - X_n)^2] \to 0$ as $m, n \to \infty$. Thus $\{X_n\}$ converges mutually in mean square, and so, by Theorem 11.6.1, has a mean square limit. $\qquad \square$

EXERCISES AND COMMENTS

1. Demonstrate, under the conditions of Theorem 11.7.2, that $E(X_m | Y_1, Y_2, \ldots, Y_n)$ converges in mean square to $E(X_\infty | Y_1, Y_2, \ldots, Y_n)$ as $m \to \infty$ (where X_∞ is the mean square limit asserted in the theorem), and hence that $X_n = E(X_\infty | Y_1, Y_2, \ldots, Y_n)$ in mean square. That is, any martingale uniformly bounded in mean square really is generated by a relation of type (25).

CHAPTER 12

Consistency and Extension: the Finite-Dimensional Case

1. The Issues

We return to our starting point: Axioms 1–5 for the expectation operator set out in Section 2.2. Consider a random vector X with components $X_1, X_2, \ldots, X_n$. The fundamental *consistency* problem is: What values of $E(X)$ are consistent with the axioms?

The axioms are framed as constraints on $E(X)$, and imply constraints that are less immediately evident. For example, if $X_2 = X_1^2$, then they imply that $E(X_2) \geq [E(X_1)]^2$. This example illustrates that we must know something about the mutual relations of the r.v.s X_j before we can come to any conclusions. These relations will normally be formulated by expressing the X_j as functions of ω, the coordinate in sample space. That is, one must ideally be able to specify what value $X(\omega)$ the vector X would have if the realization were known to be ω.

The second fundamental problem is that of *extension*. What bounds could one set on the value of $E(X)$ given that one knew the value of $E(Y)$ for another random vector Y? The problem is closely related to the first, because these bounds are to be deduced from the joint consistency conditions on $E(X)$ and $E(Y)$. Again, to obtain such conditions one should ideally be able to express X and Y as functions $X(\omega)$ and $Y(\omega)$ of the realization.

The whole of probability calculus can be regarded as the achievement of some kind of extension: given certain expectations, one wishes to calculate others, or at least set bounds on them. The idea comes explicitly to the fore in the calculation of the bounds implied by the Markov and Chebyshev inequalities. However, we are now asking as a matter of principle: What are the *exact* consistency conditions? What are the *best possible* bounds?

In pursuing these questions we shall find that convexity is the key concept. We therefore sketch the relevant parts of that theory in Section 2.

In this chapter we suppose the vectors X and Y finite dimensional. Conclusions are then quite simple and unqualified. However, there are infinite-dimensional questions, perhaps those more usually associated with the term 'extension', which take one into another class of ideas. For example, given the value of $E(e^{i\theta X})$ for a scalar r.v. X for all real θ, for what functions $H(X)$ can one in principle determine $E[H(X)]$? Given an infinite class of scalar functions of ω, is it in principle possible to assign expectation values simultaneously and consistently to all these functions? (The assignment of probabilities to sets is a special case.) The pursuit of these questions would take us too far afield, but in Chapter 15 we cover some material to which we have already made effective appeal.

2. Convex Sets

Consider the n-vector function $X(\omega)$ of ω. If we regard $x = X(\omega)$ for given ω as a point in $\mathbb{R}^n$, then, as ω varies in Ω, so x traces out a set in $\mathbb{R}^n$, which we shall denote by $X(\Omega)$. The variable ω can be regarded as labelling the points of $X(\Omega)$, although not necessarily uniquely, as $X(\omega)$ may have the same value for more than one value of ω.

A set $\mathscr{S}$ in $\mathbb{R}^n$ is said to be *convex* if it is *closed under mixing*, in that, if $x^{(1)}$ and $x^{(2)}$ belong to $\mathscr{S}$, then so does

$$x = px^{(1)} + (1 - p)x^{(2)} \tag{1}$$

for any p in $[0, 1]$. Geometrically, this can be expressed by saying that the line segment joining the points $x^{(1)}$ and $x^{(2)}$ lies wholly in $\mathscr{S}$. In Fig. 12.1 we give examples of both convex and nonconvex sets in $\mathbb{R}^n$. Note that the property of

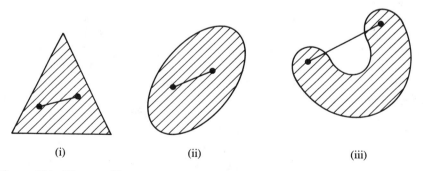

(i)   (ii)   (iii)

Figure 12.1. The sets illustrated in (i) and (ii) are convex, in that the line segment joining any two points of the set lies wholly in the set. The set illustrated in (iii) is not convex.

convexity is indeed related to probabilistic ideas, in that operation (1) represents 'averaging' of the two points, assigning them respective weights p and $q = 1 - p$.

The defining property implies the elaborated conclusion:

Theorem 12.2.1. *If* $x^{(1)}, x^{(2)}, \ldots, x^{(m)}$ *are element of* $\mathscr{S}$ *then so is*

$$x = \sum_{j=1}^{m} p_j x^{(j)} \tag{2}$$

for any m-point distribution $\{p_j\}$ $(m = 1, 2, 3, \ldots)$.

This follows immediately by repeated appeal to the definition; one can build up any m-point average from repeated two-point averages. Expression (2) is of course indeed an average of m points of $\mathscr{S}$. By considering limits of such averages one could in effect construct more general distribution on $\mathscr{S}$.

The *interior points* of $\mathscr{S}$ are those which have a neighbourhood lying wholly in $\mathscr{S}$. The others are *boundary points*. We shall in general simplify matters by supposing $\mathscr{S}$ closed, so that all limits of sequences in $\mathscr{S}$ also belong to $\mathscr{S}$.

Now we continue best by considering a set $\mathscr{S}$ in $\mathbb{R}^n$ which is *not* necessarily convex and seeing, for reasons which will transpire, how we might complete it to one which is.

Consider the two following definitions of the *extreme points* of $\mathscr{S}$, which have yet to be proved equivalent.

Definition 1. The point $\bar{x}$ of $\mathscr{S}$ is an extreme point of $\mathscr{S}$ if its only representation as a strict average of points $x^{(1)}$ and $x^{(2)}$ of $\mathscr{S}$ is that for which $x^{(1)}$ and $x^{(2)}$ both equal $\bar{x}$.

Definition 1*. The point $\bar{x}$ of $\mathscr{S}$ is an extreme point of $\mathscr{S}$ if, for some coefficient vector a, the form $a^T x$ achieves its maximum for x in $\mathscr{S}$ only at $x = \bar{x}$.

By a 'strict' average we mean that both points should receive positive weight. The geometric interpretation of Definition 1 is that $\bar{x}$ cannot be an interior point on the line segment joining any two points of $\mathscr{S}$. To interpret Definition 1*, consider a plane

$$a^T x = \sup_{x \in \mathscr{S}} (a^T x). \tag{3}$$

This has all of $\mathscr{S}$ to one side of it, but meets $\mathscr{S}$ in some point. Such a plane is said to be a *supporting hyperplane* to $\mathscr{S}$. If it meets $\mathscr{S}$ at $x^{(1)}$ then it is said to be a supporting hyperplane to $\mathscr{S}$ at $x^{(1)}$. Then Definition 1* of an extreme point $\bar{x}$ requires that there be a supporting hyperplane to $\mathscr{S}$ which meets $\mathscr{S}$ in $\bar{x}$ alone.

For example, in Fig. 12.2 we consider a set $\mathscr{S}$ consisting of the line segment

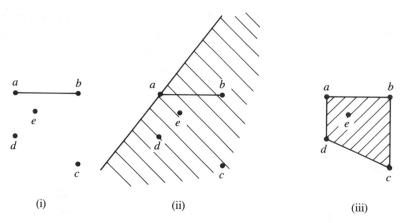

Figure 12.2. The set $\mathscr{S}$ illustrated in (i) consists of the straight line segment ab and the points c, d and e. In (ii) we see that a is an extreme point of the set under Definition 1*. Diagram (iii) indicates the conjectured form of the convex hull $[\mathscr{S}]$.

ab and the three points c, d and e. By either definition, the point e would seem not to be an extreme point. It could be represented as an average of points a, b and c, and there is no supporting hyperplane to the set which meets it. Indeed, the set of extreme points, by either definition, would seem to be the vertices of the polygon that one would obtain by drawing a string tightly around the points of $\mathscr{S}$; see Fig. 12.2(iii).

This polygon and its interior constitute a convex set, and seem to constitute the unique convex set which is generated from the elements of $\mathscr{S}$. It is called the *convex hull* of $\mathscr{S}$, denoted $[\mathscr{S}]$. We can give two alternative definitions of what we might mean by $[\mathscr{S}]$, analogous to the two definitions of the extreme points. They are both plausible, and not obviously equivalent. It is from the fact that these dual characterizations *are* in fact equivalent that much of the significant theory of convex sets stems.

Definitions 2. The convex hull $[\mathscr{S}]$ of a set $\mathscr{S}$ is the smallest convex set containing $\mathscr{S}$.

Definitions 2*. The convex hull $[\mathscr{S}]^*$ of a set $\mathscr{S}$ is the intersection of all half-spaces containing $\mathscr{S}$.

Definition 2, like Definition 1, emphasizes the 'averaging' aspect: $[\mathscr{S}]$ is the set of points obtained by averaging points of $\mathscr{S}$. (That is, averages of the form (2) and limits of them. But it is unclear what distributions on elements of $\mathscr{S}$ can be attained in this way, which is why Definition 2 is adopted as the economical characterization.)

Definition 2*, like Definition 1*, emphasizes the 'extremal' aspect. One consider the half-spaces

$$a^{\mathrm{T}}x \leq h(a), \tag{4}$$

where

$$h(a) = \sup_{x \in \mathscr{S}} (a^{\mathrm{T}}x), \tag{5}$$

and then defines $[\mathscr{S}]^*$ as the set of all points x which satisfy (4) for all a.

Theorem 12.2.2.

(i) *Definitions 1 and 1* of an extreme point are equivalent.*
(ii) *Definitions 2 and 2* of the convex hull are equivalent.*
(iii) *If $\mathscr{S}$ is a bounded set in $\mathbb{R}^n$ then any element x of $[\mathscr{S}]$ can be represented as an average*

$$x = \sum_{j=1}^{n+1} p_j x^{(j)} \tag{6}$$

of at most $n + 1$ extreme points $x^{(j)}$ of $\mathscr{S}$.

PROOF. Let us denote the sets of extreme points according to Definitions 1 and 1* by $\mathscr{A}$ and $\mathscr{A}^*$, respectively. Now, if an element $\bar{x}$ of $\mathscr{S}$ can be represented as strict average of other points $x^{(j)}$ of $\mathscr{S}$, then

$$a^{\mathrm{T}}\bar{x} = \sum_j p_j(a^{\mathrm{T}}x^{(j)}),$$

from which it is evident that, whatever a, if $a^{\mathrm{T}}x$ is maximal in $\mathscr{S}$ at $\bar{x}$ then it is also maximal at the values $x^{(j)}$. Thus $\bar{x}$ cannot be an extreme point in the sense of Definition 1*. Thus the complement of $\mathscr{A}$ lies in the complement of $\mathscr{A}^*$, so that

$$\mathscr{A}^* \subset \mathscr{A}. \tag{7}$$

Now, the set $[\mathscr{S}]^*$ certainly contains $\mathscr{S}$. It is also convex, since an average of x-values satisfying (4) also satisfies (4). Thus

$$[\mathscr{S}]^* \supset [\mathscr{S}]. \tag{8}$$

We have now to establish the inclusions reverse to (7) and (8). The elements of $\mathscr{A}^*$ are plainly boundary points (in the sense of Definition 1*) of the convex set $[\mathscr{S}]^*$. Let x be an element of $[\mathscr{S}]^*$; suppose also for the moment that it is an interior point. Choose a point $x^{(1)}$ of $\mathscr{A}^*$ and draw the line from $x^{(1)}$ to x, and continue it. It will meet the boundary of $[\mathscr{S}]^*$ in another point; x°, say. Then x is an average of $x^{(1)}$ and x°.

Now, by the construction of $[\mathscr{S}]^*$ there will be a supporting hyperplane H to $[\mathscr{S}]^*$ at x°, and the points in which H meets $[\mathscr{S}]^*$ will constitute a convex set in at most $n - 1$ dimensions, with elements of $\mathscr{A}^*$ as boundary points. We can thus continue the construction, and ultimately arrive at the representation (6). If the original x had not been an interior point then this would simply have meant that we would have started later in this dimension-reducing construction.

Assertion (iii), the validity of representation (6), is thus proved for any x in $[\mathscr{S}]^*$ and for some $x^{(j)}$ of $\mathscr{A}^*$. But the elements of $\mathscr{A}^*$ are certainly extreme points of $\mathscr{S}$, on either definition of an extreme point. We have thus proved the inclusion reverse to (8), and established assertion (ii).

Finally, relation (6) will also represent any element of $\mathscr{A}$ as an average of elements of $\mathscr{A}^*$. But since, by definition, the elements of $\mathscr{A}$ cannot be represented as a strict average of points of $\mathscr{S}$, then (6) must in fact identify each element of $\mathscr{A}$ with an element of $\mathscr{A}^*$. We thus have the inclusion reverse to (7), and assertion (i) is proved.   □

Since $[\mathscr{S}] = \mathscr{S}$ if $\mathscr{S}$ is convex, then Theorem 21.2.2 has immediate consequences for convex sets.

Corollary 12.2.3. *Let $\mathscr{S}$ be a bounded convex set in $\mathbb{R}^n$. Then:*

(i) *Any element x of $\mathscr{S}$ can be represented as an average of at most $n + 1$ extreme points of $\mathscr{S}$.*

(ii) *$\mathscr{S}$ possesses a supporting hyperplane at any point on its boundary.*

PROOF. Assertion (i) follows immediately from Theorem 12.2.2(iii). Assertion (ii) follows from the fact that $[\mathscr{S}]$, by its second definition, possesses a supporting hyperplane at all boundary points.   □

Assertion (i) of the corollary is Minkowski's Theorem. Assertion (ii), often known as the Supporting Hyperplane Theorem, is a result which is appealed to again and again. It is the property dual to the original characterization (1) of convexity.

In Fig. 12.3 we illustrate a nonconvex set $\mathscr{S}$ and its convex hull $[\mathscr{S}]$. The set $\mathscr{S}$ plainly has no supporting hyperplane at the boundary point a. In the convex hull a becomes an interior point, and the 'averages' of the contact points b and c give a new face which completes the boundary of $[\mathscr{S}]$.

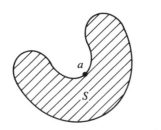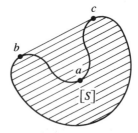

Figure 12.3. The completion of the set $\mathscr{S}$ of Fig. 12.1(iii) to its convex hull $[\mathscr{S}]$. The boundary point a of $\mathscr{S}$, which is not an extreme point and at which there is no supporting hyperplane, becomes an interior points of $[\mathscr{S}]$. The straight line segment joining the extreme points b and c becomes a face of $[\mathscr{S}]$.

EXERCISES AND COMMENTS

1. The proof of Theorem 12.2.2(iii) relied on an implicit assumption that $[\mathscr{S}]$ was bounded, so that we could assert the existence of the second intersection point $x°$ of the line with the boundary. Unbounded sets need special discussion. For instance, the set $\mathbb{R}^n$ is convex, but without boundary! Assertion (iii) of Theorem 12.2.2 (known as Minkowski's Theorem) will then certainly require modification in such unbounded cases. A weaker version which remains valid is Carathéodory's Theorem, asserting that any element of $[\mathscr{S}]$ can be represented as an average of at most $n + 1$ points of $\mathscr{S}$ (which need not then be extreme points).

3. The Consistency Condition for Expectation Values

We are now in a position to answer the questions raised in Section 1, and to characterize exactly the constraints on the vector $E(X)$ which are implicit in its definition. Note that we do not need to include the constant 1 in the vector X, to indicate that we know that $E(1) = 1$. This latter property can be taken for granted, as one of the axiomatic properties of the expectation functional.

Theorem 12.3.1. *Suppose $X(\omega)$ is a function from Ω to $\mathbb{R}^n$. Then:*

(i) *The set of possible values of $E(X)$ for varying expectation operators E is exactly $[X(\Omega)]$; the closed convex hull of the set $\{X(\omega); \omega \in \Omega\}$.*

(ii) *For any given E and $X(\cdot)$ the vector $E(X)$ can be represented as an average*

$$E(X) = \sum_{j=1}^{n+1} p_j X(\omega^{(j)}) \tag{9}$$

over at most $n + 1$ points of Ω.

PROOF. Let $\mathfrak{E}(X)$ denote the set of possible values of $E(X)$, for given $X(\cdot)$. If E_1 and E_2 are two possible expectation operators, so satisfying the axioms, then $pE_1 + (1 - p)E_2$ also satisfies the axioms. Thus the set of possible expectation operators is a convex set, and so then is $\mathfrak{E}(X)$. The set $\mathfrak{E}(X)$ also contains $X(\Omega)$, because it contains the points $X(\omega)$ for given ω in Ω, which correspond to one-point distribution. Thus

$$\mathfrak{E}(X) \supset [X(\Omega)]. \tag{10}$$

Now, if $\xi = E(X)$ and a is any real n-vector then

$$a^{\mathsf{T}}\xi = E(a^{\mathsf{T}}X) \le \sup_{\omega \in \Omega} (a^{\mathsf{T}}X(\omega)).$$

Thus ξ must lie in any half-space which contains $X(\Omega)$. But, by the second characterization of the convex hull, this implies the inclusion reverse to (10). We have thus established assertion (i). Assertion (ii) follows by appeal to Theorem 12.2.2(iii). $\square$

Assertion (i) characterizes exactly the values that $E(X)$ can take, and we shall derive the solution of the extension problem from it. Assertion (ii) is interesting in that it exhibits the expectation in the conventional form of an average with respect to a distribution. Any possible value of $E(X)$ is one that *could* have been generated by a distribution on at most $n + 1$ points of Ω. Of course, there may be infinitely many distributions $P(\cdot)$ consistent with the prescribed value of $E(X)$ in that

$$E(X) = \int X(\omega)P(d\omega), \tag{11}$$

most of them more sophisticated in form than the $(n + 1)$-point distribution yielding (9). It is a virtue of the expectation approach, however, that we do not have to face up to the possible subtleties of integrals such as (11), at least in the case of finite n.

If one considered an infinite number of expectations, perhaps even a non-denumerably infinite number, then matters would be less simple, although one would expect the assertions of Theorem 12.3.1 to survive in some sense.

4. The Extension of Expectation Values

Consider now what is in a sense the central problem: the placing of sharp bounds on $E(X)$ given that one knows the value of $E(Y)$. We shall, for simplicity, consider only the case of scalar X. Note also, as in the last section, that the relation $E(1) = 1$ is axiomatic, and need not be counted in the specification of $E(Y)$.

Let us denote by $\mathscr{Y}$ the set of scalar-valued linear functions of Y, i.e. the functions of the form

$$Z(\omega) = a^{\mathrm{T}}Y(\omega) + b \tag{12}$$

for vector a and scalar b. Let us indeed use Z to denote any such functions, i.e. any element of $\mathscr{Y}$.

Theorem 12.4.1. *Suppose that X is a random scalar and Y a random n-vector, and that the value of $E(Y)$ has been prescribed as an interior point of its feasible set $[Y(\Omega)]$. Then:*

(i) *The sharp bounds on $E(X)$ are*

$$\sup_{Z \geq X} E(X) \leq E(X) \leq \inf_{Z \geq X} E(Z), \tag{13}$$

where Z is a member of the class of linear functions (12) of Y.

(ii) *Each of the bounds (12) is attained for an extremal distribution concentrated on the values of ω for which $X(\omega) = Z(\omega)$ for the extremizing Z, and an extremal distribution exists which is concentrated on at most $n + 1$ points of Ω.*

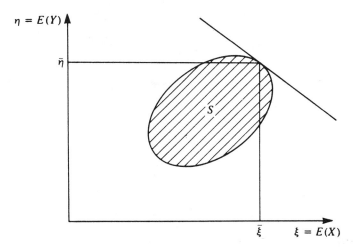

Figure 12.4. The set $\mathscr{S}$ is the set of possible values of $\xi = E(X)$ and $\eta = E(Y)$. Let $\bar{\xi}$ be the maximal value of $E(X)$ consistent with a prescribed value $\bar{\eta}$ of $E(Y)$. Then the supporting hyperplane to $\mathscr{S}$ at $(\bar{\xi}, \bar{\eta})$ defines the linear function of Y bounded below by X whose expectation equals $\bar{\xi}$.

PROOF. It is sufficient to consider the upper bound. That the upper bound of (12) is in fact valid is clear, since $Z \geq X$ implies that $E(Z) \geq E(X)$.

The set $\mathscr{S}$ of feasible values (ξ, η) of $(E(X), E(Y))$ is the convex hull $[X(\Omega), Y(\Omega)]$. We seek bounds on ξ for given η; say, $\bar{\eta}$. The point $(\bar{\xi}, \bar{\eta})$ at which ξ takes its maximal value consistent with $\eta = \bar{\eta}$ is a boundary point of $\mathscr{S}$. Since $\mathscr{S}$ is convex then a supporting hyperplane exists at this point; see Fig. 12.4. That is, there exist coefficients a, b and c such that

$$c\xi \leq a^{\mathsf{T}}\eta + b \qquad (14)$$

for (ξ, η) in $\mathscr{S}$, with equality at $(\bar{\xi}, \bar{\eta})$. Now, the coefficient c cannot be zero, otherwise $\bar{\eta}$ would yield equality in $0 \leq a^{\mathsf{T}}\eta + b$, and this would imply that $\bar{\eta}$ was a boundary point of $[Y(\Omega)]$, against hypothesis. The fact that equality holds in (14) at $(\bar{\xi}, \bar{\eta})$ and strict inequality for $(\xi, \bar{\eta})$ with $\xi < \bar{\xi}$ implies that c is positive. We can thus normalize c to unity, when (14) becomes

$$\xi \leq a^{\mathsf{T}}\eta + b \qquad (15)$$

If we define Z as in (12) with this choice of a and b, then the fact that (15) holds for all (ξ, η) in $\mathscr{S}$ with equality at $(\bar{\xi}, \bar{\eta})$ implies that $X \leq Z$, but that the maximal value $\bar{\xi}$ of $E(X)$ equals $a^{\mathsf{T}}\bar{\eta} + b = a^{\mathsf{T}}E(Y) + b = E(Z)$. The upper bound of (13) is thus attainable, and consequently sharp.

The distribution for which the bound is attained must be confined to ω-values for which $X(\omega) = Z(\omega)$, if equality of $E(X)$ and $E(Z)$ is to be compatible with $X \leq Z$. The remainder of assertion (ii) follows from the fact that $(\bar{\xi}, \bar{\eta})$ is a boundary point of the $(n + 1)$-dimensional convex set $\mathscr{S}$. It thus lies in a convex set of dimension n at most (a face of $\mathscr{S}$, of some dimension) and so is representable as an average of at most $n + 1$ points of the set. That is, as an average over at most $n + 1$ points of Ω.   □

To solve the extremal problem of determining a form (12), whose expectation is minimal consistent with $Z(\omega) \geq X(\omega)$ for all ω in Ω, is not necessarily an easy matter, and may defeat all ingenuity in some cases. However, there is at least a simple test of whether one has succeeded.

Theorem 12.4.2. *Suppose one has found a form $Z = a^{\mathrm{T}}Y + b$ for which $X \leq Z$ and a distribution over the values of ω for which $X(\omega) = Z(\omega)$ that gives $E(Y)$ its prescribed value. Then $a^{\mathrm{T}}E(Y) + b$ is the sharp upper bound to $E(X)$.*

The assertion follows simply from the fact that $E(Z)$ is an upper bound to $E(X)$ which is attained under the particular distribution.

EXERCISES AND COMMENTS

1. For those familiar with linear programming: the problem of finding the form Z for which $E(Z)$ is minimal subject to $Z \geq X$ is the dual of the problem of finding a probability measure P which maximizes $\int X(\omega)P(d\omega)$ subject to prescription of $\int Y(\omega)P(d\omega)$.

2. The case when $E(Y)$ is a boundary point of its permitted set may need special discussion. For example, in the graph of Fig. 12.4, if $E(Y)$ were given its maximal possible value, then the supporting hyperplane we have drawn would be vertical, which means that $c = 0$ in (14). We cannot then make the required deduction (15). In this particular case there is only one value that $E(X)$ can take, consistent with the prescribed $E(Y)$.

5. Examples of Extension

When it comes to actual cases one tends to consider a r.v. X and to pose the problem as one of setting bounds on an expectation $E[h(X)]$ given that one knows expectations $E[g_j(X)]$ $(j = 1, 2, \ldots, n)$. That is, one replaces ω, $X(\omega)$ and $Y_j(\omega)$ by X, $h(X)$ and $g_j(X)$, respectively, and the set Ω by the set $\mathscr{X}$ of possible x-values.

   The Markov inequality of Section 2.9 is the simplest case, with $g(x) = x$, $h(x) = I(x - a)$ and $\mathscr{X}$ the positive half-axis. It is indeed simple, but we shall give the full-dress treatment, to set the pattern for more demanding cases.

Theorem 12.5.1. *The Markov inequality in the form*

$$P(X \geq a) \leq \min(1, E(X)/a) \qquad (16)$$

is sharp as a bound in terms of $E(X)$. If $E(X) \leq a$ then the unique distribution *giving equality is that with probability mass $E(X)/a$ at a and the rest at 0. If $E(X) > a$ then there are infinitely many extremal distributions; one is that which places all mass at $x = E(X)$.*

PROOF. We require to find coefficients a, b such that

$$I(x - a) \le ax + b, \tag{17}$$

and such that the margin of inequality is as small as possible in some sense. The two cases one might consider tightest are: $I(x - a) \le x/a$ (when equality holds at $x = 0$, a) and $I(x - a) \le 1$ (when equality holds on $x \ge a$). Taking expectations in these two inequalities we see that the two bounds expressed in (16) hold separately, whence the combined bound holds. We verify also that the bounds are attained by the distributions indicated. ☐

We quote the obvious analogue of these assertions for the Chebyshev inequality, and leave verification by the same methods to the reader.

Theorem 12.5.2. *The Chebyshev inequality in the form*

$$P(|X - E(X)| \ge a) \le \min(1, \text{var}(X)/a^2)$$

is sharp as a bound in terms of $E(X)$ and $E(X^2)$. If $\text{var}(X) \le a^2$ then the unique distribution giving equality is that with probability mass $\text{var}(X)/(2a^2)$ at the two points $E(X) \pm a$ and the rest at $E(X)$. If $\text{var}(X) > a^2$ then there are infinitely many extremal distributions; one is that which places mass $\frac{1}{2}$ at each of the two points $E(X) \pm \sqrt{\text{var}(X)}$.

These two example are very straightforward, although they do illustrate how the analytic form of the sharp bound can change as the values of the prescribed expectations change. Some of the exercises are more demanding. However, for the real tour de force in this direction one must return to Chebyshev, who obtained sharp bounds on the distribution function $F(x)$ of X in terms of the first $2r$ moments: $E(X^j)$ for $j = 1, 2, \ldots, 2r$. The aim was to establish the central limit theorem; something which we now achieve by the c.f. methods of Section 7.3. However, the analysis is dazzling in its ingenuity and in the elegance of its conclusions. To give the analysis would take us too far; we refer the interested reader to Whittle (1971), pp. 110–118. However, if we quote the results then the reader may realize that there are depths which could not have been suspected from the simple examples above.

There are two cases. Suppose that a distribution on r or fewer points can be found consistent with the given moments; let this have distribution function $G(x)$. Then the distribution is essentially unique and determined, in that the sharp bounds on $F(a)$ are $G(a-) \le F(a) \le G(a+)$. That is, the only indeterminacy in the distribution is that corresponding to an infinitesimal perturbation of the probability mass, which will not affect the expectation of any continuous function.

Suppose that one is not in the above case. Then there is a unique distribution giving mass to a and to r other points. If this has distribution function $G_a(x)$ then the sharp bounds on $F(a)$ are $G_a(a-) \leq F(a) \leq G_a(a+)$. Another way to express this is to say that the distribution function $F(x)$ interlaces $G_a(x)$, because one will have the same distribution function $G_a(x)$ if one takes a equal to one of the other points of increase of $G_a(x)$.

EXERCISES AND COMMENTS

1. Suppose the r.v. X is known to lie in the interval (a, b) and to have expectation μ. Show that

$$P(X \geq a) \geq \max\left[0, \frac{\mu - a}{b - a}\right]$$

and that this inequality is sharp.

2. Suppose that the values of $E(X)$ and $E(X)^2$ are known. Show that

$$P(X \geq a) \leq \frac{\text{var}(X)}{\text{var}(X) + [a - E(X)]^2},$$

for $a \geq E(X)$, and that the inequality is sharp in this range.

3. Consider the profit function $g_N(X)$ defined by the first displayed formula of Section 2.6. Suppose all one knows of the distribution of demand X is that $E(X) = \mu$ and $\text{var}(X) = \sigma^2$. Show then that a sharp lower bound for $G_N = E[g_N(X)]$ is

$$G_N \geq a\mu + \tfrac{1}{2}(a - b + c)(N - \mu) - \tfrac{1}{2}(a + b + c)\sqrt{(N - \mu)^2 + \sigma^2}$$

or

$$G_N \geq -c\mu + \frac{(a + c)\mu^2 - b\sigma^2}{\mu^2 + \sigma^2} N$$

according as $E(X^2)$ is less than or greater than $2NE(X)$.

   If one had to optimize stocks on the basis of this limited information it would be reasonable to choose the value of N which maximizes this lower bound. Show that this is $N = \mu + [(a - b + c)\sigma]/2\sqrt{b(a + c)}$ if $\sigma/\mu < \sqrt{(a + c)/b}$ and zero otherwise, with corresponding values $a\mu - \sigma\sqrt{b(a + c)}$ and $-c\mu$ for the bound. The discontinuous behaviour of N is interesting, when compared with that for the case of full information.

4. The example of the Markov inequality shows that, strictly, a distinction should be made between a maximum and a supremum. Consider the case $\mu = E(X) < a$, when the bound (16) is μ/a. This constitutes a maximum for $P(X \geq a)$ (attained for a distribution concentrated on 0 and a) but merely a supremum for $P(X > a)$ (approached by a distribution concentrated on 0 and $a + \varepsilon$ as $\varepsilon \downarrow 0$). The effect is due to lack of continuity of $h(x) = I(x - a)$.

5. Suppose that (X_1, X_2) takes values on the square $(0 \leq x_1, x_2 \leq M)$ and that $E(X_1)$ and $E(X_2)$ are known. Show that $E(X_1 X_2) \leq M \min[E(X_1), E(X_2)]$, and that this inequality is sharp. [One looks for a relation $x_1 x_2 \leq a_1 x_1 + a_2 x_2 + b$ valid on the square. If $E(X_1) = E(X_2)$ then presumably $a_1 = a_2$ and one finds an extremal distribution concentrated on $(0, 0)$ and (M, M). If $E(X_1) < E(X_2)$ then consider

$x_1 x_2 \le M x_1$, with an extremal distribution concentrated on the two sides of the square: $x_1 = 0$ and $x_2 = M$. In fact, consider a distribution on $(0, 0)$, $(0, c)$ and (M, M) for appropriate c.]

6. Suppose that r.v.s X_i have individual distribution functions $F_i(x)$ $(i = 1, 2)$ and that these are continuous. Show that the sharp upper bound for the joint distribution function $F(x_1, x_2)$ is

$$F(a_1, a_2) \le \min[F_1(a_1), F_2(a_2)].$$

[Use the probability integral transformation to achieve the reduction $F_i(x_i) = x_i$. Then consider an extremizing distribution which is constant on each of the four rectangles into which the unit square is divided by the lines $x_i = a_i$ $(i = 1, 2)$.]

6. Dependence Information: Chernoff Bounds

If one knows that certain r.v.s are independent or are identically distributed then this is information of quite another character than the specification of a few expectations. The extremal problem is then no longer linear (or, if one likes, it is subject to infinitely many linear constraints of the type of (5.18) or of $E[H(X_1)] = E[H(X_2)]$ for all H). Great ingenuity was spent earlier in this century on the evaluation of close bounds for the distribution of a sum of IID r.v.s (prescribed mean and nonnegative, or prescribed mean and variance). The most natural type of result in the case when the whole distribution of the summand is specified now seems to be the Chernoff bound, which is not sharp, but which is asymptotically sharp in a sense to be explained.

Suppose that $S_n = \sum_{j=1}^n X_j$ is a sum of IID r.v.s X_j whose moment generating function (m.g.f.)

$$M(\theta) = E(e^{\theta X})$$

exists in some real interval with the origin as an interior point. Then an application of Markov's inequality and an optimization of the parameter θ gives the bound

$$P(S_n \ge na) \le \inf_{\theta \ge 0} M(\theta)^n e^{-n\theta a}. \tag{18}$$

We could write this as

$$P(S_n \ge na) \le e^{-nR(a)}, \tag{19}$$

where

$$R(a) = \sup_{\theta \ge 0} [\theta a - \log M(\theta)] \tag{20}$$

is known as the *rate function*. Bound (18) is the *Chernoff bound*. The rewriting of it in the form (19) emphasizes its exponential dependence on n and its place in a much wider theory; the theory of large deviations.

Although derived by simple means, the Chernoff bound is surprisingly good. It is asymptotically sharp in that one can refine (19) to

$$P(S_n \ge na) = e^{-nR(a) + o(n)}, \tag{21}$$

an assertion close to *Cramér's Theorem*, whose proof we sketch in the exercises. It is so useful that we have in fact already appealed to it three times: in Exercises 2.9.5 and 2.9.6, in Exercise 5.7.3 and in Exercise 7.5.10.

The natural character of the bound also reveals itself in the close relationship it demonstrates to the X distribution itself. We give the following examples for the reader to verify, assuming in every case that $a \geq E(X)$.

Distribution	Moment generating function	Chernoff bound
Standard normal	$e^{\theta^2/2}$	$e^{-na^2/2}$
Poisson	$\exp[\lambda(e^\theta - 1)]$	$[e^{a-\lambda}(\lambda/a)^a]^n$
0/1	$pe^\theta + q$	$\left[\dfrac{p}{a}\right]^{na}\left[\dfrac{q}{1-a}\right]^{n(1-a)}$

The distribution of S_n in the last case is, of course, binomial. The bound for the normal case is normal in form; the bounds for the Poisson and binomial cases are again close to the known expressions for $P(S_n = na)$, with Stirling approximations for the factorials.

EXERCISES AND COMMENTS

1. Consider the 'tilted' X distribution whose expectation operator $\tilde{E}$ is defined by

$$\tilde{E}[H(X)] = \frac{E[H(X)e^{\theta X}]}{M(\theta)}.$$

Suppose that θ is given the value which achieves the extreme in (20). Then show that $\tilde{E}(X) = a$.

2. Under the tilted distribution S_n/n will converge in distribution to a. That is, $\tilde{E}[H(S_n/n)] \to H(a)$ as $n \to \infty$, for any bounded continuous H. Use this result to demonstrate (21).

CHAPTER 13

Stochastic Convergence

1. The Characterization of Convergence

Probability theory is founded on an empirical limit concept, and its most characteristic conclusions take the form of limit theorems. Thus, a sequence of r.v.s $\{X_n\}$ which one suspects has some kind of limit property for large n is a familiar object. For example, the convergence of the sample average $\overline{X}_n$ to a common expected value $E(X)$ (in mean square, Exercise 2.8.6; in probability, Exercise 2.9.14 or in distribution, Section 7.3) has been a recurrent theme. Other unforced examples are provided by the convergence of estimates or conditional expectations with increasing size of the observation set upon which they are based (Chapter 11), and the convergence of the standardize sum u_n to normality (Section 7.4). Any infinite sum of r.v.s which we encounter should be construed as a limit, in some sense, of a finite sum. Consider, for instance, the sum $\sum_{t=0}^{\infty} R_t z^t$ of Section 6.1, or the formal solution $X_t = \sum_{s=0}^{\infty} \alpha^s \varepsilon_{t-s}$ of the stochastic difference equation $X_t = \alpha X_{t-1} + \varepsilon_t$.

The question of convergence of a sequence of r.v.s $\{X_n\}$ is rather less straightforward than that of a sequence of constants $\{a_n\}$. One understands quite clearly what is meant by convergence in his latter case: that there exists a constant a such that $a_n - a \to 0$, or, more precisely, that for any positive ε there exists a finite number $n(\varepsilon)$ such that $|a_n - a| \le \varepsilon$ for all n greater than $n(\varepsilon)$. A necessary and sufficient condition for such convergence is that the sequence should be *mutually convergent*. That is, that $|a_m - a_n| \to 0$ as m and n tend to infinity independently. The usefulness of this second criterion is that it does not suppose knowledge of the limit value a. A sequence having the mutual convergence property is sometimes called a *Cauchy sequence*.

However, when it comes to a sequence of r.v.s $\{X_n\}$ (which is really a sequence of functions $\{X_n(\omega)\}$ on a general space Ω) to a limit X then one can conceive of many types of convergence. One might first make the requirement of *pointwise convergence*: that $X_n(\omega)$ should converge to $X(\omega)$ for any given ω as n increases. That is, $X_n \to X$ for all realizations. However, convergence in such a strong sense is almost completely uncharacteristic of probability models. For example, the proportion of heads in n tosses of a fair coin does not converge to $\frac{1}{2}$ for *all* sequences, although the infinite sequences for which it does not converge have zero probability (the strong law of large numbers; Section 14.2).

So, already it is natural to introduce a weaker concept of convergence: that $X_n(\omega)$ should converge to $X(\omega)$ for all ω except a set of zero probability (*almost sure convergence*). However, in some physical situations, even this is too strong a concept, for there are cases where one can establish only the convergence of certain expectations rather than of the sequence of r.v.s itself.

We might thus weaken requirements to a demand that

$$E[H(X_n)] \to E(H(X)) \tag{1}$$

for all function H of some suitable class. This is exactly the course already followed in Chapter 7 to define the valuable concept of *convergence in distribution* or *weak convergence*, corresponding to the requirement that (1) should hold for bounded continuous H.

However, convergence in this sense merely implies that the distribution of X_n approximates, at least partially, to that of X, and not that X_n itself approximates to X. To ensure virtual identity of 'lim X_n' and X one would have to require something like

$$E[H(X_n - X)] \to 0 \tag{2}$$

for suitable H. A generalization of this would be to require that

$$E[H(X_n, Y)] \to E[H(X, Y)],$$

where Y is another r.v. This would would ensure that 'lim X_n' bore the same relationship to the reference r.v. Y as did X itself.

This does not exhaust the possibilities; one might make convergence depend on the whole *tail* of the sequence:

$$E[H(X_n, X_{n+1}, \ldots ; Y)] \to E[H(X, X, \ldots ; Y)] \tag{3}$$

for suitable H. Almost sure convergence constitutes just such a demand.

There are obviously many possibilities, and by playing with them one can construct an extensive and somewhat sterile theory. In a given physical situation one must always be guided by operational considerations: What kind of convergence is it natural to demand?

In this chapter we shall briefly consider some of the formal implications. In the next we shall study the class of sequences which more than any other can be said to occur naturally and converge naturally: the martingales.

1. Determine the functions H which is are invoked for mean square convergence, convergence in probability and almost sure convergence.

2. Types of Convergence

Let us list some of the types of convergence in standard use. We have already mentioned the concept of weak convergence, or convergence in distribution, which requires that (1) should hold for bounded continuous H. One writes the convergence as $X_n \xrightarrow{D} X$, although, as we have already emphasized, it is the distribution rather than the variable itself which converges.

Suppose we demand relation (2) with $H(u) = |u|^r$, where r is a prescribed positive index. We are thus requiring that

$$E[|X_n - X|^r] \to 0. \tag{4}$$

This is *convergence in rth mean*, or L_r-*convergence*, written $X_n \xrightarrow{r} X$. The particular case $r = 2$ is that we already know as *mean square convergence*, written $X_n \xrightarrow{m.s.} X$.

Suppose we require (2) to hold for H the indicator function of the complement of the interval $(-\varepsilon, \varepsilon)$, for arbitrarily small positive ε. That is, we require

$$P(|X_n - X| > \varepsilon) \to 0 \tag{5}$$

for any fixed positive ε. Then X_n is said to *converge in probability* to X, written $X_n \xrightarrow{P} X$.

Finally, we obtain a special type of the tail-convergence (3) by choosing H so that

$$H(X_n, X_{n+1}, \ldots) = \begin{cases} 1 & \text{if } |X_m - X| \le \varepsilon \text{ for } m \ge n, \\ 0 & \text{otherwise,} \end{cases}$$

and so requiring that

$$P(|X_m - X| \le \varepsilon; m \ge n) \to 1, \tag{6}$$

for arbitrarily small positive ε. This is *almost sure convergence*, written $X_n \xrightarrow{a.s.} X$. It is identical with the almost sure convergence mentioned in the previous section, because it implies that X_n converges to X in the conventional sense with probability one. We shall often speak of it simply as a.s. convergence.

The types of convergence (5) and (6) are sometimes referred to as *weak* and *strong convergence in probability*, respectively. This is quite a good nomenclature, but is perhaps best avoided, because of possible confusion with the concept of weak convergence (i.e. convergence in distribution).

In general, there are a number of implications between the various criteria, summarized by the arrows of the following diagram:

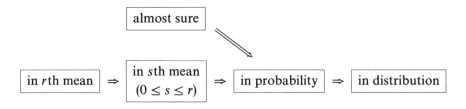

We leave verification of these implications to the reader; some hints are given in the exercises. It is a rather less direct matter to prove that there is in fact no implication where none is indicated on the diagram; appropriate counter-examples are also sketched in the exercises.

Any mode of convergence has a *mutual* analogue. That is, if we write convergence to X in sense S as $X_n \overset{S}{\to} X$, then mutual convergence in the same sense requires that $X_m - X_n \overset{S}{\to} 0$ as m and n go to infinity independently. For every mode listed above, except convergence in distribution, convergence (to some X) and mutual convergence are equivalent, as we shall show where this is not evident.

EXERCISES AND COMMENTS

1. Use the fact that $[E(|X|^r)]^{1/r}$ is an increasing of r (Exercise 2.9.12) to establish the first implication in the bottom row of the diagram.

2. Use Markov's inequality to establish the second.

3. Neither convergence in probability nor in rth mean imply a.s. convergence. Consider a sequence of independent r.v.s $\{X_n\}$ for which X_n is 1 or 0 with respective probabilities $1/n$ and $1 - 1/n$.

4. Convergence in probability does not imply convergence in rth mean. Consider an independent sequence for which X_n equals $n^{2/r}$ or zero with respective probabilities $1/n$ and $1 - 1/n$.

5. Almost sure convergence does not imply convergence in rth mean. Consider $X_n = a_n Y$, where $\{a_n\}$ is a convergent sequence of constants and $E(|Y|^r)$ is infinite.

6. Not even pointwise convergence implies corresponding convergence of expectations. Consider a sample space $\omega \geq 0$ with $X_n = n^2 \omega e^{-n\omega}$ and ω exponentially distributed. Then $X_n \to 0$ pointwise, but $E(X_n) \to 1$.

7. Consider $u_n = n^{-1/2} \sum_{j=1}^{n} X_j$, where the X_j are IID standardized variables. Then the sequence $\{u_n\}$ converges in distribution to normality, as we known from Section 7.4. Show, however, that the sequence is not mean square mutually convergent, and so not mean square convergent.

3. Some Consequences

In this section we shall consider some simple conditions which ensure various types of convergence, or which permit one to make some of the reverse implications not included in the diagram of Section 2 (being not generally valid).

Let us first follow up the helpful fact that, since the event 'mutual convergence' (in the deterministic sense) is equivalent to the event 'convergence', then a.s. mutual convergence is equivalent to a.s. convergence.

Theorem 13.3.1. *If* $\sum_n P(|X_{n+1} - X_n| \geq \varepsilon_n) \leq \infty$*, where* $\sum_n \varepsilon_n$ *is a convergent sum of positive terms, then* $\{X_n\}$ *is a.s. convergent.*

PROOF. Define $\delta_n = \sum_{j=n}^{\infty} \varepsilon_j$. Then

$$P(|X_j - X_k| < \delta_n; k \geq n) \geq P(|X_{j+1} - X_j| < \varepsilon_j; j \geq n)$$

$$\geq 1 - \sum_{j=n}^{\infty} P(|X_{j+1} - X_j| \geq \varepsilon_j) \to 1,$$

the second step following by Boole's inequality. The sequence $\{X_n\}$ is thus a.s. mutually convergent, and so a.s. convergent. □

Theorem 13.3.2. *If* $\{X_n\}$ *is mutually convergent in rth mean then it contains a subsequence* $\{X_n'\}$ *which is a.s. convergent.*

PROOF. Since $E(|X_m - X_n|^r) \to 0$ we can extract a subsequence $\{X_n'\}$ for which

$$E(|X_{n+1}' - X_n'|^r) \leq \eta_n,$$

where η_n tends to zero sufficiently rapidly with increasing n that $\sum_n (\eta_n/\varepsilon_n^r) < \infty$. Here $\{\varepsilon_n\}$ is the sequence of the previous theorem. An application of Markov's inequality then yields

$$\sum_n P(|X_{n+1}' - X_n'| \geq \varepsilon_n) \leq \sum_n (\eta_n/\varepsilon_n^r) < \infty,$$

so that $\{X_n'\}$ is a.s. convergent, by the previous theorem. □

For example, consider the partial sum $X_n = \sum_{t=0}^{n} R_t z^t$ where R_t is the number of renewals at time t, introduced in Section 6.1. We have

$$E(|X_{n+1} - X_n|) = |z|^{n+1} E(R_{n+1}) \leq \frac{|z|^{n+1}}{1 - p_0},$$

(see Exercise 6.1.2). Thus, if $|z| < 1$ we have

$$\sum_n P(|X_{n+1} - X_n| \geq |z|^{n/2}) \leq (1 - p_0)^{-1} \sum_n |z|^{n/2} < \infty,$$

so that $\{X_n\}$ is a.s. convergent, as is then the infinite sum $\sum_t R_t z^t$.

A final result, which we give for interest, is essentially a probabilistic version of the dominated convergence theorem (Exercise 15.5.4), but framed under much weaker assumptions.

Theorem 13.3.3. *Suppose that* $X_n \xrightarrow{P} X$ *and* $Y \leq X_n \leq Z$ *for all n, where* $E(|Y|)$ *and* $E(|Z|)$ *are both finite. Then* $X_n \xrightarrow{1} X$ *and* $E(X) = \lim E(X_n)$.

PROOF. We have $0 \leq E(Z - Y) < \infty$. Let A_n denote the event $|X - X_n| > \varepsilon$, so that $P(A_n) \to 0$. We have

$$E(|X - X_n|) = P(\bar{A}_n)E(|X - X_n|\,|\,\bar{A}_n) + P(A_n)E(|X - X_n|\,|\,A_n)$$

$$\leq \varepsilon + P(A_n)E(|Y - Z|) \to \varepsilon.$$

Since ε is arbitrarily small, the first result is proved. The second follows from $|E(X) - E(X_n)| \leq E(|X - X_n|)$. $\square$

EXERCISES AND COMMENTS

1. Show that a sequence convergent in probability is mutually convergent in probability and contains an a.s. convergent subsequence.

2. Suppose that $\sum_n E(X_n^2) < \infty$. Show that $X_n \overset{\text{a.s.}}{=} 0$.

3. Construct a version of Theorem 13.3.3 which establishes that $X_n \xrightarrow{r} X$.

4. Convergence in rth Mean

The notion of convergence in rth mean is very useful one. The case $r = 2$ is a particularly natural one to work with, as we have seen right from Section 2.8 and more explicitly from Chapter 11. The case $r = 1$ also presents itself readily; e.g. for the extension problem of Chapter 15.

First, a few definitions. We shall say that X belongs to L_r if $E(|X|^r) < \infty$. Thus, the extent of L_r depends upon the particular process (i.e. the sample space and the distribution of realizations on it) that one is considering. It is often convenient to deal with the *norm* of X

$$\|X\| = [E(X|^r)]^{1/r}.$$

This depends upon r, but if we work with a fixed r there is rarely any need to indicate the dependence.

Two r.v.s X and X', for which $E(|X - X'|^r) = 0$, are said to be L_r-*equivalent*. There are two useful inequalities. One is the c_r *inequality*

$$|X + Y|^r \leq c_r[|X|^r + |Y|^r], \tag{7}$$

where $c_r = 1$ for $0 \leq r \leq 1$ and $c_r = 2^{r-1}$ for $r \geq 1$. The other is the *Minkowski inequality*

$$\left\| \sum_{j=1}^{n} X_j \right\| \leq \sum_{j=1}^{n} \|X_j\| \qquad (r \geq 1), \tag{8}$$

(see Exercise 1).

Theorem 13.4.1. *If* $X_n \overset{r}{\to} X$ *then* X *belongs to* L_r *and* $E(|X|^r) = \lim E(|X_n|^r)$.

PROOF. If $r \leq 1$ then, by the c_r inequality,

$$|E(|X_n|^r) - E(|X|^r)| \leq E(|X_n - X|^r) \to 0,$$

and, if $r > 1$, then, by the Minkowski inequality,

$$|\|X_n\| - \|X\|| \leq \|X_n - X\| \to 0. \qquad \square$$

However, the key theorem is the following.

Theorem 13.4.2. $\{X_n\}$ *is* L_r-*convergent for a given* $r > 0$ *if and only if it is* L_r-*mutually convergent.*

PROOF. The second assertion follows directly from the c_r inequality:

$$E(|X_m - X_n|^r) \leq c_r E(|X_m - X|^r) + c_r E(|X_n - X|^r).$$

It is the direct assertion which is rather more tricky to prove, but extremely useful: that mutual convergence implies convergence.

We have already proved the direct result for the case $r = 2$ in Theorem 11.6.1. This proof rested on the validity of the triangular inequality (11.24). However, between them the c_r inequality and Minkowski's inequality provide sufficient equivalents of the triangular inequality for all $r \geq 0$. The forms in which they enable a direct appeal to the earlier proof are

$$\|X - Y\| \leq \|X\| + \|Y\| \qquad (r \geq 1)$$

and

$$\|X - Y\|^r \leq \|X\|^r + \|Y\|^r \qquad (0 < r \leq 1).$$

The proof of Theorem 11.6.1 now adapts directly to the general case $r > 0$.

$\square$

EXERCISES AND COMMENTS

1. *Proof of Minkowski's inequality.* Show by direct minimization of the first member with respect to X that

$$|X|^r - r\theta|XY| \geq (1 - r)|\theta Y|^s,$$

where $r > 1$ and $r^{-1} + s^{-1} = 1$. Taking expectations over X and Y and minimizing the upper bound for $E(|XY|)$ thus obtained with respect to θ, show that

$$E(|XY|) \leq [E(|X|^r)]^{1/r}[E(|Y|^s)]^{1/s}.$$

This is Hölder's inequality, a generalization of the Cauchy inequality. Show that equality is attainable.

Alternatively, this result can be written

$$[E(|X|^r)]^{1/r} = \max_{Y} E(|XY|),$$

where the r.v. Y is restricted by $E(|Y|^s) = 1$. Show that Minkowski's inequality follows from this identity and the relation

$$\max_{Y} E\left(\left|\sum_{j} X_j Y\right|\right) \le \sum_{j} \max_{Y} E(|X_j Y|).$$

CHAPTER 14
_____

Martingales

1. The Martingale Property

There is a particular structure whose seemingly weak defining property is so significant that the structure pervades the theory of probability and leads to powerful conclusions. This is the *martingale structure*, already encountered in Sections 9.7, 10.7 and 11.7.

Let us assume for the moment that all sequences have the subscript set $n = 1, 2, 3, \ldots$, although occasionally we also attach significance to $n = 0$. A sequence of r.v.s $\{X_n\}$ is said to be a *martingale* with respect to a sequence of r.v.s $\{Y_n\}$ if

$$E(X_{n+1} | Y_1, Y_2, \ldots, Y_n) = X_n. \tag{1}$$

This implies that X_n must itself be a function of $Y_1, Y_2, \ldots, Y_n$.

The significance of the property (1) is that any conditional expectation

$$X_n = E(Z | Y_1, Y_2, \ldots, Y_n) \tag{2}$$

is a martingale; relation (1) in this case expresses just the iterative property (5.17) of conditional expectations. Moreover, it just seems to be the case that a sequence $\{X_n\}$ generated by the mechanism (2) is a powerful object, naturally occurring and naturally convergent. Indeed, we began our whole discussion of conditioning in Chapter 5 with the picture of a doctor who carried out a sequence of observations on a patient, accumulating ever more information and converging on an ever-sharpening diagnosis for the particular patient. The patient and observation error constitute the realization ω, the observations constitute the $\{Y_n\}$ sequence, and the conditional expectations of interest are $E(I_j | Y_1, Y_2, \ldots, Y_n)$, where I_j is the indicator function of the event 'the patient suffers from condition j'. These conditional probabilities (r.v.s, in that they reflect the randomness of both patient and observation error) will converge with increasing n, and the doctor will make his diagnosis on the basis of them.

We have two formal aims. One is to establish that a sequence (2) generated as a conditional expectation on an increasing observation set does indeed converge, in some stochastic sense. The other is to demonstrate that a martingale, defined by (1), has the properties that (a) it could have been generated by a relation (2), and (b) it also converges. Indeed, properties (a) and (b) are almost equivalent, in that (1) implies that

$$X_n = E(X_{m+n}|Y_1, Y_2, \ldots, Y_n) \qquad (m \geq 0) \tag{3}$$

so that if $\{X_n\}$ had a limit X_∞, then (2) would presumably hold with $Z = X_\infty$.

In Section 11.7 we proved just these conjectures, for convergence in the mean square sense and under the assumption of uniformly bounded mean squares. We shall see in later sections that we can also assert almost sure (a.s.) convergence.

The martingale concept has its origin in gaming; and relation (1) can be seen as representing the notion of a fair game. Suppose that Y_n represents the run of play in the nth round, and X_n the size of Alex's capital after that round. He may have a system, so that his mode of play (size of stake, etc.) in the next round, the $(n + 1)$th, depends upon the run of play hitherto. Relation (1) then expresses the fact that, whatever he does, his expected net winnings in that round are zero. Of course, his system might lead him to play for a random time, in that it tells him to withdraw from the game first when one of a target set of outcomes has been attained. The optional stopping theorem of Section 4 states that he can draw no advantage from that either.

There are various specializations. One can very well have X and Y identical, in which case (1) reduces to

$$E(X_{n+1}|X_1, X_2, \ldots, X_n) = X_n, \tag{4}$$

expressing a property of the sequence when conditioned by its own past.

For another, suppose that $\{X_n\}$ is generated by (2), that $\{Y_n\}$ is time-homogeneous Markov and that Z is a r.v. defined on the 'distant future' in that it is a function of $\{Y_m; m \geq n\}$ for any n. Then, by the Markov property and time-homogeneity, (2) reduces to

$$X_n = E(Z|Y_n) = \psi(Y_n),$$

say. $\{X_n\}$ is certainly a martingale, and the martingale condition (1) reduces to

$$E[\psi(Y_{n+1})|Y_n] = \psi(Y_n).$$

That is, ψ is an invariant under the action of the transition operator P of the Y process:

$$P\psi = \psi. \tag{5}$$

The continuous-time analogue requires that $\Lambda\psi = 0$, as we have seen in Section 10.7.

As far as examples go, there are processes which are martingales by nature, processes which have martingale character thrust on them by construction, and processes which just seem to happen to be martingales.

For processes which are martingale by construction: any process produced by the conditional expectation specification (2) is automatically a martingale, and can be regarded as a 'solution' of the martingale equation (1). For an example, suppose that $\{Y_n\}$ is a Markov process with an absorbing set A, and let X_n be the probability of absorption in the set A by time m conditional on the history of the process at time n. Then

$$X_n = E[I(X_m \in A)| Y_n, Y_{n-1}, \ldots] = E[I(X_m \in A)| Y_n] \qquad (n \le m), \qquad (6)$$

the second equality following from the Markov character of $\{Y_n\}$. The definition (6) of $\{X_n\}$ implies that it is a martingale; see Exercise 4 for more detailed discussion.

Another way of manufacturing a martingale is just to look for solutions ψ of (5). For example, we know from Section 9.7 that if $\{Y_n\}$ is a random walk on the integers whose increment has p.g.f. $\Pi(z)$ then $X_n = z^{Y_n}$ is a martingale if z satisfies $\Pi(z) = 1$.

The 'fair game' we discussed serves as an example of a process which is a martingale by nature. Whether it has a representation (2) and whether it converges are dependent on the nature of the stopping rule: the rules which specify when and how the game will finish. The material of Section 4 is relevant.

For a process (in continuous time), which just seems to happen to be a martingale, we could note the epidemic example of Section 10.7. This could be written

$$X(t) = \binom{Y_1(t)}{k} \left[\frac{b}{ak + b} \right]^{Y_1(t) + Y_2(t)},$$

where $Y_1(t)$ and $Y_2(t)$ are, respectively, the number of susceptibles and the number of infecteds at time t. For another apparently chance example, consider the *likelihood ratio*

$$X_n = \prod_{j=1}^{n} [f_2(Y_j)/f_1(Y_j)], \qquad (7)$$

where, under hypothesis i ($i = 1, 2$), the r.v.s Y_j are IID with probability density $f_i(y)$ relative to a measure μ. This is a martingale under hypothesis 1 because

$$E[X_n| Y_1, Y_2, \ldots, Y_n] = X_n \int [f_2(y)/f_1(y)]f_1(y)\mu(dy) = X_n.$$

In fact, there is also a very good reason that it should be so; it is a disguised random walk (see Exercises 5 and 6).

In the more abstract accounts of the theory, the notion of conditioning with respect to an increasing observation set is replaced by that of conditioning with respect to a σ-field which increases (i.e. grows finer) with increasing n. The notion of a function of current observations is then replaced by that of a r.v. measurable on the current σ-field. We shall not go so far, but we shall adopt the convenient practice of denoting the set of observations $Y_1, Y_2, \ldots,$ Y_n available at time n collectively by W_n. The martingale property (1) then becomes simply

$$E(X_{n+1}|W_n) = X_n. \tag{8}$$

EXERCISES AND COMMENTS

1. Note that representation (2) need not be unique for a given martingale. For example, we could add Z' to Z, where Z' is any r.v of zero mean which is independent of $\{Y_n\}$.

2. Note a consequence of the martingale property (8): that

$$X_{n+1} - X_n \perp H(W_n)$$

   for any H.

3. Consider $\Phi(X_n)$ where $\{X_n\}$ is a martingale and Φ convex. Then an appeal to (3) and Jensen's inequality shows that

$$E[\Phi(X_{m+n})|W_n] \geq \Phi(X_n).$$

   (This is the so-called *submartingale* property for the sequence $\{\Phi(X_n)\}$.) In the case $\Phi(X) = X^2$, one has the stronger assertion

$$E[\Phi(X_{m+n})|W_n] = E[\Phi(X_{m+n} - X_n)|W_n] + \Phi(X_n),$$

   (see Exercise 2).

4. The absorption probability X_n defined by (6) is a function $\rho(Y_n, n)$. One would intepret $\rho(j, n)$ as the probability of absorption in A by time m conditional on a start from state j at time n ($\leq m$). It will obey a backward equation $\rho(j, n) = \sum_j p_{jk}\rho(k, n + 1)$, which is just a more explicit way of writing the martingale equation $X_n = E(X_{n+1}|Y_n)$. The latter equation is written in r.v.s, the former in particular values of those r.v.s. The 'limit' of the sequence $\{X_n\}$ is attained already at $n = m$. It is a r.v. which, conditional on a start from Y_0 at time 0, takes values 1 and 0 with respective probabilities $\rho(Y_0, 0)$ and $1 - \rho(Y_0, 0)$. In the limit $m \to \infty$, the function $\rho(j, n)$ becomes independent of n, and X_n does not attain its limit value for any finite n.

5. Denote the logarithm of the likelihood ratio defined in (7) by U_n. Then this is a sum of IID r.v.s u_j, and so describes a random walk whose increment has moment generating function

$$M(v) = E(e^{vu}) = \int (f_2/f_1)^v f_1 \mu(dy).$$

   It follows then, as for the random walk of Sections 9.6 and 9.7, that $X_n = e^{vU_n}$ will be a martingale if v is such that $M(v) = 1$. This equation has just two real roots: $v = 0$ (giving the trivial martingale $X_n = 1$) and $v = 1$, giving the martingale (7).

6. More generally, suppose that there are m exclusive and exhaustive hypotheses, where hypothesis i has prior probability π_i. The posterior probability of hypothesis i after n observations is then $P_{in} \propto \zeta_{in}$, where

$$\zeta_{in} = \pi_i \prod_{j=1}^{n} f_i(Y_j).$$

Note that $\log \zeta_{in}$ will follow a random walk under any fixed hypothesis. Denote an expectation conditional on the truth of hypothesis i by E_i. Furthermore, define

$$M(v) = \int \left(\prod_i f_i(y)^{v_i} \right) \mu(dy).$$

Show that if v is such that $M(v) = 1$ and $\sum_i v_i = 1$, then

$$X_n = \Phi(P_n) = \prod_i P_{in}^{v_i}$$

is a martingale under $E = \sum_i \pi_i E_i$ and

$$X_{in} = \frac{\partial \Phi(P_n)}{\partial P_{in}}$$

is a martingale under E_i (with respect to $\{Y_n\}$ in both cases). These ideas have implications for the sequential testing of hypotheses (Whittle, 1964).

2. Kolmogorov's Inequality: the Law of Large Numbers

Let $\{X_n; n \geq 1\}$ be a martingale with respect to $\{Y_n; n \geq 1\}$. Then we can always normalize this to a martingale $\{X_n; n \geq 0\}$ with $X_0 = 0$ by considering $X_n - E(X_n)$ instead of X_n.

Note that the martingale property (with the conditioning observations denoted collectively by W_n, as in (8)) will imply that

$$E[(X_n - X_j)|W_j] = 0 \qquad (j \leq n),$$

and so that

$$E[(X_n - X_j)Z_j|W_j] = 0 \qquad (j \leq n),$$

where Z_j is any function of W_j. In particular, taking $Z_j = X_j$ we find that $X_n - X_j \perp X_j$ conditional on W_j, and so

$$E(X_n^2|W_j) = E[(X_n - X_j)^2|W_j] + E(X_j^2|W_j) \qquad (j \leq n). \tag{9}$$

If we assume the normalization $X_0 = 0$ and set

$$\zeta_j = X_j - X_{j-1} \qquad (j = 1, 2, 3, \ldots),$$

then we deduce from (9) that

$$E(X_n^2) = \sum_{j=1}^{n} E(\zeta_j^2). \tag{10}$$

We now come to the classic result which provides the bridge from mean square convergence to almost sure convergence.

Theorem 14.2.1 (Kolmogorov's Inequality). *Let $\{X_n\}$ be a martingale for which $X_0 = 0$ and u a positive number. Then*

$$P(|X_j| \leq u; j = 1, 2, \ldots, n) \geq 1 - E(X_n^2)/u^2. \tag{11}$$

If the expression were a bound simply for $P(|X_n| \le u)$ then (11) would amount exactly to Chebyshev's inequality. What is new is that (11) gives a lower bound for the probability that $X_1, X_2, \ldots, X_n$ are simultaneously bounded in modulus, this bound seemingly made possible by the martingale character of $\{X_n\}$.

PROOF. Let A be the event $|X_j| \le u_j$ for $j = 1, 2, \ldots, n$, and let B_j be the event that $|X_i| > u_i$ first for $i = j$ $(j = 1, 2, \ldots, n)$. These events are exhaustive and exclusive. We have

$$E(X_n^2|A) \ge 0,$$

$$E(X_n^2|B_j) = E[(X_n - X_j)^2|B_j] + E(X_j^2|B_j) \ge u^2.$$

Thus

$$E(X_n^2) = P(A)E(X_n^2|A) + \sum_{j=1}^{n} P(B_j)E(X_n^2|B_j) \ge u^2 \sum_{j=1}^{n} P(B_j) = u^2[1 - P(A)],$$

whence (11) follows. $\square$

However, there is a generalization of inequality (11) due to Hajek and Rényi which will prove useful. The following proof (Whittle, 1969) also covers the material of the exercises.

Theorem 14.2.2 (The Kolmogorov–Hajek–Rényi Inequality). *Let $\{X_n\}$ be a martingale for which $X_0 = 0$, and let $\{u_n\}$ be a sequence of constants such that $0 = u_0 \le u_1 \le u_2 \le \cdots$. Then*

$$P(|X_j| \le u_j; j = 1, 2, \ldots, n) \ge 1 - \sum_{j=1}^{n} E(\zeta_j^2)/u_j^2. \tag{12}$$

We see from (10) that this reduces to the Kolmogorov inequality in the case $u_j = u$ $(j > 0)$.

PROOF. Denote the event $(|X_j| \le u_j; j = 1, 2, \ldots, n)$ by A_n. Then

$$P(A_n) = E[I(A_n)] = E[I(A_{n-1})I(|X_n| \le u_n)]$$
$$\ge E[I(A_{n-1})(1 - X_n^2/u_n^2)] \tag{13}$$
$$= E[I(A_{n-1})(1 - X_{n-1}^2/u_n^2 - \zeta_n^2/u_n^2)]$$
$$\ge E[I(A_{n-2})(1 - X_{n-1}^2/u_{n-1}^2)] - E(\zeta_n^2)/u_n^2. \tag{14}$$

The equality after (13) follows from (9), and the following inequality from the facts that $u_n \le u_{n-1}$ and $I(|X| \le u)(1 - X^2/u^2) \ge (1 - X^2/u^2)$. Iteration of the reduction (14) of (13) yields (12). $\square$

The interest in the KHR inequality (as we shall abbreviate it) is that it leads to an immediate proof of the strong law of large numbers under conditions weaker than that the summands should be independent.

Theorem 14.2.3. *Let* $\{Y_n\}$ *be a sequence of r.v.s for which*

$$E(Y_n | Y_1, Y_2, \ldots, Y_{n-1}) = \mu, \qquad E(Y_n - \mu)^2 = \sigma_n^2. \tag{15}$$

Define the sum

$$S_n = \sum_{j=1}^{n} (Y_j - \mu).$$

Then:

(i) $\{S_n\}$ *is a martingale with* $S_0 = 0$, *and for any nonnegative, nondecreasing sequence of constants* $\{a_n\}$ *and positive constant* ε

$$P(|S_j/a_j| \leq \varepsilon; j \geq n) \geq 1 - \varepsilon^2 \sum_{j=1}^{n} \sigma_j^2/a_n^2 - \varepsilon^2 \sum_{j=n+1}^{\infty} \sigma_j^2/a_j^2. \tag{16}$$

(ii) *If, moreover,* $\sigma_n^2 \equiv \sigma^2 < \infty$ *and* $\sum_n a_n^{-2} < \infty$ *then*

$$S_n/a_n \overset{\text{a.s.}}{\to} 0. \tag{17}$$

In particular, the strong law of large numbers holds:

$$\frac{1}{n} \sum_{j=1}^{n} Y_j \overset{\text{a.s.}}{\to} \mu \tag{18}$$

and

$$\frac{S_n}{\sqrt{n(\log n)^\alpha}} \overset{\text{a.s.}}{\to} 0 \tag{19}$$

for any $\alpha > 1$.

PROOF. One readily verifies that $\{S_n\}$ is a normalized martingale relative to $\{Y_n\}$ as asserted, and that $Y_n - \mu$ can be identified with the increments $\zeta_n = S_n - S_{n-1}$. The sequence $(0, S_n, S_{n+1}, \ldots)$ is then also a martingale. Applying Theorem 14.2.2 to the whole course of this martingale we deduce inequality (16).

The consequence (17) will follow if the right-hand member of (16) tends to unity with increasing n. Under the assumptions of assertion (ii) it will do this if $\sum_{n+1}^{\infty} a_j^{-2}$ and na_n^{-2} both converge to zero with increasing n. The condition $\sum_1^{\infty} a_j^{-2} < \infty$ will imply both; the first by definition, and the second because

$$na_{2n}^{-2} \leq \sum_{n+1}^{2n} a_j^{-2} \to 0.$$

Assertion (ii) then follows from the particular choices $a_n = n$ and $a_n = \sqrt{n(\log n)^\alpha}$. $\qquad\square$

In (18) we have a proof of the strong law of large numbers, obtained under the assumptions (15) and $\sigma_n^2 = \sigma^2 < \infty$. These are weaker than the IID assumption, although stronger than necessary in that $E(Y_n^2)$ is required to be finite, rather than $E(|Y_n|)$. The second conclusion (19) is even stronger. It is a result half-way to the law of the iterated logarithm, which states that $\max_{j \leq n}|S_j|$ essentially grows as $\sqrt{2\pi\sigma^2 n \log\log n}$ for n large.

While the restriction to square-integrable variables is regrettable, it does permit an economical and powerful treatment. Further weakening of conditions can certainly be achieved, but only at the expense of a considerable amount of special argument.

EXERCISES AND COMMENTS

The following exercises are both based on the assumption that $\Phi(x)$ is a nonnegative symmetric function of a scalar x increasing in $|x|$, and that $\{X_n\}$ is a sequence of r.v.s with the properties $X_0 = 0$ and

$$0 \leq E[\Phi(X_{n+1})|W_n] - \Phi(X_n) \leq \Delta_{n+1} \qquad (n = 0, 1, 2, \ldots).$$

In particular, the first inequality will hold if $\{X_n\}$ is a martingale and Φ is convex. We also denote the event $(|X_j| \leq u_j; j = 1, 2, \ldots, n)$ by A_n, and define $\zeta_n = X_n - X_{n-1}$.

1. Show that if $\{u_n\}$ is positive and nondecreasing, then

$$P(A_n) \geq 1 - \sum_{j=1}^{n} \frac{E[\Phi(X_j)] - E[\Phi(X_{j-1})]}{\Phi(u_j)}.$$

This generalizes the KHR inequality.

2. Show that if the u_j are positive and $\Phi(u_j) \geq \Phi(u_{j-1}) + \Delta_j$ $(j = 1, 2, \ldots, n)$, then

$$P(A_n) \geq \prod_{j=1}^{n} [1 - \Delta_j/\Phi(u_j)].$$

This has the multiplicative form one might reasonably expect. In particular, in the case $\Phi(x) = x^2$, it yields *Dufresnoy's inequality*

$$P(A_n) \geq \prod_{j=1}^{n} [1 - E(\zeta_j^2)/u_j^2],$$

valid if $u_j^2 \geq u_{j-1}^2 + E(\zeta_j^2)$ $(j = 1, 2, \ldots, n)$.

3. Martingale Convergence: Applications

Theorem 14.3.1 (The Martingale Convergence Theorem). *Suppose that $\{X_n\}$ is a martingale for which $E(X_n^2)$ is bounded uniformly in n. Then $\{X_n\}$ converges almost surely to a limit X_∞ which equals its means square limit with probability one.*

PROOF. Assume the martingale normalized so that $X_0 = 0$ (see the beginning of the last section) and define $\zeta_n = X_n - X_{n-1}$. It follows then from (10) and the assumption on $E(X_n^2)$ that

$$\sum_{j=1}^{\infty} E(\zeta_j^2) < \infty. \qquad (20)$$

Now, for given n, the sequence $\{X'_k\} = \{X_{n+k} - X_n; k \geq 0\}$ is also a martingale with $X'_0 = 0$. It then follows from Kolmogorov's inequality that

$$P(|X_j - X_n| \leq \varepsilon; n < j \leq m) \geq 1 - (1/\varepsilon^2) \sum_{n+1}^{m} E(\zeta_j^2)$$

and so that

$$P(|X_m - X_n| \leq \varepsilon; m > n) \geq 1 - (1/\varepsilon^2) \sum_{n+1}^{\infty} E(\zeta_j^2).$$

Relation (20) implies that this bound tends to unity with increasing n. The sequence $\{X_n\}$ is thus a.s. mutually convergent, and so a.s. convergent.

The sequence must then converge in probability to its a.s. limit. But it also converges in probability to its m.s. limit, so the two are equal in probability.
$\square$

Theorem 14.3.1 is the celebrated martingale convergence theorem. The assumption that the X_n are square-integrable (indeed, uniformly so) can be weakened by the use of truncation arguments. That is, one restricts the r.v.s to a range $(-R, R)$ and then shows that one can obtain results valid for the untruncated case by letting R tend to infinity. However, although one obtains stronger results in this way, one loses the essential simplicity of the square-integrable case; a simplicity grounded in the intepretation of a martingale as a sequence of least square estimates.

The martingale convergence theorem has, of course, an immediate corollary for this latter case.

Theorem 14.3.2. *Suppose $\{X_n\}$ generated as the sequence of conditional expectations (2) and that $E(Z^2) < \infty$. Then $\{X_n\}$ has an a.s.. limit, identifiable with its m.s. limit with probability one.*

The proof follows simply from the fact that $\{X_n\}$ is a martingale and $E(X_n^2) \leq E(Z^2)$.

We have already seen in Section 1 that an absorption probability for a Markov process (regarded as a r.v., in that it is a function of current state, which is a r.v.) is a martingale. It is intrinsically bounded and so, by the theorem, converges. The limit r.v. takes only a single value, which must be either 0 or 1.

The capital of a player in a fair game is a martingale. It will not in general be a bounded one, however, unless one adds stopping rules which state that the game terminates (so that X_n is constant thereafter) as soon as X_n leaves some bounded set. A sum

$$S_n = \sum_{j=1}^{n} [Y_j - E(Y_j)]$$

with independent Y_j is also a martingale, although again not one that is bounded in any sense. We have nevertheless seen from the previous section that martingale-related methods enable one to set bounds on the rate of growth of S_n.

The martingale convergence theorem does not yield very interesting results when applied to processes which ultimately become absorbed in some single state (such as the absorption probability itself discussed above, the gambler's ruin problem or random walks subject to absorption) because the limit r.v. is always trivial in these cases. The optional stopping theorem, to be discussed in the next section, is more significant in such cases, because it tells one something of the ultimate distribution over absorbing states to be expected. The martingale convergence theorem is of more interest in situations where the limit r.v. is nontrivial. One such case is indeed that of estimation or conditional expectation (2), when one ends up with an estimate based on the whole observation history which is nevertheless genuinely random, because it is the estimate of a r.v.

Another example which is interesting, because the process is one that is marked by its early history, is that of the branching process considered in Section 6.4. We assume that the population is intrinsically an increasing one in that $\alpha > 1$, where α is the expected number of sons per father. Denote the population size at generation n by Y_n, and suppose fixed initial conditions $Y_0 = K$. Define

$$X_n = Y_n/E(Y_n) = Y_n/(K\alpha^n).$$

We leave the reader to verify that $\{X_n\}$ is a martingale and that

$$E(X_n^2) \le \frac{\beta}{K\alpha(\alpha - 1)},$$

where β is the variance of the number of sons per father. If we assume β finite then X_n will be a.s. convergent.

The r.v. to which it converges is nontrivial; a normalized population size. The intuitive reason for such convergence is that the population size varies erratically while it is small, but the growth rate stabilizes as the population becomes larger. So the variation of X_n stems principally from the variation of the population while it is small and has not yet stabilized its growth rate, either by becoming very large or by becoming extinct. Graphs of sample paths demonstrate this behaviour very clearly (see Harris, 1963, p. 12).

We can actually derive more definite information on the distribution of X_∞; see Exercise 3.

EXERCISES AND COMMENTS

1. Consider the likelihood ratio example (7). Note that $\log X_n$ follows a random walk with expected increment $\int f_1 \log(f_2/f_1)\mu(dy)$. By a classic inequality (the Gibbs inequality) this is negative, and strictly so if f_1 and f_2 differ on a set of positive μ-measure. Thus, by the strong law of large numbers, $\log X_n \overset{\text{a.s.}}{\to} -\infty$ and $X_n \overset{\text{a.s.}}{\to} 0$.

2. The X_n of the branching process is so strongly convergent that we can use much cruder methods. Show that $\sum_n \gamma^{2n} E(|X_{n+1} - X_n|^2) < \infty$, and hence that $\gamma^n X_n$ converges mutually a.s. if $|\gamma| < \sqrt{\alpha}$, without appeal to the martingale property.

3. Let X be the limit r.v. for the branching process under the supposition $Y_0 = 1$. Show that its c.f. $\phi(\theta)$ must satisfy

$$\phi(\alpha\theta) = G[\phi(\theta)], \tag{21}$$

where $G(z)$ is the progeny p.g.f. Consider the particular case

$$G(z) = \frac{a + bz}{1 - cz},$$

where a, b and c are necessarily nonnegative and add up to unity. Show that (21) has the solution

$$\phi(\theta) = \rho + \frac{1 - \rho}{1 - i\theta d}, \tag{22}$$

where ρ is the probability of ultimate extinction for the process with $Y_0 = 1$, and d is arbitrary (but fixed by the normalization $E(X) = 1$ implicit in the definition of X_n). Interpret solution (22).

4. Consider the continuous-time simple birth and death process of Section 10.4. Show that the analogue of (21) in this case is

$$(\lambda - \mu)\theta \frac{\partial \phi}{\partial \theta} = (\lambda\phi - \mu)(\phi - 1),$$

and show that this has a general solution of the form (22) with $\rho = \mu/\lambda$.

4. The Optional Stopping Theorem

Many stochastic processes come to a stop, not at some predetermined time, but at a time when some predetermined set of conditions has been fulfilled. So, the doctor with whom we began our study of conditioning in Chapter 5 will stop his tests once he feels he has accumulated enough information to make a confident diagnosis. The gambler will stop when he is compelled to or when he feels that further play is not worthwhile. The inspector examining a batch in Section 5.8 may do so on a sequential basis (i.e. item by item) and stop when he has seen enough, one way or another.

These are all situations in which one feels that stopping has been a matter of choice on the part of some agent. There are others where clearly it is not. For example, the extinction of a population or the completion of an epidemic (i.e. the extinction of the infecteds) are events intrinsic to the model: entry into a state which is absorbing by its nature.

'Stopping time' is a technical term; it means a stopping rule which is expressed in terms of current observables only (perhaps ideal observables; see

Exercise 2). That is, if W_n denotes collectively all observations available at time n (including specification of n itself), then stopping will occur when W_n first enters a prescribed *stopping set* $\mathscr{D}$. The time at which W first enters the stopping set determines the moment of termination, which we shall denote by τ. The variable τ will be a r.v., unless the stopping rule simply specifies a value of n at which the process is to terminate. Specification of a stopping rule is equivalent to specification of the stopping set $\mathscr{D}$. When probabilists speak of 'a stopping time', they do not mean the time τ itself but rather the rule (as specified by $\mathscr{D}$) which determines it—confusing! The complement of $\mathscr{D}$ in W-space is the *continuation set* $\mathscr{C}$. Sometimes one is dealing with a Markov process and the stopping rule is couched in terms solely of the current value of state. In that case, $\mathscr{C}$ and $\mathscr{D}$ are effectively complementary sets in state space.

Our opening examples indicate that a martingale may well be associated with a stopping rule. We write the martingale condition as

$$E(X_{n+1}|W_n) = X_n \qquad (n = 0, 1, 2, \ldots). \tag{23}$$

We suppose that W_0 (the information available at time 0) is simply X_0 itself, and that unconditioned expectations are those which are conditioned by this information alone. If the martingale can continue freely (so that (23) holds for all W_n) then (23) implies that

$$X_0 = E(X_n). \tag{24}$$

However, suppose that the process is subject to a stopping rule, so that (23) holds only for W_n in $\mathscr{C}$. One would then like to strengthen the conclusion (24) to

$$X_0 = E(X_\tau). \tag{25}$$

We have already seen in Sections 9.7 and 10.7 how useful identity (25) (if valid) is, in determining the distribution of stopping time τ and stopping coordinate W_τ. It implies Wald's identity, for example.

Relation (25) is not valid without qualifications, however, and clarification of these is the purpose of

Theorem 14.4.1 (The Optional Stopping Theorem). *Let $\{X_n\}$ be a martingale with respect to the information sequence $\{W_n\}$ and let τ be the moment when W first enters the stopping set. Then the following conditions are jointly sufficient for the validity of identity* (25):

(a) $P(\tau < \infty) = 1$.
(b) $E(|X_\tau|) < \infty$.
(c) $E[X_n I(\tau \geq n)] \to 0$ *as* $n \to \infty$.

PROOF. Note first that, for $n > j$,

$$E[X_n I(\tau = j)] = E\{E[X_n I(\tau = j)|W_j]\} = E[X_j I(\tau = j)]$$
$$= E(X_j|\tau = j)P(\tau = j).$$

We then have, for any n,

$$X_0 = E(X_n) = E\left\{X_n\left[I(\tau \geq n) + \sum_{j=1}^{n-1} I(\tau = j)\right]\right\}$$

$$= E[X_n I(\tau \geq n)] + \sum_{j=1}^{n-1} E(X_j|\tau = j)P(\tau = j),$$

and so

$$E(X_\tau) - X_0 = E[(X_\tau - X_n)I(\tau \geq n)]. \tag{26}$$

Furthermore, since

$$\sum_j E[|X_j|I(\tau = j)] = E(|X_\tau|) < \infty$$

then

$$|E[X_\tau I(\tau \geq n)]| = \left|\sum_{j=n}^{\infty} E[X_j I(\tau = j)]\right| \leq \sum_{j=n}^{\infty} E[|X_j|I(\tau = j)] \to 0$$

as $n \to \infty$. The hypotheses thus imply that both expectations in the right-hand member of (26) converge to zero with increasing n, and the validity of (25) is established.   □

Note that the conditions of the theorem are convenient sufficient conditions; certainly not the best possible and possibly not the best suited.

EXERCISES AND COMMENTS

1. Suppose that $X_n = \psi(Y_n)$ is a martingale with respect to $\{Y_n\}$, where $\{Y_n\}$ is a Markov process with finite state space. Show that, if the Markov process is irreducible, then X_n must be constant. In other cases, it is then constant within any closed class. Suppose that τ is the moment of first entry to a closed class. Then (25) is certainly valid.

2. A stopping time determined by the observer could certainly depend only on current observables. A stopping time determined by the process could depend only on current process history, which, by definition, constitutes the maximal possible observable. For example, a missile terminates when it strikes or otherwise comes to rest—events perhaps not accessible to any human observer, but accessible to an ideal observer who could follow the course of the process in real time. The point is that the definition of a stopping time must preclude anticipatory behaviour, in which action is based on events whose outcome cannot be predicted at the time. For example, to say 'I will kill myself when you finally leave me' is, strictly speaking, not a well-defined stopping rule.

5. Examples of Stopped Martingales

Equation (10.38) represents an appeal to (25) for the epidemic process. We see from Theorem 14.4.1 that the appeal is justified. Passage to one of the termination states is certain, and the process is Markov with finite state space, so that all r.v.s are bounded. This represents a particular case of the situation envisaged in Exercise 4.1.

On the other hand, we know that relation (25) can fail for the martingale that one would regard as the simplest possible: the capital X_n of a player in a fair game with unit stakes. This is indeed a martingale, which is just another way of saying that $\psi(x) = x$ is a solution of

$$\psi(x) = \tfrac{1}{2}[\psi(x+1) + \psi(x-1)].$$

Suppose that the player starts with $X_0 = k$ and continues until he has attained $X = k + 1$. This can be attained with probability one, so that $E(X_\tau) = k + 1 \neq X_0$. The condition of the optional stopping theorem that is violated is condition (c): the process can show infinite excursions (necessarily in a negative direction) before the stopping set $X = k + 1$ is attained. In other words, the player has positive probability of being infinitely deep in debt before he finally makes his net unit gain. If debt were restricted in some way then (25) would be valid (either because the player would be kept within a bounded region of state space or because the game could terminate in other ways; e.g. by his ruin).

Suppose we consider the example when $\{Y_n\}$ is a random walk (in several dimensions) whose increment $\Delta_n = Y_n - Y_{n-1}$ has m.g.f. $M(\alpha) = E(e^{\alpha^T\Delta})$. Then we know from the discussion of Sections 9.6 and 9.7 that

$$X_n = M(\alpha)^{-n} e^{\alpha^T Y_n}$$

is a martingale with respect to $\{Y_n\}$ for any α for which $M(\alpha)$ is defined. Suppose a stopping set is defined in the state space $\mathscr{Y}$, and that the process begins at the origin, supposed a point of the continuation set. Thus $Y_0 = 0$ and $X_0 = 1$, and the identity (25) in this case would yield

$$1 = E[M(\alpha)^{-\tau} e^{\alpha^T Y_\tau}], \tag{27}$$

which is just Wald's identity. However, rather than establishing the conditions of validity for (27) by appealing to the optional stopping theorem, it is better to use an argument specific to the case.

Note that, for this random walk,

$$\frac{E(e^{i\theta^T Y_n} X_n)}{E(X_n)} = E(e^{i\theta^T Y_n} X_n) = E[M(\alpha)^{-n} e^{(\alpha + i\theta)^T Y_n}] = \left[\frac{M(\alpha + i\theta)}{M(\alpha)}\right]^n \tag{28}$$

which would be the c.f. of Y_n for a random walk whose increment had c.f. $E(e^{i\theta^T \Delta}) = M(\alpha + i\theta)/M(\alpha)$. Let us denote such a random walk by $RW(\alpha)$; the original walk is then $RW(0)$. Then relation (28) implies

Theorem 14.5.1. *The random walk $RW(0)$, when weighted by X_n as in (28), is equivalent to the random walk $RW(\alpha)$. In particular, validity of the statement (27) of Wald's identity for $RW(0)$ is equivalent to validity of the statement that $RW(\alpha)$ terminates in the prescribed stopping set with probability one.*

This is because $E(X_\tau)$ for $RW(0)$ is just the integral of the stopping distribution for $RW(\alpha)$. Since it is rather easy to determine conditions for certainty of termination, one can then readily determine a region of validity for (27). Let us do this for the one-dimensional case.

Theorem 14.5.2. *Consider the one-dimensional case. Suppose that $M(\alpha)$ is defined in the real interval (α_1, α_2) and in this is minimal at α_0. Suppose also that the increment Δ is not identically constant, and that the continuation set is an interval $a \le Y \le b$, with the initial value $Y_0 = 0$ an interior point.*

(i) *In the case when a and b are both finite Wald's identity (27) is valid for any α in (α_1, α_2).*

(ii) *In the case a finite and $b = +\infty$ the identity is valid for any α in the interval (α_1, α_0).*

PROOF. Recall that $M(\alpha)$ is convex for real α. In case (ii) $RW(\alpha)$ has negative drift (i.e. $E(\Delta) = M'(\alpha)/M(\alpha)$ is negative) if $\alpha < \alpha_0$. Thus crossing of the boundary $Y = a$ is certain for $RW(\alpha)$ for α in this range.

In case (i) drift of either sign will assure termination for $RW(\alpha)$. In the case of zero drift ($\alpha = \alpha_0$) $RW(\alpha)$ will still terminate. This is because the case $\Delta \equiv 0$ is excluded by hypothesis, and the spread in distribution of Y_n with increasing n will ultimately ensure termination (see Exercise 2). $\qquad\square$

The modification of $RW(0)$ to $RW(\alpha)$ is an example of the tilting of a distribution we have already encountered in Exercises 2.9.13 and 12.6.1.

EXERCISES AND COMMENTS

1. Suppose that $\{Y_n\}$ is a Markov process with transition operator P, that $\{X_n\} = \{\psi(Y_n)\}$ is a martingale with respect to it and that τ is the time of first passage into a stopping set $Y \in \mathscr{D}$. Show that validity of (25) is equivalent to the statement that passage from Y_0 into $\mathscr{D}$ is certain for the Markov process with transition operator $(1/\psi)P\psi$.

2. Consider the one-dimensional random walk $\{Y_n\}$ with increment Δ. To say that Δ is not identically zero is to say that positive constants ε, δ exist, such that $P(|\Delta| \ge \varepsilon) \ge \delta$. We can suppose ourselves in the case $P(\Delta \ge \varepsilon) \ge \delta$. Then $P(Y_n \in \mathscr{D}) \ge \delta^n$ for any Y_0 in $\mathscr{C}$ if $n > (a + b)/\varepsilon$. Thus passage to $\mathscr{D}$ has positive probability from any starting point in $\mathscr{C}$, and so is ultimately certain.

CHAPTER 15

Extension: Examples of the Infinite-Dimensional Case

1. Generalities on the Infinite-Dimensional Case

In this chapter we consider to some extent the transfer of the material of Chapter 12 to the infinite-dimensional case, i.e. to the case in which infinitely many expectation values are specified. The two issues to be faced remain those of consistency and extension. These issues are too large for us to treat systematically, and we shall in fact consider only some particular cases of extension, of interest either in that they make a point or in that they have already loomed into sight. So, in Sections 2–4 we indicate how the expectation approach ties in with the usual one, based on the concepts of σ-fields of subsets of Ω. Interestingly, this idea of a σ-field generalizes to the more attractive concept of a linear lattice of r.v.s.

The interest of these sections might be regarded as abstract; in Section 5 it becomes concrete. Here we treat the question of what expectations are determinable from knowledge of the c.f. of a r.v. This is immediately related to the fact that convergence of a sequence of c.f.s to a c.f. is equivalent to weak convergence of the corresponding sequence of r.v.s—a fact which we have repeatedly taken on faith (in Sections 7.3 and 7.4, for example). It is equivalences of this type that make the powerful concept of weak convergence also a powerful tool.

Let us return to the formulation of Section 12.4. We shall suppose all r.v.s scalar, so that specification of a vector expectation means specification of several scalar expectations. In general, the set of r.v.s Y whose expectation is specified will be infinite. Let us extend this to the class $\mathscr{Y}$ of r.v.s whose expectations are immediately determinable from those given by appeal to Axioms 1–5 of Section 2.2. That is, $\mathscr{Y}$ is closed under the application of finite linear operations and monotone limits. Other extensions may be possible, but this we have yet to determine.

It is the possibility of considering limits that introduces a new feature. Let us note that if the prescribed expectation values are consistent with Axioms 1–4, then Axiom 5 is self-consistent. In other words, if one has two monotone sequences tending to the same limit, then the two corresponding sequences of expectations will also tend to the same limit.

Suppose that two monotone nondecreasing sequences $\{Y_n\}$ and $\{Y_n'\}$ have a common limit Y. Then, for fixed m and variable n, $Y_n - Y_m'$ is a nondecreasing sequence with a nonnegative limit, so that, by Axioms 1 and 5, $\lim_n E(Y_n - Y_m') \geq 0$ or $\lim E(Y_n) \geq E(Y_m')$. Taking the limit of large m in this last relation we deduce that $\lim E(Y_n) \geq \lim E(Y_n')$, and the reverse inequality follows by the same argument. The two limiting expectations are then equal, and can be unequivocally identified with $E(Y)$. It is because we have restricted ourselves to monotone limits that such identification is possible, but we would like to be able to draw similar conclusions for more general limit sequences.

We give the analogies of those assertions of Theorem 12.4.1 which have an immediate analogue.

Theorem 15.1.1. *Suppose that the infinite vector of specified expectations lies in the interior of its feasible set. Denote by Y a member of the set $\mathcal{Y}$ of r.v.s generated from those of specified expectation by finite linear operation and monotone limit. Then*:

(i) *The sharp bounds on $E(X)$ are*

$$\sup_{Y \leq X} E(Y) \leq E(X) \leq \inf_{Y \geq X} E(Y). \tag{1}$$

(ii) *$E(X)$ is completely determined just for those r.v.s X for which sequences $\{Y_n\}$ and $\{Y_n'\}$ exist in $\mathcal{Y}$ such that $Y_n \leq X \leq Y_n'$ and $E(Y_n' - Y_n) \to 0$. The r.v. X is then an L_1 limit of either sequence.*

Assertion (i) follows by the supporting-hyperplane proof as before (although a 'hyperplane' now becomes a linear functional). The validity of (ii) is evident. Note that the X envisaged in (ii) is an L_1-limit of the sequences indicated for *all* distributions on Ω consistent with the given expectations. The addition of these L_1-limits to $\mathcal{Y}$ gives the maximal set of r.v.s whose expectation is completely determined.

One can scarcely progress further without making explicit assumptions concerning the class $\mathcal{Y}$. We shall discuss some particular cases.

2. Fields and σ-Fields of Events

The probabilistic analogue of a family of r.v.s of known expectation is a set of events $\{A_i\}$ of known probability. That is, a family of subsets of Ω, which we shall denote by $\mathscr{F}$.

In order to make progress one must assume that $\mathscr{F}$ has some structure. The structure advanced by Kolmogorov (1933) is that $\mathcal{F}$ should be a *field*, by which is meant that:

(i) $\varnothing \in \mathcal{F}$.
(ii) $A \in \mathcal{F}$ implies that $\bar{A} \in \mathcal{F}$.
(iii) $A, B \in \mathcal{F}$ implies that $A \cap B \in \mathcal{F}$.

That is, the empty set belongs to $\mathcal{F}$, and $\mathcal{F}$ is closed under the operations of complementation and intersection of its members. Thus, Ω itself belongs to $\mathcal{F}$, by (i) and (ii). So also do unions and differences of sets, because $A \cup B$ is the complement of $\bar{A} \cap \bar{B}$ and $A \Delta B = A\bar{B} + \bar{A}B$.

The advantage of a field is then that all the compound events which one could naturally generate from a field of events (sets) also lie within the field. So, if A and B belong to $\mathcal{F}$ then so do the four events AB, $\bar{A}B$, $A\bar{B}$ and $\bar{A}\bar{B}$ (which form a decomposition of Ω) and all possible unions of these. In fact, this collection of $2^4 = 16$ possible unions constitutes the smallest field containing both A and B.

If the field $\mathcal{F}$ is infinite then one can compound events indefinitely; whether limit events thus generated belong to $\mathcal{F}$ is a matter of definition. The extended field obtained by adjoining all such limits to $\mathcal{F}$ is the Borel extension of $\mathcal{F}$, denoted $B\mathcal{F}$.

Otherwise expressed, a field which obeys the stronger version of (iii):

(iii)′ If $A_1, A_2, A_3, \ldots$ belong to $\mathcal{F}$ then so does $\bigcap_{i=1}^{\infty} A_i$,

is termed a *Borel field* or a *σ-field*. The Borel extension $B\mathcal{F}$ can be characterized as the smallest σ-field containing $\mathcal{F}$.

Kolmogorov's principal theorem asserted that a probability measure (for which properties (i)–(iv) of Theorem 3.2.1 are taken as axiomatic) on a field $\mathcal{F}$ has a unique extension to $B\mathcal{F}$. That is, if $P(A)$ is consistently specified for $A \in \mathcal{F}$ then $P(A)$ is uniquely determined for $A \in B\mathcal{F}$. We shall consider a r.v. verison of this result in the next section.

EXERCISES AND COMMENTS

1. Let $\mathcal{F}$ be the smallest field containing the sets $x \leq a$ on the real line for varying a. Then $\mathcal{F}$ contains all intervals $b < x \leq a$ and finite unions of such intervals. The extension $B\mathcal{F}$ contains monotone limits of sequences of such sets.

3. Extension on a Linear Lattice

Let us return to the discussion of a family $\mathcal{Y}$ of scalar r.v.s of known expectation. As in Section 1, we suppose this closed under linear operations and monotone limits. However, let us add the assumption that it is also closed under the taking of moduli, so that $|Y|$ belongs to $\mathcal{Y}$ if Y does. Such a family of functions is termed a *linear lattice*.

The additional assumption implies that if Y_1 and Y_2 belong to $\mathcal{Y}$, then so do $Y_1 \vee Y_2 = \max(Y_1, Y_2)$ and $Y_1 \wedge Y_2 = \min(Y_1, Y_2)$. This follows because

$$Y_1 \vee Y_2 = \tfrac{1}{2}(Y_1 + Y_2 + |Y_1 - Y_2|), \qquad Y_1 \wedge Y_2 = \tfrac{1}{2}(Y_1 + Y_2 - |Y_1 - Y_2|).$$

The assumption that $|Y|$ belongs to $\mathcal{Y}$ if Y does is certainly convenient, and in certain cases reasonable. Its main use is that it provides us with a distance function

$$\|X - Y\| = E(|X - Y|), \tag{2}$$

with which we can measure the effective separation of two elements of $\mathcal{Y}$. This in turn enables us to be explicit about the limit behaviour of more general sequences than the monotone sequences.

As an example, suppose that the two indicator functions $I(A)$ and $I(B)$ belong to $\mathcal{Y}$, so that we know the values of $P(A)$ and $P(B)$. Then, since $I(AB) = I(A) \wedge I(B)$, we also know the value of $P(AB)$. The additional assumption is thus the analogue for r.v.s of assumption (iii) of the last section, which gave the family of sets the character of a field.

In fact, the case discussed in Exercise 1 of the last section is just that which gives us our principal application of the results of this section; see Section 5. If we know the distribution function $F(x) = E[I(X \leq x)]$ for all x, then for what functions $H(x)$ can one determine $E[H(X)]$?

Let us return to the general problem. The treatment becomes simpler if we reduce $\mathcal{Y}$ to $\mathcal{Y}'$, the set of members Y of $\mathcal{Y}$ for which $E(|Y|) < \infty$. What we shall now prove is that the maximal extension (in the sense of complete determination of an expectation value) of $\mathcal{Y}'$ is obtained by appending to $\mathcal{Y}'$ the L_1 limits of L_1 Cauchy sequences in $\mathcal{Y}'$. That is, of sequences $\{Y_n\}$ such that $E(|Y_m - Y_n|) \to 0$ as m and n become infinite independently.

Theorem 15.3.1. *Any sequence $\{Y_n\}$ in $\mathcal{Y}'$ which is L_1 Cauchy is L_1 convergent to a class of L_1-equivalent r.v.s Y for which $E(Y) = \lim E(Y_n)$. By appending such limits to $\mathcal{Y}'$ we obtain a consistent extension of $\mathcal{Y}'$ which is also maximal.*

PROOF. All these assertions have already been proved one way or another. We know from Theorems 13.4.1 and 13.4.2 that an L_1 Cauchy sequence has an L_1 limit with expectation $\lim E(Y_n)$. As this theorem emphasizes, and as is evident from the constructive proof of Theorem 11.6.1, this 'limit' is really a class of L_1-equivalent r.v.s. Terms such as 'L_1-convergent' and 'L_1-equivalent' are valid for all distributions on Ω consistent with the given expectations.

The extension is maximal, since we know from Theorem 15.1.1 that any element in the maximal extension is an L_1 limit of a sequence in $\mathcal{Y}'$. Being maximal, it is then itself closed under all the operations we have used: linear combination, the taking of moduli and the taking of L_1 limits.

To prove consistency, we must demonstrate facts such as that the sum of L_1 limits of sequences is the L_1 limit of the sequence of sums, and that, if two sequences have the same L_1 limit, then the limits of the expectations for the two sequences then also agree. This is straightforward, and we leave verification to the reader.   □

Theorem 15.3.2. *Theorem* 15.3.1 *still holds if one does not know the r.v.s Y as functions* $Y(\omega)$ *of* ω, *but simply the expectation values* $E(Y)$ $(Y \in \mathscr{Y}')$.

This is surprising, because, in the deduction of the basic extension Theorems 12.3.1 and 12.4.1, it was essential that we should know the r.v.s as explicit functions on Ω. However, the fact that a linear lattice is such a rich class means that we can 'relate' two r.v.s Y_1 and Y_2 of $\mathscr{Y}$ (see Exercise 2), and so can do just as well as if we knew them as functions of ω. On the other hand, we must know that a representation of an r.v. as a function on Ω exists in principle.

Proof of the theorem is immediate: since we can evaluate $E(|Y_m - Y_n|)$ we can construct or recognize Cauchy sequences. We can thus achieve the same extension as before, and, as this is now based on less information, it is *a fortiori* maximal.

However, the two cases differ when it comes to identification of limits. If the r.v.s are known as functions of ω then the limit of a Cauchy sequence can always be recognized as an r.v (or class of r.v.s) Y by the construction of Theorem 15.1.1. However, for the case of Theorem 15.3.2 one cannot always make such an identification. The limit of an L_1 Cauchy sequence in $\mathscr{Y}'$ is always meaningful, but can be identified with a specific r.v. (i.e. a specific function of ω) only if the limit also lies in $\mathscr{Y}'$.

EXERCISES AND COMMENTS

1. Suppose that the linear lattice $\mathscr{Y}$ is generated from the single scalar r.v. Y. (The constant 1 is always taken as being an element.) Show then that $(y - Y)_+ = (y - Y) \vee 0$ belongs to $\mathscr{Y}$. So then does $1 \wedge K(y - Y)_+$, which is a continuous approximation to the indicator function $I(Y \le y)$, approaching this as K becomes large. One can thus evaluate the distribution function of Y at its continuity points.

2. Suppose that the linear lattice $\mathscr{Y}$ is generated from two scalar r.v.s Y_1 and Y_2. By the same argument as for Exercise 1, one can essentially determine the joint distribution function $F(y_1, y_2)$, and so detect a possible functional relationship (in a sufficiently smooth class) between Y_1 and Y_2.

3. The linear lattice is very much an L_1-structure. As one might expect, there is a somewhat corresponding L_2-structure, which really does ask least in the way of information. Suppose that the value of $E(Y_1 Y_2)$ is known for all elements Y_1 and Y_2 of a class of r.v.s $\mathscr{G}$. This is the sole information; it is not assumed that the r.v.s are known as functions of ω. One can, of course, calculate mean products of finite linear combinations of the elements of $\mathscr{G}$, and we shall assume $\mathscr{G}$ closed under the formation of such combinations. Monotone sequences cannot be recognized, since one cannot recognize if one element is larger than another. It is convenient to reduce $\mathscr{G}$ to $\mathscr{G}'$, the set of elements of $\mathscr{G}$ of finite mean square.

   One can extend $\mathscr{G}'$ to $\mathscr{G}''$, say, by adjoining to it all mean square limits of sequences in $\mathscr{G}$. Show that this extension is consistent in that $E(Y_\infty^2)$ and $E(Y_\infty Z)$ $(Z \in \mathscr{G}')$ are the limits of corresponding expectations, and that it is closed under the taking of further mean square limits. Show that the extension is maximal in that, if one can determine the value of $E(X^2)$ from knowledge of $E(XY)$ $(Y \in \mathcal{G})$, then X must be in $\mathcal{G}''$. [Project X on to $\mathcal{G}''$, and show that if X is not in $\mathcal{G}''$, then there are other r.v.s with the same projection which are not mean square equivalent to X.]

4. Integrable Functions of a Scalar Random Variable

Consider the problem raised in the last section: if the distribution function $F(x)$ of a scalar r.v. X is prescribed, then for what functions $H(x)$ is the value of the expectation $E[H(X)]$ thereby determined? In particular, for what sets A is the probability $P(A)$ determined? Such functions are termed *integrable* and such sets *measurable*. This is obviously an extension problem, starting from the given expectations $F(x)$. We shall take the problem in this classic form, although a more natural version of it might be to consider extension on a linear lattice generated from the single r.v. X which, as we saw in Exercise 3.1, amounts to specification of $F(x)$ only at continuity points.

The application of countably many linear operations to the indicator functions will generate the class of *simple functions*; that is, those functions for which the x-axis can be divided up into a countable number of intervals, on each of which the function is constant. We then know from Theorem 15.3.1 that $H(X)$ is integrable if and only if we can deduce from the axioms that $E(|H(X) - H_n(X)|) \to 0$, where $\{H_n(x)\}$ is a sequence of simple functions.

Let us restrict our attention to nonnegative functions, allowing the possibility of infinite expectations. The general case then follows, provided we avoid the case which leads to the indeterminate evaluation $+\infty - \infty$.

Theorem 15.4.1. $E[H(X)]$ *is determined by $F(x)$ and equal to $\int H(x)\,dF(x)$ if H is a continuous function, or a monotone limit of simple functions (a Borel function).*

PROOF. Consider first the case of simple H; suppose $H(x)$ takes the value h_j on the interval $(a_{j-1}, a_j]$, where these intervals constitute a decomposition of th real axis. It then follows from the axioms that

$$E(H) = \sum_j h_j[F(a_j) - F(a_{j-1})]. \tag{3}$$

Turning now to continuous H, let A_j be the x-set satisfying

$$\frac{j-1}{n} < H(x) \le \frac{j}{n}, \tag{4}$$

where n is a fixed positive integer. Then A_j is a countable union of intervals, so that we can evaluate $P(A_j)$. If $H_n(x)$ is the simple function taking the value j/n on A_j, we have then

$$H_n(X) - 1/n \le H(X) \le H_n(X)$$

so that

$$E(H_n) - 1/n \le E(H) \le E(H_n),\tag{5}$$

where

$$E(H_n) = \sum_j (j/n)P(A_j).\tag{6}$$

The expectation of $E(H)$ is thus evaluated to within $1/n$ by formulae (4) and (5). As we let n increase the lower and upper bounds on $E(H)$ will, respectively, increase and decrease, and will converge to a common value which must be the value of $E(H)$. Obviously, $E(|H - H_n|) \le 1/n$.

The final assertion of the theorem is a direct application of Axiom 5. □

Our initial set of r.v.s was essentially the class of simple functions of X; we have extended it to the class of Borel functions of X. This is a great deal less than the maximal extension which is possible; a fact demonstrated by the observation that the maximal extension depends upon the given expectations (i.e. the distribution of X) while the extension of the theorem is one that is possible for *all* distributions. Of course, it is useful to have a 'universal' extension of this kind, and the Borel functions are adequate for many purposes, but one should keep in mind that a much larger extension is possible: to the class of all functions H which, for the given distribution, are L_1 limits of simple functions.

The convergence theorems of classic measure and integration theory are rather different in character to those of probability theory, because they emphasize more the study of $X(\omega)$ itself as a function than the study of its 'statistical' behaviour under various averaging operations. A classic problem is this: if $\{X_n\}$ is a sequence of integrable functions of ω converging pointwise to a limit X, under what conditions can one assert that X is integrable and $E(X) = \lim E(X_n)$? One sufficient condition (by axiom) is that $\{X_n\}$ be monotone. Another is that $Y \le X_n \le Z$ where Y and Z are integrable (or, as a special case, $|X_n| \le Y$ with integrable Y). This is the *dominated convergence theorem* which we have proved in a probabilistic version in Theorem 13.3.3. That proof would not be acceptable in the present context, because we cannot *a priori* regard $E(|X - X_n|)$ and $P(|X - X_n| > \varepsilon)$ as well-defined quantities. The treatment can be made satisfactory, however; we indicate the conventional direct proof in Exercises 3 and 4.

EXERCISES AND COMMENTS

1. Suppose X positive. Show that $E(X)$ is finite if and only if $\int x \, dF(x)$ is, and that the two are equal. (Use equations (3) and (4).)

2. Consider a sequence of r.v.s $\{X_n\}$ and define $\bar{X}_n = \sup_{m \ge n} X_m$ and $\underline{X}_n = \inf_{m \ge n} X_m$. Show that if $\{X_n\}$ converges pointwise (i.e. for each given ω) to a *finite* limit X then $\{\bar{X}_n\}$ and $\{\underline{X}_n\}$ converge monotonely to X (from above and below, respectively).

3. *Fatou's lemma.* One defines lim sup X_n and lim inf X_n as lim $\bar{X}_n$ and $\underline{X}_n$, respectively. Show from $X_n \geq \underline{X}_n$ that lim inf $E(X_n) \geq E(\text{lim inf } X_n)$.

4. *The dominated convergence theorem.* Prove this theorem (see the text above) by applying Fatou's lemma to the sequences $\{X_n - Y\}$ and $\{Z - X_n\}$.

5. Expectations Derivable from the Characteristic Function: Weak Convergence

So many of our results have been expressed in terms of c.f.s that it is natural to ask: For what functions $H(x)$ is $E[H(X)]$ determinable from knowledge of $\phi(\theta) = E(e^{i\theta X})$ for real θ? This is an extension problem, and related to it are the questions that arose particularly in Chapter 7: If $\phi_n(\theta)$ is the c.f. of X_n and $\{\phi_n(\theta)\}$ has a pointwise limit $\phi(\theta)$, then is $\phi(\theta)$ a c.f. (of a r.v. X, say)? If so, for what H can one assert that $E[H(X_n)] \to E[H(X)]$?

The definition of $\phi(\theta)$ might make it natural to regard this problem as one of extension on a quadratic field (see Exercise 15.3.3) since, if $Y(\theta) = e^{i\theta X}$, then we are given the expectations

$$E[Y(\theta)\overline{Y(\theta')}] = \phi(\theta - \theta').$$

The basic field $\mathcal{G}$ thus consists of the trigonometric sums $\sum_j a_j e^{i\theta_j X}$, and we know from Exercise 15.3.3 that we can evaluate $E[H(X)]$, and indeed $E[H(X)e^{-i\theta X}]$ and $E(|H|^2)$, for any H which is the mean square limit of such sums. The fact that the r.v.s are complex is no problem; see Exercise 11.1.6.

However, in taking this mean square approach we are neglecting an important piece of information: that the r.v.s $Y(\theta)$ are known as functions of X. This knowledge should enable us to reach stronger results; for example, that $E(H)$ is the limit of a sequence $E(H_n)$ even if H is not the mean square limit of $\{H_n\}$.

As an example of the pure mean square approach, however, we give the following.

Theorem 15.5.1. *If $\phi''(0)$ exists then $E(X^2) = -\phi''(0)$ and $E(Xe^{i\theta X}) = -i\phi'(\theta)$ and all these quantities are finite.*

We know from Theorem 7.2.4 that if $E(X^2)$ is finite then $-\phi''(0)$ exists and is equal to it, so the theorem provides a partial converse to this result. The proof is outlined in Exercises 1 and 2.

We shall now turn our attention to the more general results which exploit the functional form of $e^{i\theta X}$. We shall assume that all specified r.v.s are finite with probability one; the corresponding c.f.s are then uniformly continuous, by Theorem 7.2.3. Whether this holds true for limits of sequences is a matter to be determined.

Theorem 15.5.2. *Suppose that $H(x)$ has the Fourier representation*

$$H(x) = \frac{1}{2\pi} \int h(\theta)e^{i\theta x} \, d\theta. \tag{7}$$

where $h(\theta)$ is absolutely integrable. Then

$$E[H(X)] = \frac{1}{2\pi} \int h(\theta)\phi(\theta) \, d\theta. \tag{8}$$

PROOF. Equation (8) follows formally from (7) if one takes expectations under the integral sign. This commutation of integral and expectation will be valid if $h(\theta)e^{i\theta X}$ is absolutely integrable as a function of θ (Fubini's theorem; see Kingman and Taylor, 1966, p. 144), which it will be if $h(\theta)$ is. □

However, this theorem puts conditions on H which are excessively strong, and not very explicit. One would like simple and fairly undemanding conditions directly on H itself which would ensure the validity of some version of (8). The finding of these necessarily involves Fourier theory to some extent.

We can note a few helpful points. First, if H is absolutely integrable (i.e. $\int |H(x)| \, dx < \infty$) then

$$h(\theta) = \int e^{-i\theta x} H(x) \, dx \tag{9}$$

exists, and is bounded by $\int |H(x)| \, dx$. Second, $h(\theta)$ will converge to zero as $\theta \to \pm\infty$ at a rate determined by the degree of continuity of H. For example, if H and its derivatives up to order p exist and are absolutely integrable then repeated partial integration of (9) shows that $h(\theta) = O(\theta^{-p})$ for large θ. For $p = 2$ this would ensure absolute integrability of h. Third, if both H and h are absolutely integrable then the inversion of (9) to (7) is valid (Goldberg, 1961, p. 16).

Thus, if we can tailor a given H so that it and its first two derivatives are absolutely integrable, without changing its expectation by more than a pre-assigned amount, then we can certainly calculate its expectation.

Theorem 15.5.3. *Suppose that $H(x)$ is bounded and continuous. Then $E[H(X)]$ is determinable in terms of the c.f. of X.*

PROOF. Define $H_N(x)$ as $(1 - |x|/N)H(x)$ if $|x| \le N$, zero otherwise. Because H is bounded and X finite, then $E(H_N)$ will differ arbitrarily little from $E(H)$ if N is chosen large enough. However, H_N is absolutely integrable for given finite N.

Modify this again to

$$H_{\sigma N}(x) = \int H_N(x + \sigma u)f(u) \, du,$$

where f is a standard normal density, so that $H_{\sigma N}$ is a smoothed version of H_N. If H continuous then H_N is uniformly continuous, and $H_{\sigma N} - H_N$ can be made arbitrarily and uniformly small if σ is chosen small enough. However, for σ positive, $H_{\sigma N}$ possess derivatives of all orders, each of these also being absolutely integrable.

We see from the discussion before the theorem that $H_{\sigma N}$ obeys the conditions of Theorem 15.5.2 and that $E[H_{\sigma N}(X)]$ is given by the σN-version of (8). However, by choosing N large enough and σ small enough we can make $|E(H) - E(H_{\sigma N})|$ arbitrarily small. The expectation $E(H)$ is thus determinable from knowledge of $\phi(\theta)$. □

The theorem can be further refined, but is general enough for many purposes as it stands. We should now like to transfer the assertion of the theorem to the case when we have a convergent sequence of c.f.s. First, we need a partial converse to Theorem 7.2.3: the assertion of an upper bound for $P(|X| > N)$ in terms of the degree of smoothness of the c.f. of X.

Theorem 15.5.4. *Suppose that the scalar r.v. X has c.f. $\phi(\theta)$. Then*

$$P(|X| \geq N) \leq \frac{N}{2\alpha\beta} \int_{-\beta/N}^{\beta/N} [1 - \phi(\theta)]\, d\theta, \tag{10}$$

where α, β are universal positive constants. Thus, if $\phi(\theta)$ is continuous at the origin, then X is finite with probability one.

PROOF. The values of α and β are those derived from the graph of $g(x) = 1 - (\sin x)/x$ in Fig. 15.1. That is, α is the value of $g(x)$ at its first minimum

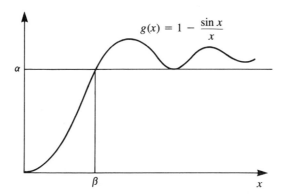

Figure 15.1. A graph of the function $g(x) = 1 - (\sin x)/x$, illustrating the determination of the derived constant α and β.

and β is the smallest positive root of $g(x) = \alpha$. Note the two relations

$$g(x) \geq \alpha I(|x| \geq \beta),$$

$$g(ax) = \frac{1}{2a} \int_{-a}^{a} (1 - e^{i\theta x}) \, d\theta. \tag{11}$$

We then have

$$P(|X| \geq N) = E[I(|X| \geq N)] \leq \alpha^{-1} E[g(\beta X/N)]$$

and insertion of representation (11) into this last relation yields the bound (10).
If $\delta(\varepsilon)$ is the maximal value of $|1 - \phi(\theta)|$ for $|\theta| \leq \varepsilon$, then it follows from (10) that

$$P(|X| \geq N) \leq \frac{\delta(\beta/N)}{\alpha},$$

and this will tend to zero with increasing N if $\phi(\theta)$ is continuous at the origin.
□

We now come to the result of principal interest.

Theorem 15.5.5. *Let $\{X_n\}$ be a sequence of scalar r.v.s with corresponding c.f.s $\phi_n(\theta)$, and suppose that $\phi_n(\theta)$ converges with increasing n to a limit $\phi(\theta)$ for every real θ. Then:*

(i) *If $\phi(\theta)$ is a c.f. then $\{X_n\}$ converges weakly to the r.v. X which has c.f. $\phi(\theta)$.*
(ii) *If $\phi(\theta)$ is continuous at the origin then it is a c.f. and assertion (i) holds.*

By weak convergence we mean of course that

$$E[H(X_n)] \to E[H(X)] \tag{12}$$

for all bounded continuous H, and the point of the theorem is that this convergence is assured if it holds for the particular cases $H(x) = e^{i\theta x}$ (θ real). The theorem is often termed the *continuity theorem* because of the condition expressed in (ii). In the general context of weak convergence the conditions expressed in (i) and (ii) are *tightness conditions*, which constrain X to be a finite r.v.

PROOF. Suppose provisionally that all r.v.s X_n are finite with probability one, uniformly in n for n sufficiently large. Then we know from the proof of Theorem 15.5.2 that we can establish the convergence (12) for bounded continuous H if we can establish it for absolutely integrable H with absolutely integrable Fourier transform $h(\theta)$. We restrict ourselves then to this latter case.
By the hypothesis of either (i) or (ii) $\phi(\theta)$ is continuous at the origin. It is then uniformly continuous, being the limit of functions whose degree of continuity is determined by that at the origin. The convergence $\phi_n(\theta) \to \phi(\theta)$ is then uniform over any finite θ interval. This fact and the absolute integrability

of $h(\theta)$ imply that

$$E[H(X_n)] = \frac{1}{2\pi} \int h(\theta)\phi_n(\theta)\, d\theta \to \frac{1}{2\pi} \int h(\theta)\phi(\theta)\, d\theta. \qquad (13)$$

If $\phi(\theta)$ is a c.f. we have then proved the convergence (12), and the finiteness of X justifies our provisional initial assumption (that the X_n are uniformly bounded in probability for sufficiently large n). Assertion (i) is thus proved.

The convergence (13) still holds under the hypothesis of (ii). In particular, it will hold for

$$h(\theta) = e^{-\sigma^2\theta^2/2} \frac{e^{-i\theta b} - e^{-i\theta a}}{-i\theta},$$

the Fourier transform of the smoothed indicator function of the interval (a, b). In other words,

$$\tilde{F}_n(b) - \tilde{F}_n(a) \to \tilde{F}(b) - \tilde{F}(a) = \frac{1}{2\pi} \int_{-\infty}^{\infty} \frac{e^{-i\theta b} - e^{-i\theta a}}{-i\theta} e^{-\sigma^2\theta^2/2}\phi(\theta)\, d\theta, \quad (14)$$

for any a, b. Here $\tilde{F}_n(x)$ is the distribution function of $X_n + \sigma u$, where u is a standard normal variable independent of X_n, and $\tilde{F}(x)$ is defined by the final equality. In view of the convergence (14) we see that $\tilde{F}(x)$ must be a distribution function at least in that $0 \le \tilde{F}(b) - \tilde{F}(a) \le 1$ for $a \le b$. The corresponding c.f. is then $\tilde{\phi}(\theta) = e^{-\sigma^2\theta^2/2}\phi(\theta)$. To show that it is a proper distribution function, and so ϕ a proper c.f., we must show that $\tilde{F}(N) - \tilde{F}(-N)$ tends to unity with increasing N. But if ϕ is continuous at the origin then so is $\tilde{\phi}$, and it follows then by appeal to Theorem 15.5.4 that $1 - [\tilde{F}(N) - \tilde{F}(-N)]$ tends to zero with increasing N.

Letting σ now tend to zero, we complete the proof of assertion (ii).   □

EXERCISES AND COMMENTS

1. Consider the sequence of r.v.s $H_n(X) = (i/n)(e^{iX/n} - 1)$. Show that this is mean square mutually convergent and so convergent if and only if $\phi''(0)$ exists.

2. *Continuation.* Use the inequality $(\sin \theta)/\theta > (2/\pi)$ for $|\theta| < \pi/2$ to show that $E(X_n^2)$ exists and is equal to $\lim E(H_n^2)$, and hence that the mean square limit of $\{H_n\}$ is equal to its pointwise limit X and can be identified with it. Theorem 15.5.1 follows from this identification.

3. Show more generally that if $i^{2j}\phi^{(2j)}(0)$ exists, then $E(X^{2j})$ exists and is equal to it.

CHAPTER 16

Some Interesting Processes

In this final chapter we consider three independent topics, each of substantial interest in its own right: information theory, dynamic optimization and quantum theory. The first two constitute applications of the theory as we have developed it. In the case of quantum theory it is a question of studying the radical variant of probability which the quantum world seems to require. However, for all its fundamental differences, this theory is also based on expectation axioms very close to those with which we began our exposition of the classical theory in Chapter 2.

1. Information Theory: Block Coding

Suppose one has a message source which emits data in the form of a sequence $\{u_1, u_2, u_3, \ldots\}$. Here u_t is the 'letter' which is emitted at time t; we suppose that it can take values in an 'alphabet' of m elements. Then, over a time interval $1 \leq t \leq n$, one has a block of data $U_n = (u_1, u_2, \ldots, u_n)$. Suppose also that the 'source statistics' are known, in that one regards U_n as a r.v. which has a known distribution (over the set consisting of n-tuples from the alphabet) for any n.

For example, u_t might be the tth letter in a text which is spelled out in 'telegraph English', which has the $m = 27$ characters comprising the 26 letters of the English alphabet plus the character 'word space'. One knows the statistics of such texts (let us be specific: of a quality English newspaper, say) in that one knows the frequencies of letters, of letter-pairs, etc.

The block U_n can take $m^n = 2^{n \log m}$ values, where the logarithm is to base 2. This is the same number of values as if the source had expressed the information in a *binary* (i.e. a two-letter) alphabet, but had emitted $n \log m + o(n)$ letters over a long time interval of length n, or $\log m$ letters per unit time.

270

One expresses this by saying that the source *generates data at a rate of* log *m bits per unit time* ('bit' being the recognized abbreviation of 'binary digit').

One might now ask whether the output of the source might not be *compacted* to one of lower rate by neglecting values of U_n which are relatively improbable. Specifically, if A is a set of values of U_n then there are two measures of this set which are of interest. One is its probability

$$P(A) = \sum_{U_n \in A} P(U_n)$$

and the other is its 'size'

$$\#(A) = \text{number of elements in } A.$$

Suppose that for each n one can find a set A_n of U_n values constrained by

$$\#(A_n) \leq 2^{Rn} \tag{1}$$

for all n, but such that

$$P(A_n) \to 1 \tag{2}$$

as $n \to \infty$. The idea is that one takes the data in large blocks of size n. If U_n falls in A_n then it is forwarded faithfully. If it falls in the complementary set $\bar{A}_n$ then it is simply not forwarded, the loss of data being regarded as 'error'. Condition (1) then states that one is forwarding data at an effective rate of R bits per unit time, because one can relabel the blocks of the reduced set A_n using at most nR binary digits. We assume that this reduction of the data is acceptable provided condition (2) is satisfied: that $P(\bar{A}_n) \to 0$, so that the probability of error tends to zero as block size increases.

If conditions (1) and (2) can both be met then it is said that one can *encode the source reliably at a rate of R bits per unit time.* We define

$$H = \inf R \tag{3}$$

as the *information rate* of the source, where the infimum is over reliable encoding rates. That is, one can encode reliably at no smaller rate than H, and can encode reliably at rate $H + \varepsilon$ for any $\varepsilon > 0$. The value of H depends upon the statistical properties of the source; it has a very clear operational significance. Even if the encoding of large blocks is not considered practicable, the information rate H gives one a measure of what can ideally be achieved.

Let us now consider the simplest statistics of all: those of a *Bernoulli source*, for which the letters are regarded as IID r.v.s. This is not a very realistic case, but constitutes a natural first model.

Theorem 16.1.1. *Suppose the digits u_t of the message are IID, taking the jth letter value with probability p_j ($j = 1, 2, \ldots, m$). Then the source has information rate*

$$H = h(u) = -\sum_j p_j \log p_j \tag{4}$$

bits per unit time.

We have used $h(u)$ to denote the final expression in (4), the famous Shannon measure of information. (More specifically, in this case, the measure of the information in a single digit.) This is a quantity which occurs in many contexts, is invoked in many more, and is given justifications ranging from the axiomatic to the metaphysical. It is enough for us that the measure has emerged spontaneously from the mathematical treatment of an operational problem, and with a clear interpretation.

PROOF. Regard $\zeta_t = -\log P(u_t)$ as a r.v. That is, this is a r.v. which takes the value $-\log p_j$ with probability p_j. Similarly, define the r.v. $\xi_n = -\log P(U_n)$. By virtue of the independence of the digits we have $P(U_n) = \prod_{t=1}^{n} P(u_t)$, so that

$$\xi_n = \sum_{t=1}^{n} \zeta_t,$$

where the ζ_t are IID with expectation

$$E(\zeta) = h(u).$$

Now, suppose A_n specified as the set of U_n for which

$$2^{-n(h+\delta)} \le P(U_n) \le 2^{-n(h-\delta)},$$

i.e. for which

$$h - \delta \le \xi_n/n \le h + \delta.$$

Here δ is a prescribed positive number, and we have written $h(u)$ as h for simplicity. By the law of large numbers ξ_n/n will converge to $E(\zeta) = h$ in probability, so that certainly $P(A_n) \to 1$ with increasing n. Furthermore

$$\#(A_n) \le 2^{n(h+\delta)},$$

since $P(A_n) \le 1$ and each member of A_n has probability not less than $2^{-n(h+\delta)}$. Reliable encoding at rate $h + \delta$ is thus possible, and, since δ is arbitrary, we have demonstrated that $H \le h$.

To demonstrate that we can do no better, suppose we tried encoding at a rate R by using sets A'_n of U_n. Then

$$P(A'_n) = P(A'_n \cap A_n) + P(A'_n \cap \bar{A}_n) \le P(A'_n \cap A_n) + P(\bar{A}_n)$$
$$\le 2^{nR} 2^{-n(h-\delta)} + P(\bar{A}_n),$$

which tends to zero with increasing n unless $R > h - \delta$. So, reliability will require that $R > h - \delta$. Since δ is arbitrary, we thus have the reverse inequality $H \ge h$, whence assertion (4) follows.   □

The following strengthening of Theorem 16.1.1 follows immediately, by the same proof.

Theorem 16.1.2. *Suppose that the r.v. $-(1/n) \log P(U_n)$ converges in probability to a constant H_∞ with increasing n. Then the information rate H of the source is just H_∞.*

So, suppose that the source were *Markov* in that $\{u_t\}$ was an irreducible Markov process with transition probabilities p_{jk}. Then appeal to Theorem 16.1.2, and verification of the required convergence (which takes some calculation) yields the evaluation

$$H = -\sum_j \sum_k \pi_j p_{jk} \log p_{jk} \tag{5}$$

as the information rate of the source. Here $\{\pi_j\}$ is the equilibrium distribution over letter values.

EXERCISES AND COMMENTS

1. It follows from the definition of H that, for a source in an alphabet of size m,

   $$0 \le H \le \log m.$$

   Show that these bounds can be attained. (Consider cases in which U_n can take only one value and in which all m^n values are equally likely.)
      It follows then from (4) that $h(u)$ must be subject to the same bounds. What are the letter distributions $\{p_j\}$ which achieve these bounds?

2. If the conditions of Theorem 16.1.2 are met, then one can state that

   $$P[2^{-n(H+\delta)} \le P(U_n) \le 2^{-n(H-\delta)}] \to 1$$

   as $n \to \infty$, where H is the information rate of the source. This is *roughly* expressed by saying that 2^{nH} values of U_n each have probability 2^{-nH} and the rest have probability zero. The fact that all the significant block values have (in the above sense) asymptotically the same probability is spoken of as the *asymptotic equipartition property*.

3. We have spoken of *data compaction*, which is not the same as *data compression*. In compaction one passes on faithfully that part of the data which is not discarded; the infidelity then takes the form of outright error on a set of small probability. In compression one represents all messages in terms of a smaller number of 'typical messages', and approximates any given message by the typical message to which it is closest. The infidelity then takes the form of a 'distortion' which affects most messages, although by no more than an acceptable amount, it is hoped.

2. Information Theory: More on the Shannon Measure

The Shannon information measure $h(u)$ has many properties. We shall briefly indicate just those which we need to see us through the next section.
   The interpretation of h as an information rate H (for the Bernoulli case) implies the rough interpretation of the measure that one should carry in one's mind: the number of bits required to represent the r.v. u. (More exactly, the number of bits per character required to represent a long sequence of IID

u-values.) The interpretation as an information rate for a sequence in an alphabet of size m implies the bounds

$$0 \le h(u) \le \log m, \tag{6}$$

with equality in the cases of a degenerate (one-point) and a uniform u-distribution. However, these assertions can be proved directly; see the exercises.

A compact way of writing the definition (4) is

$$h(u) = -E[\log P(u)],$$

where $P(u)$ is regarded as a r.v.; the variable that takes value p_j with probability p_j. Note that 'u' is not a functional argument in $h(u)$, but a label; an indication that the information measure concerns the r.v. u.

If u and v are a pair of r.v.s then their joint information measure is

$$h(u, v) = -E[\log P(u, v)],$$

with the expectation taken over both u and v. This is not a new definition; simply the old one with u replaced by (u, v). Note that $h(u, v) = h(u)$ in the case when v is in fact determined by u, so that specification of v conveys no further information once u is known.

A useful concept is that of the *conditional information measure*

$$h(u|v) = -E[\log P(u|v)],$$

where the expectation is again over *both* u and v. So, one calculates the Shannon measure for u conditional on a prescribed value of v, and then averages it over v. The notation again requires clarification: the variables u and v in $P(u|v)$ are actual arguments; in $h(u|v)$ they are simply *labels* of the conditioned and conditioning variables. The two usages should be distinguished in notation, but are now both accepted by convention.

It follows from the relation $P(u, v) = P(v)P(u|v)$ that

$$h(u, v) = h(v) + h(u|v).$$

The relation makes sense: the number of bits required to specify u and v is the number of bits required to specify v plus the number of bits required to specify u once v is known.

EXERCISES AND COMMENTS

1. Note that $-p \log p \ge 0$ for $0 \le p \le 1$, with equality only at p equal to 0 or 1. It follows from this that $h(u) \ge 0$ with equality if and only if the u-distribution is degenerate.

2. Let $\{p_j\}$ and $\{q_j\}$ be two alternative distributions on the same m-element set. Show that

$$-\sum_j p_j \log(p_j/q_j) \le 0 \tag{7}$$

with equality if and only if the two distributions are identical. [Appeal to Jensen's inequality or the inequality $\log x \leq x - 1$.] This is *Gibbs' lemma*, appealed to constantly in this context, and certainly appealed to in the next three exercises.

3. By taking q_j as the uniform distribution, show that $h(u) \leq \log m$, with equality if and only if the u-distribution is uniform.

4. Show that $h(u, v) \geq h(u)$. What are the cases of equality?

5. Show that expression (5) for the information rate of a Markov source does not exceed $-\sum_j \pi_j \log \pi_j$. The presence of statistical dependence between letter values thus decreases the information rate. Such dependence makes the sequence more predictable, so that it carries less information.

6. The quantity $\sum_j p_j \log(p_j/q_j)$ is sometimes called the *Kullback–Liebler information number*, and is a measure of distance between the two distributions (although an asymmetric one). If we consider the multinomial distribution

$$P(\mathbf{n}) = n! \prod_j (q_j^{n_j}/n_j!) \tag{8}$$

and define the empirical proportion $p_j = n_j/n$, then

$$n^{-1} \log P(\mathbf{n}) = \text{constant} - \sum_j p_j \log(p_j/q_j) + o(1)$$

for large n. Assertion of Gibbs' lemma is thus equivalent to saying that the probability distribution is maximal in $\mathbf{n}$ for $n_j/n \sim q_j$ (all j). Alternatively, if we consider the distribution as a function of the q_j rather than of the n_j, then n_j/n is the maximum likelihood estimator of q_j.

3. Information Theory: Sequential Interrogation and Questionnaires

Consider a problem which may be related to the data-compaction problem of Section 1, but is certainly posed differently. Suppose that a quantity u can take m different values, and that one must determine its actual value in a particular case by posing a sequence of questions. The questions are restricted; they may, for example, be constrained so as to admit only Yes/No answers. How should one choose the questions so as to determine u as quickly as possible? More specifically, if u is regarded as a r.v. of known distribution, how should one proceed so as to minimize the expected number of questions needed?

The question that one asks at a given point will depend upon previous answers. The problem can then indeed be characterized as one of 'sequential interrogation'. The design of a questionnaire is just such a problem, the aim of the questionnaire being to categorize the subject (i.e. determine u) with as few questions as possible. The same problem occurs in diagnostics: if a system has a fault, then one wishes to determine the nature of the fault with as few tests as possible. The variable u is 'the nature of the fault' and the tests correspond to the questions. There will be physical constraints on the tests: they will not locate the fault immediately, but merely indicate into which of a number of classes it falls. The supposition that u is a r.v. of known distribution corresponds to the assumption that one knows the frequencies of occurrence of the different type of fault from previous experience.

Let us suppose that all the questions have a possible answers, and that these answers are truthful, mutually exclusive and exhaustive. (That is, one and only one of the answers will be correct in a given case, and that answer will be given.) We shall also suppose that the questions which are actually used are informative under all circumstances, in that, whatever the answer, it will have the effect of strictly reducing the set of values within which u is known to lie.

Let $s(u)$ denote the minimal expected number of questions needed to determine u. We shall also use the notation $s(u|v)$; if u and v are a pair of r.v.s then $s(u|v)$ is the minimal expected number of questions required to determine u given that v is known to have a particular value, averaged over this v-value.

Theorem 16.3.1. *Suppose that all questions have a answers and are informative. Then $s(u)$, the minimal expected number of questions needed to determine u, has the lower bound*

$$s(u) \geq \frac{h(u)}{\log a}, \tag{9}$$

with equality if and only if every question can be so posed that the a possible answers are equiprobable.

The Shannon information measure thus again makes an unforced appearance. However, the optimization rule has at least a different expression from that which it had in Section 1. In Section 1 it was optimal to restrict attention to the (roughly) 2^{nH} most probable values of U_n. Here the suggested rule has a sequential form: at every step one tries to frame the next question so that, conditional on the answers already received, the a possible answers to the question are equally likely. There seems to be no appeal to asymptotic notions for this problem. There is a concealed aspect, however. The more values u can take (of comparable probability) the easier it will be to find a question whose answers are approximately equiprobable, and the smaller will be the relative margin of inequality in (9).

PROOF. We shall prove the result by induction on m, the number of values that u can adopt. If $m = 1$ then u has been completely determined, so both $s(u)$ and $h(u)$ are zero, and equality holds in (9).

Suppose that u takes m values and that we have established the theorem for r.v.s taking fewer than m values. Let v be the answer to the first question posed. If this question is optimal then

$$s(u) = 1 + s(u|v). \tag{10}$$

The comparable equation for $h(u)$ is

$$h(u) = h(u, v) = h(v) + h(u|v),\qquad (11)$$

the first equality holding because v is a function of u. We thus deduce that

$$s(u) - \frac{h(u)}{\log a} = \left(1 - \frac{h(v)}{\log a}\right) + \left(s(u|v) - \frac{h(u|v)}{\log a}\right)$$

$$\geq 1 - \frac{h(v)}{\log a} \geq 0.\qquad (12)$$

The first inequality in (12) follows from the inductive hypothesis: since the question will locate v in a set of smaller size than m, one has $s(u|v) \geq h(u|v)/(\log m)$, with equality if and only if all subsequent questions yield equiprobable outcomes. The second inequality corresponds to the known upper bound on $h(v)$, with equality if and only if all answers v are equiprobable. The induction is thus complete. ☐

EXERCISES AND COMMENTS

1. Although the problem is presented as sequential (in that one chooses a question on the basis of answers already received) one could in fact prepare a scheme of questioning in advance, which would prescribe the question to be asked in any given circumstance. Then, for any given u, one knows what the sequence of answers will be: $v_1, v_2, \ldots, v_\tau$, say, where τ is the (u-dependent) stage at which u is fully determined. The determination of the questions is thus equivalent to the determination of a coding in which u is represented by an a-ary sequence $(v_1, v_2, \ldots, v_\tau)$, and one wishes the expected length of sequence to be as small as possible.

   Note that this coding will necessarily obey a *prefix condition*: if one particular 'codeword' $(v_1, v_2, \ldots)$ terminates at τ then there cannot be another codeword which begins with this same block of τ symbols and then continues. For, by hypothesis, observation of the first codeword has determined u.

2. Suppose the questions posed admit a variable number of answers, that being asked at stage i having a_i possible answers. Suppose we redefine $s(u)$ as the minimal value of $E[\log(a_1 a_2 \ldots a_\tau)]$, where τ is the stage at which the value of u is determined. That is, the criterion 'number of questions needed' has been replaced by the criterion 'number of bits needed'. Show that Theorem 16.3.1 still holds, in that $s(u) \geq h(u)$, with equality if and only if all answers are equiprobable at every stage.

4. Dynamic Optimization

In Section 5.6 we considered statistical decision theory: the optimal choice of a decision in the face of uncertainty. The situation was static, in that it was a question of taking a decision at a given point in time on the basis of given information. However, if one has a sequence of decisions to take, then one has a problem in dynamic optimization. This brings in at least two new features. First, decisions will interact, in that one's decisions at a given time and under given circumstances must depend upon the decision rules that will be used *later*. Second, the decisions will in general be made on different information, because in general one will acquire more information as time progresses.

Since we live in time and try to do the best we can, examples are not far to seek. On a descending scale: the conduct of life, the control of systems, economic control, investment decisions, gambling.

We can afford no more than the briefest of glimpses at what is by now a very large subject, but shall do enough to see that a number of now-familiar threads are pulled together. In particular, we shall see that expectations over future events conditional on current information play a fundamental role, as do the induced backward equations.

Just as in Section 5.6, there will be a loss function L, whose expectation is to be minimized. In this context one tends to speak more of a *cost function*, and we shall do so. The cost will in general depend upon the whole course of the process: that is, upon both the values of the process variables and of the decisions taken over the course of time. Consider first the situation when there are in fact no decisions to be taken. Then the best assessment one can make at time t of the cost that will be incurred is the conditional expectation

$$G(W_t) = E(L|W_t), \tag{13}$$

where, as in Chapter 14, W_t represents the total information available at time t. This turns out to be significant, and is called the *value function*. We assume that no information is discarded (so that knowledge of W_{t+1} implies knowledge of W_t). In this case the value function obeys the recursion (a backward equation)

$$G(W_t) = E[G(W_{t+1})|W_t]. \tag{14}$$

Now suppose that there are decisions to be taken, the decision taken at time t being denoted by u_t. These decisions will in general affect the course of the process, affect costs, and possibly also observations. The decision taken at time t can depend only on current information, W_t. A rule which prescribes the decision to be taken under all circumstances, i.e. for all t and all W_t, is called a *policy*. The policy is denoted π, and the expectation operator induced by policy π denoted by E_π.

If the policy is to be optimized, then it is appropriate to redefine the value function as the *minimal* expected cost, conditional on information at time t:

$$G(W_t) = \inf_\pi E_\pi(L|W_t). \tag{15}$$

We shall suppose that W_t includes information on past decisions; the infimum in (15) can then affect only the current decision u_t and decisions after that. If one took an arbitrary decision u_t at time t and took optimal decisions after that, then the expected cost conditional on W_t would be $E[G(W_{t+1}|W_t, u_t],$

where the inclusion of u_t as a conditioning variable really corresponds to the fact that the constant u_t acts as a parameter of the stochastic dynamics of the system. It is plausible, then, that the optimal value of u_t is that which minimizes this expression, in which case one has the relation

$$G(W_t) = \inf_{u_t} E[G(W_{t+1})|W_t, u_t]. \tag{16}$$

This is the *dynamic programming equation,* or *Bellman optimality equation.* It generalizes relation (14) as a recursion for the value function. However, the significant fact is that it also determines the optimal value of u_t; in this it can be regarded as a dynamic version of Theorem 5.6.1. These assertions require proof, but the reader will perhaps accept them as plausible.

We can now make matters a good deal more explicit if we assume Markov structure. For notational economy we shall use lowercase letters to denote both the r.v.s and their observed values; the context will make clear which is intended. Suppose we regard the course of the system in time as a stochastic process $\{x_t\}$ which is Markov and fully observed in the following senses:

(a) The distribution of x_{t+1} conditional on previous values of state and control $\{x_s, u_s; s \le t\}$ is in fact conditioned only by the immediately previous values (x_t, u_t).
(b) The value of the current state is observable, so that W_t consists of x_t and $\{x_s, u_s; s < t\}$.
(c) The cost function has the additive form

$$L = \sum_t \beta^t c(x_t, u_t),$$

where β is a discount factor.

If one defines the minimal *future* cost, conditional on current information

$$F(W_t) = \inf_{\pi} E_{\pi}\left(\sum_{s \ge t} \beta^{s-t} c(x_s, u_s)|W_t\right),$$

then one finds by recursive argument that this is in fact a function of x_t and t alone, $F(x_t, t)$, say, and that the optimality equation (16) reduces to

$$F(x_t, t) = \inf_{u_t} (c(x_t, u_t) + \beta E[F(x_{t+1}, t+1)|x_t, u_t]). \tag{17}$$

The minimizing value of u_t in (17) is again the optimal value. (There are many questions of regularity, and the less academic ones of terminal conditions; these we by-pass). The merit of this reduction to 'state structure' (as it is often termed) is that one need not carry the whole observational history W_t; it is optimal to base decisions solely on the current state variable x_t. This reveals itself in that the optimality equation (16) reduces to (17), in which both the future value function F and the optimal control are determined in terms of x_t and t alone.

As an example, we can consider a problem which has exercised us throughout, the text, and which is properly seen as a dynamic optimization problem. This is the question of an optimal policy for our friend Alex, the gambler seeking ways of guaranteeing himself a return. He has a current capital of x, and can either stop play with a sum of x in his possession, or can continue play. If he continues play then he places a unit stake, and doubles it or loses with probabilities p and q. Suppose we take expected capital on retirement as the criterion. It is plain that x, the current capital, is the effective state variable; the decision at each stage is whether to continue or retire. Let $F(x)$ be the future value function: the maximal expected return conditional on a start from an initial capital x. (It is plain that we can as well consider maximization of a return as minimization of a cost. The value function will be independent of t if we are prepared to consider indefinite play, when the whole problem becomes invariant to a time-translation.)

An increase in generality which appears slight but which proves significant is to introduce a discount factor β and a cost c for every round of play. This latter can be regarded as the value that the player places upon his own time, if nothing else. The optimality equation then becomes

$$F(x) = \max\{x, \ -c + \beta[pF(x + 1) + qF(x - 1)]\}, \tag{18}$$

where the two alternatives under the maximization correspond to the choices of retirement or continuation.

Theorem 16.4.1. *Let α denote the larger root*

$$\alpha = \frac{1 + \sqrt{1 - 4\beta^2 pq}}{2p}$$

of the equation

$$\beta p\alpha^2 - \alpha + \beta q = 0. \tag{19}$$

Then it is optimal to retire if $x \geq a$, where a is the smallest integer such that

$$a \geq \frac{1}{\alpha - 1} - \frac{c}{1 - \beta}, \tag{20}$$

and to continue if $x < a$. The future value function has the evaluation

$$F(x) = \begin{cases} x & (x \geq a), \\ -\dfrac{c}{1 - \beta} + \dfrac{\alpha^{x-a}}{a + c/(1 - \beta)} & (x < a). \end{cases} \tag{21}$$

PROOF. The optimal policy will obviously have the form indicated for some a. We have then, from (18), that $F(x) = x$ for $x \geq a$ and

$$F(x) = -c + \beta[pF(x + 1) + qF(x - 1)] \qquad (x < a). \tag{22}$$

The solution of (22) is

$$F(x) = \phi(x) = -\frac{c}{1 - \beta} + d\alpha^x + d'(\alpha')^x, \tag{23}$$

where α and α' are the two roots of (19) and d and d' are constants to be determined. But the roots of (19) are both positive, the smaller being less than unity. Since $F(x) \geq x$ and $F(x)$ is monotone nondecreasing we must then have $d' = 0$.

We use $\phi(x)$ to denote expression (23) for all x; it equals $F(x)$ only for $x \leq a$. The boundary and optimality conditions implicit in (18) have as consequence

$$\phi(a) = a, \qquad \phi(a + 1) \leq a + 1, \qquad \phi(a - 1) \leq a - 1, \qquad (24)$$

(see Exercise 2), so that one has the discrete equivalent of continuity of $F(x)$ and $dF(x)/dx$ at the transition point $x = a$. The first condition of (24) implies the lower evaluation of $F(x)$ given in (21); the others imply that

$$a + \frac{c}{1 - \beta} \geq \frac{1}{\alpha - 1} \qquad \text{and} \qquad a + \frac{c}{1 - \beta} < \frac{\alpha}{\alpha - 1} = 1 + \frac{1}{\alpha - 1},$$

whence we deduce the determination of a asserted. $\square$

The course of $F(x)$ is sketched in Fig. 16.1. The optimal policy indeed takes the form that the player retires when his holdings (*not* his winnings) reach the threshold value

$$a \approx \frac{1}{\alpha - 1} - \frac{c}{1 - \beta}.$$

For $(1 - \beta)$ small and positive one finds that a is given, to within a term $o(1)$, by $p/(q - p) - c/(1 - \beta)$ for $p < q$, by $1/\sqrt{2(1 - \beta)} - c/(1 - \beta)$ for $p = q$ and by $(p - q - c)/(1 - \beta)$ for $p > q$. So, in the case of an unfavourable game the threshold at which Alex should cease play slips away to $-\infty$ (i.e. he should never play) as the discount factor β approaches unity.

The case of a fair game is remarkable in its instability; seemingly slight perturbations of the problem cause wild variations in the optimal policy. As

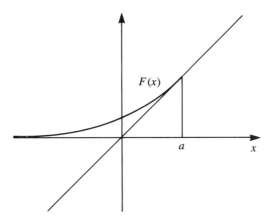

Figure 16.1. A graph of the value function $F(x)$ for the gambling problem: the maximal expected (discounted net) return for a gambler starting with a holding of x.

$\beta \uparrow 1$ then a tends to $+\infty$ or $-\infty$ according as c is zero or not. That is, if Alex places no value on his time he should play indefinitely; if he places the slightest value on his time then he should never play (in the limit of zero discounting). This unstable character of the fair game explains why the question of a good policy for this simplest of games has always proved so troublesome.

In the case of a favourable game a will tend to $+\infty$ or $-\infty$ according as $p - q$ is greater than or less than c; i.e. according as to whether the game is 'net favourable' or not.

One can say that the gambling problem has held its historic fascination, not merely as one of the early exercises in probability, but because it is in fact a dynamic optimization problem, and one which, even in the simplest version, shows quirks in its optimal policy. For a deeper discussion of these matters we refer to the excellent text by Dubins and Savage (1965).

The reader will perhaps agree that study of the dynamic optimization problem has brought together several of our previous themes in an inevitable fashion. The notion of an ever-increasing observation set is evident. Optimization compels one to consider expectations conditional on these observations, and to consider in particular the generalization of the backward equation constituted by the optimality equation (16). The invocation of Markov structure produces its own special simplifications.

Another theme which also enters naturally is that of second-order theory. If we suppose that both the process variable and the observations are generated by linear equations driven by Gaussian 'noise' and that the cost functions are quadratic in the vector variables x and u, then we obtain a very natural and complete second-order theory (the so-called LQG theory) with many pleasing features. For example, one has the certainty equivalence principle (a conclusion, not an assumption), which states that, in an optimal policy, one should behave as if unobservables were in fact known and equal to their current LLS estimates. For a discussion of these matters, we refer to Whittle (1982, 1983).

EXERCISES AND COMMENTS

1. Note that equation (10) for the interrogation problem was one that held only if the question whose answer is v had been posed optimally. The equation should really be seen as

$$s(u) = 1 + \inf_{v} s(u|v),$$

where by the infimum we mean that the question should be chosen so as to minimize $s(u|v)$. Note that this equation is an optimality equation: $s(u)$ is the value function, the *distribution* of u is the state variable, and the question is the decision.

2. Suppose we have an equation $F(x) = \max[K(x), TF(x)]$, where $TF(x) = \sum_{j=-1}^{1} c_j F(x + j)$. Suppose that $F = K \geq TF$ holds for $x \geq a$ and $F = TF > K$ for $x < a$. If ϕ is the free solution of $\phi = T\phi$ which agrees with F in $x < a$, then it also agrees with F at $x = a$, because $F = TF$ ($x < a$) yields $\phi(a) = F(a)$ as an effective

boundary condition. Subtraction of the relation $\phi = T\phi$ from the relation $K \geq TF$ at $x = a$ yields $K(a + 1) \geq \phi(a + 1)$. Correspondingly, the relation $TF > K$ at $x = a - 1$ yields $K(a - 1) > \phi(a - 1)$.

3. Note that to play just one round of the game will never be worthwhile if $\beta(p - q) < c$. Nevertheless, we see from Theorem 16.4.1 that, even under these conditions, continuation of play can be optimal.

5. Quantum Mechanics: the Static Case

It is remarkable that one of the most basic and impressive structures of theoretical physics, quantum mechanics, should be intrinsically probabilistic, and yet with a logic deviating essentially from that of the 'classic' probability theory we have studied. The quantum mechanical approach is quite similar to that which we took to classic probability in Chapter 2, in its selection of expectation as the basic concept, and in the detailed form of the axioms to which the expectation functional is subject (and was thus formulated by Dirac (1930) and von Neumann (1932)). However, certain differences in the initial formulation produce profound differences in structure and conclusions.

We shall have to accept a certain number of basic ideas derived from physical arguments. Thus, once done, we shall find that the theory develops in a self-contained and striking manner.

P1. To any variable a of physical interest there corresponds a Hermitian linear operator A. (We shall refer to such a variable as an *observable*, the point being that it is only quantities which are principle observable which are regarded as having physical interest.)

P2. The operator corresponding to the function $f(a)$ of a is $f(A)$.

P3. The operator corresponding to a linear function $\sum_j c_j a_j$ of observables is the same function $\sum_j c_j A_j$ of the corresponding operators.

For formal simplicity we shall assume that all the operators A are $n \times n$ matrices, although in actual physical contexts one must consider operators on more general spaces. If the matrix A has spectral representation

$$A = \sum_{j=1}^n \alpha_j \phi_j \phi_j^\dagger$$

(where $\phi^\dagger$ is the transpose of the complex conjugate of ϕ), then we can define $f(A)$ in a self-consistent fashion by

$$f(A) = \sum_{j=1}^n f(\alpha_j) \phi_j \phi_j^\dagger. \tag{25}$$

The expectation value of a is assumed to be a functional $\mathfrak{E}$ of the corresponding operator:

$$E(a) = \mathfrak{E}(A). \tag{26}$$

The functional $\mathfrak{E}$ is assumed to have the following properties:

E1. $\mathfrak{E}$ is a real scalar.
E2. $\mathfrak{E}$ is linear:

$$\mathfrak{E}(c_1 A_1 + c_2 A_2) = c_1 \mathfrak{E}(A_1) + c_2 \mathfrak{E}(A_2). \tag{27}$$

E3. $\mathfrak{E}(A)$ is nonnegative if A is nonnegative definite.
E4. $\mathfrak{E}(I) = 1$ (where I is the identity matrix).

Axioms E1–E4 are plainly very similar to Axioms 1–4 of Section 2.2. One can also add a continuity axiom, similar to Axiom 5, but we shall not need it if we consider only the finite structure of the theory. However, the fact that $\mathfrak{E}$ has a matrix as an argument, rather than a scalar function $X(\omega)$, makes a great difference. Note, in connection with E3, that any Hermitian matrix can be written as a difference of nonnegative definite matrices, in analogy with the relation $X = X_+ - X_-$ for scalars.

Our first assertion is the following

Theorem 16.5.1. *The expectation functional $\mathfrak{E}$ has the form*

$$\mathfrak{E}(A) = \operatorname{tr}(UA), \tag{28}$$

where U is a nonnegative definite Hermitian matrix of unit trace.

Here we have used tr( ) to denote the trace of a matrix; the sum of its diagonal elements. Statement (27) corresponds to the assertion in the classical case with finite state space that $E(X)$ must be of the form (2.6). The matrix U, termed the *density matrix*, specifies the statistics of the system, just as did the distribution $\{p_k\}$ in the classical case.

PROOF. Since $\mathfrak{E}$ is linear, $\mathfrak{E}(A)$ can be written as a linear function of the elements of A, and relation (28) does just this in matrix form. Since A is always Hermitian then U can be normalized to Hermitian form. Setting $A = I$ in (28) we deduce the unit trace property from E4; setting $A = \phi\phi^\dagger$ we deduce from E3 that $\phi^\dagger U\phi \geq 0$, and so the nonnegative definite character of U. The conditions of the theorem are also sufficient; the functional $\mathfrak{E}(A)$ constructed there has all the properties E1–E4. $\qquad\square$

From (25) and (28) we deduce that

$$E[f(a)] = \sum_{j=1}^{n} f(\alpha_j)(\phi_j^\dagger U\phi_j). \tag{29}$$

Since f is arbitrary, relation (29) must have the following interpretation.

Theorem 16.5.2. *Suppose that the operator A corresponding to the observable a has eigenvalues α_j and corresponding normalized eigenvectors ϕ_j. Then a can* *adopt only the values* α_j, *and if these are distinct, then*

$$P(a = \alpha_j) = \phi_j^\dagger U \phi_j \qquad (j = 1, 2, \ldots, n). \tag{30}$$

This is one of the most interesting and radical conclusions of quantum theory. As we see, it follows quite quickly from the axioms.

If the α_j are not distinct then (29) would imply rather that

$$P(a = \alpha) = {\sum_j}' \phi_j^\dagger U \phi_j,$$

where the summation Σ' covers all j for which $\alpha_j = \alpha$. It is then better to interpret $\phi_j^\dagger U \phi_j$ as the probability that *the system is in eigenstate ϕ_j for the observable a*. We shall denote this by $P(A_j)$: the probability that the system is in the jth eigenstate of the observable a.

It could be said that we are appealing to classic probability concepts, in that we are concluding from relation (29) that the coefficient of $f(\alpha)$ in the sum can be identified with the probability that a takes the value α. However, one could regard this identification simply as providing a *definition* of probability, a concept that has not hitherto been introduced.

We cannot in general extend formula (29) to expectations involving more than one observable. If a and b are two such observables with corresponding operators A and B, then $f(A, B)$ is not well defined (for polynomial f, for example) unless A and B commute. In this case they have identical eigenvectors, with corresponding eigenvalues α_j and β_j, say, and one can define

$$f(A, B) = \sum_{j=1}^n f(\alpha_j, \beta_j) \phi_j \phi_j^\dagger.$$

The conclusion is then that, when the system is in the common eigenstate ϕ_j, the observables a and b take values α_j and β_j, respectively.

Observables a and b are said to be *simultaneously observable* when their operators commute. In some approaches one demands the linearity condition E2 only for commuting A_i on the grounds that the linear combination is not otherwise meaningful. Remarkably, Theorem 16.5.1 remains vaid under this weakened axiom, as was demonstrated in an ingenious paper by Gleason (1957).

In discussing general observables a and b we shall, of course, not assume simultaneous observability. Theorem 16.5.2 still has a very strong implication.

Theorem 16.5.3. *Suppose that the observables a and b have operators with spectral representations*

$$A = \sum_{j=1}^n \alpha_j \phi_j \phi_j^\dagger, \qquad B = \sum_{j=1}^n \beta_j \psi_j \psi_j^\dagger,$$

and that b is known to have the value β_k, where β_k is a simple eigenvalue of B.

Then necessarily

$$U = \psi_k \psi_k^\dagger,\tag{31}$$

so that

$$E(a) = \psi_k^\dagger A \psi_k$$

and

$$P(A_j) = |\phi_j^\dagger \psi_k|^2.$$

PROOF. Suppose that U has spectral representation

$$U = \sum_{j=1}^n \lambda_j \xi_j \xi_j^\dagger,\tag{32}$$

where the λ_j are necessarily real, nonnegative and add to unity. Since b is known to take the value β_k then we know from Theorem 16.5.2 that $\eta^\dagger U \eta = 0$ for every vector η orthogonal to ψ_k, so that

$$\sum_j \lambda_j |\xi_j^\dagger \eta|^2 = 0.$$

Since all terms in this sum are nonnegative they must be individually zero. That is, if $\lambda_j \neq 0$ then ξ_j must be proportional to ψ_k. This can hold for only one j-value, say $j = 1$, so that then $\lambda_j = 0$ $(j > 1)$ and $\lambda_1 = 1$. Assertion (31) thus holds. □

This is again one of the basic ideas: that the taking of an observation affects all expectations, which are thereby 'conditioned'. When U has the form $\psi\psi^\dagger$ we shall say that the system is in a *pure state*, and that the elements of ψ constitute the *wave function* describing the state. For convenience we shall describe ψ itself as the wave function. The general case (32) is regarded as one of *mixed states*, in which the system has wave function ξ_j with probability λ_j $(j = 1, 2, \ldots, n)$.

The case of a pure state might seem to correspond to that of a degenerate distribution in the classical case, when a r.v. can take only a single value. However, it is a peculiarity of quantum mechanics that there is no situation corresponding to complete absence of randomness.

Theorem 16.5.4. *A deterministic situation, in which the values of all observables are known, is impossible if $n > 1$.*

PROOF. In such a case U would certainly have the form $\psi\psi^\dagger$, with ψ an eigenvector corresponding to an observable of known value. But it is impossible that a given ψ can simultaneously be an eigenvector of *all* Hermitian operators. □

Indeed, one can obviously make the stronger statement, that one cannot simultaneously know the values of two observables whose operators do not share at least one eigenvector. Thus some degree of uncertainty is essential in a quantum-mechanical situation—an idea which finds another expression later in the uncertainty relation (38).

The uncertainty manifests itself also in that the taking of an observation does more than reveal something of the system: it actually changes the state of the system. We see this if we carry out successive observations on a, b and a; the second observation on a will in general not be concordant with the first. Suppose, for definiteness, that we observe $a = \alpha_1$. If we repeated the measurement immediately we would again observe $a = \alpha_1$. However, suppose that after the first observation we measured the value of b as β_1, throwing the system into eigenstate B_1 if β_1 is a simple eigenvalue of B. Then a second subsequent observation on a would yield the value $a = \alpha_j$ with probability $|\psi_1^\dagger \phi_j|^2$ ($j = 1, 2, \ldots, n$).

One can speak of a conditional probability in that one could regard $|\psi_k^\dagger \phi_j|^2$ as $P(A_j | B_k)$. From the symmetry of this expression in j and k we deduce

Theorem 16.5.5. *Conditional probabilities obey the symmetry relation*

$$P(A_j | B_k) = P(B_k | A_j) \quad (= |\phi_j^\dagger \psi_k|^2). \tag{33}$$

The quantity

$$(A_j, B_k) = \phi_j^\dagger \psi_k, \tag{34}$$

known as the *probability amplitude*, turns out to play a special role.

Let us slightly generalize the experiment carried out above, in that we carry out successive observations on observables a, b and c. Suppose that all eigenvalues are simple, so that these three observations successively throw the system into eigenstates A_j, B_k and C_l, respectively, say. The definition (34) implies the identity

$$(A_j, C_l) = \sum_k (A_j, B_k)(B_k, C_l),$$

which we would regard as a 'Chapman–Kolmogorov equation' in probability amplitudes. The amplitude associated with the observation history B_k, C_l after the initial A_j is $(A_j, B_k)(B_k, C_l)$ and, if b is not in fact observed, then one obtains the correct probability amplitude simply by summing B_k out.

However, the corresponding Chapman–Kolmogorov equation in *probabilities* does not hold. One has, in general,

$$|(A_j, C_l)|^2 \neq \sum_k |(A_j, B_k)|^2 |(B_k, C_l)|^2. \tag{35}$$

The sum in the right-hand member is in fact interpretable as the probability that an observation on c will throw a system initially in eigenstate A_j into eigenstate C_l if, between these two observations, the value of b *has been measured but not recorded*. It has been experimentally demonstrated that the taking of an observation can change the state of a system even if the result of the observation is not revealed to the experimenter (see Feynman, 1985, p. 80).

However, there has been an unspoken assumption in our discussion. In considering the sequential taking of observations, we have admitted the passage of time. We have, however, assumed that the system did not change in time, except in response to the taking of an observation. One would like to know what the quantum analogue of a dynamic system (i.e. of a stochastic process) would be, and whether such a dynamic model could allow equilibrium behaviour.

EXERCISES AND COMMENTS

1. Deduce conditions under equality *would* hold in (35).

2. Suppose one tries to construct a joint distribution over eigenstates A_j and B_k consistent with the conditional probabilities (33). Show that in general this can only be achieved by the choice $U = n^{-1}I$, when all eigenstates for any observable are equiprobable. This does seem to constitute a quantum version of a 'universal prior distribution' representing 'complete ignorance', a concept which has been sought in classical statistics, but does not there exist.

3. Consider a pair of observables a and b and denote the commutator $AB - BA$ of their operators by $2iD$. From that fact that, for real λ,

$$0 \le \mathfrak{E}[(A + i\lambda B)(A - i\lambda B)] = \mathfrak{E}(A^2 + 2\lambda D + \lambda^2 B^2)$$

deduce the Cauchy inequality

$$E(a^2)E(b^2) \ge [\mathrm{tr}(UD)]^2.$$

By replacing a by $a - E(a)$, and so A by $A - E(a)I$, and similarly for b, one can strengthen this to

$$\mathrm{var}(a)\,\mathrm{var}(b) \ge [\mathrm{tr}(UD)]^2, \tag{36}$$

which indicates that a and b cannot be simultaneously known if D is nonzero.

4. Certain pairs of observables have a commutator proportional to the identity:

$$AB - BA = 2i\kappa I, \tag{37}$$

where κ is a real scalar. These are termed *conjugate observables*. For such pairs inequality (36) takes the form

$$\mathrm{var}(a)\,\mathrm{var}(b) \ge \kappa^2, \tag{38}$$

independent of U. This is the so-called *uncertainty principle*, expressing the fact that under no circumstances can a and b be simultaneously determined.

We see also that a relation such as (37) cannot hold for finite n, because (38) would then imply that $\mathrm{var}(b)$ would be infinite if a were known, which is impossible in a finite system. However, for an infinite system the conjugacy relation (37) is possible, as we now see.

5. Consider the operators A and B whose action on a function $\phi(x)$ of scalar argument is $A\phi(x) = d\phi(x)/dx$ and $B\phi(x) = x\phi(x)$. If ϕ is square integrable on the real axis then these are Hermitian and $AB - BA = I$. They are proportional to the operators representing momentum and position, respectively, which are consequently conjugate.

6. Show that inequality (36) becomes an equality if $U = \phi\phi^\dagger$, where ϕ satisfies $(A - i\lambda B)\phi = \mu\phi$ for some scalar λ and μ. Hence show that in the preceding example one has equality if and only if $\phi(x)$ has the 'normal' form $\phi(x) = \exp(\alpha + \beta x + \gamma x^2)$, where γ has negative real part.

6. Quantum Mechanics: the Dynamic Case

It can be said that, in the classical case, one thinks of a stochastic process as being determined by specification of the joint distribution of the process variable at all relevant moments of time. Alternatively, one describes the stochastic dynamics of the process by specifying the distribution of the process variable at time t conditional on its values at all previous times. However, such characterizations must now fail, because the concepts of joint distribution and conditional distribution both fail. We must then find another way of expressing the dynamics.

Suppose we now allow that quantities such as the density matrix U and the wave function ϕ can be time-dependent, written $U(t)$ and $\phi(t)$, and make the following assumptions concerning them:

D1. The density matrix U provides a complete description of the system, in that the value of $U(t)$ for $t \geq 0$ can be deduced from that of $U(0)$.

D2. A system which begins in a pure state remains in one.

D3. The transformation which generates the wave function $\phi(t)$ from its initial prescription has an infinitesimal generator.

So let $\mathscr{D}$ denote the set of possible density matrices (nonnegative definite Hermitan matrices of unit trace) and let $\mathscr{W}$ denote the set of possible wave functions (vectors of unit norm). Then assumption D1 states that $U(t)$ can represented as a time-dependent function $G[U(0), t]$ of $U(0)$, this transformation necessarily being from $\mathscr{D}$ to $\mathscr{D}$. Assumption D2 states that if $U(0)$ is of the form $\phi(0)\phi(0)^\dagger$ then $U(t)$ is necessarily of the form $\phi(t)\phi(t)^\dagger$; the transformation G will then induce a transformation $\phi(t) = F[\phi(0), t]$ from $\mathscr{W}$ to $\mathscr{W}$.

Assumptions D1 and D2 constitute structural requirements, with a fairly evident motivation. Assumption D3 has more the status of a regularity condition. It states that there is a function f with the property

$$\lim_{t \downarrow 0} \frac{F(\phi, t) - \phi}{t} = if(\phi).$$

The presence of the 'i' implies no assumption, but merely a definition of f which turns out to be convenient.

We need one further regularity condition if we are to reach our conclusions. It could doubtless be relaxed, but is convenient:

D4. As a function of the complex variables $\phi^{(r)}$ constituted by the components of ϕ, the function $f(\phi)$ is analytic in some annulus $\delta_1 < |\phi^{(r)}| < \delta_2$ $(r = 1, 2, \ldots, n)$.

These assumptions do not seem unduly strong, but have a powerful implication.

Theorem 16.6.1. *Assumptions D1–D4 imply that the wave function ϕ obeys a linear equation*

$$\frac{\partial \phi}{\partial t} = iH\phi, \tag{39}$$

where the matrix H is Hermitian.

Equation (39) is just the *wave equation* or *Schrödinger equation*, with H interpretable as the Hamiltonian operator of the system. What is surprising is that the strong property of linearity emerges from the apparently mild assumptions.

PROOF. Assumptions D1–D3 imply that the wave function $\phi(t)$ for a system starting in a pure state obeys the equation

$$\frac{\partial \phi}{\partial t} = if(\phi).$$

The rate of change of $|\phi|^2$ is

$$\frac{\partial |\phi|^2}{\partial t} = \dot{\phi}^\dagger \phi + \phi^\dagger \dot{\phi} = i(f^\dagger \phi - \phi^\dagger f).$$

Since we know that the norm $|\phi|$ is preserved we must thus have

$$\phi^\dagger f - f^\dagger \phi = 0. \tag{40}$$

In view of D4 the elements of $f(\phi)$ can be expressed as a Laurent power series in the components $\phi^{(r)}$. Let us set $\phi^{(r)} = \exp(\zeta_r) = \exp(\xi_r + i\eta_r)$, so that the power series representing the rth element of $f(\phi)$ is

$$f_r = \sum_m f_{rm} \exp[(\xi + i\eta)^\mathrm{T} m],$$

where m varies over all n-tuples of signed integers. This representation is valid for all real η and for all real ξ in an interval. We then have

$$\phi^\dagger f = \sum_r \sum_m \exp(\xi_r - \eta_r) f_{rm} \exp[(\xi + i\eta)^\mathrm{T} m]$$

$$= \sum_r \sum_m f_{rm} \exp[\xi^\mathrm{T}(m + e_r) + i\eta^\mathrm{T}(m - e_r)],$$

where e_r is the n-vector with a unit in the rth place and zeros elsewhere. Holding ξ fixed and varying η we see that the coefficient of $\exp(i\eta^\mathrm{T} m)$ in the expansion of the left-hand member of (40) must be zero; this leads to the equation

$$\sum_r \{f_{r,m+e_r} \exp[\xi^\mathrm{T}(m + 2e_r)] - \bar{f}_{r,e_r-m} \exp[\xi^\mathrm{T}(2e_r - m)]\} = 0, \tag{41}$$

valid for all m. The coefficient in this expression of $\exp(\xi^T m)$ for each distinct m must be zero. If m is not of the form $e_j - e_k$ for some j, k then all exponents in the sum (41) are distinct, and the consequent f-coefficients must be individually zero. If $m = e_j - e_k$ then we deduce that

$$f_{j,e_k} = \bar{f}_{k,e_j} \tag{42}$$

and that all other f_{rm} are zero. Thus $f(\phi)$ reduces to a linear function $H\phi$ with the matrix $H = (f_{j,e_k})$ being Hermitian, by (42).   □

The regularity condition D4 could be weakened, but not dispensed with altogether if we are to reach the conclusion of the theorem; see Exercise 2.

Suppose we add yet a further assumption:

D5. In the case of a mixed state

$$U = \sum_j \lambda_j \phi_j \phi_j^\dagger,$$

each ϕ_j evolves independently as it would in the pure state case and the λ_j remain unchanged.

This is a stronger form of D2, implying that there is no interference between the alternatives in a mixture. That is, whatever pure state in fact holds, then evolution continues (in a pure state) irrespective of the alternative situations which might have held, or of their prior probabilities λ_j.

Theorem 16.6.2. *Assumptions* D1–D5 *imply that the density matrix* U *obeys the equation*

$$\frac{\partial U}{\partial t} = i(HU - UH), \tag{43}$$

where H *is the Hamiltonian operator of* (39).

PROOF. We readily verify from (39) that (43) holds in the pure case $U = \phi\phi^\dagger$. Assumption D5 then implies that it holds also in the mixed case.   □

Note that assumption D5 is necessary; the validity of (43) for the pure case does not imply its validity for the mixed case.

It is equation (43) which is the analogue of the Kolmogorov forward equation, the analogy being more evident if we write this latter in the form

$$\frac{\partial \pi_j}{\partial t} = \sum_k (\pi_k \lambda_{kj} - \pi_j \lambda_{jk}).$$

Equations (39) and (43) have the formal solutions

$$\phi(t) = e^{iHt}\phi(0), \qquad U(t) = e^{iHt}U(0)e^{-iHt}.$$

We pursue some of the immediate consequences of Theorem 16.6.1 in Exercise 1.

EXERCISES AND COMMENTS

1. The Hamiltonian operator H is associated with the observable 'energy' and its eigenvalues λ are regarded as possible energy levels. U is an equilibrium solution of (43) if it commutes with H, and so represents a mixture of energy eigenstates. If the system is in the pure eigenstate corresponding to energy λ_j and eigenvector ϕ_j, then the wave function $\phi(t)$ must be proportional to ϕ_j. We see then from (39) that

$$\frac{\partial \phi}{\partial t} = iH\phi = i\lambda_j\phi,$$

so that $\phi(t) = e^{i\lambda_j t}\phi(0)$. The fixed eigenvector ϕ_j is then modulated in time by a rotation in complex space at a speed proportional to the energy in that state.

2. Any dynamic equation $\dot{\phi} = g(\phi)$ can be made to preserve the norm of ϕ if we supplement the transformation $\phi \rightarrow \phi + sg$ over time s by a renormalization of ϕ. Show, in this way, that the modified equation

$$\dot{\phi} = g - \frac{1}{2}\frac{(\phi^\dagger g + g^\dagger \phi)}{|\phi|^2}\phi \qquad [= h(\phi), \text{ say}]$$

always preserves the norm. However, if $g(\phi)$ is analytic, and if $h(\phi)$ differs from it, then h is not analytic (because $\phi^\dagger$ is not an analytic function of ϕ).

References

References are to actual sources and, for example, systematic reference to the periodical literature has not been attempted. I have, of course, benefited from reading some of the standard texts. Among these I would single out Breiman (1968) for its individual and direct style, Ross (1970) for its synthesis of probability and optimization, and Grimmett and Stirzaker (1982), which stylishly succeeds in the task which must be tackled every decade or so: of slimming the swelling body of theory down to a leaner and livelier form.

Apostol, T.M. (1957): *Mathematical Analysis*. Addison-Wesley, Reading, Mass.

Breiman, L. (1968): *Probability*. Addison-Wesley, Reading, Mass.

De Finetti, B. (1970): *Teoria delle probabilita*, Vols. 1 and 2. Einaudi, Turin. English translation, Wiley, New York, 1974, 1975.

Dirac, P.A.M. (1930): *The Principles of Quantum Mechanics*. Clarendon, Oxford.

Dubins, L. and Savage, L. (1965): *How to Gamble If You Must*. McGraw-Hill, New York.

Feller, W. (1966): *An Introduction to Probability Theory and Its Applications*, Vol. 2, Wiley, New York.

Feynman, R.P. (1985): *QED*. Penguin, Harmondsworth, U.K.

Gleason, A.M. (1957): Measures on the closed subspaces of a Hilbert space. *J. Math. Mech.*, **6**, 885–894.

Goldberg, R.R. (1961): *Fourier Transforms*. Cambridge University Press, Cambridge, U.K.

Goldie, C.P. and Pinch, R. (1991): *Communication Theory*. Cambridge University Press, Cambridge, U.K.

Grimmett, G.R. and Stirzaker, D.R. (1982): *Probability and Random Processes*. Clarendon, Oxford.

Hald, A. (1990): *A History of Probability and Statistics and Their Applications before 1750*. Wiley, New York.

Harris, T.E. (1963): *The Theory of Branching Processes*. Springer-Verlag, Berlin; Prentice-Hall, Englewood Cliffs, N.J.

Kerrich, J.E. (1956): *An Experimental Introduction to the Theory of Probability*. Einar Munksgaard, Copenhagen.

Kingman, J.F.C. and Taylor, S.J. (1966): *Introduction to Measure and Probability*. Cambridge University Press, Cambridge, U.K.

Kolmogorov, A.N. (1933): *Grundbegriffe der Wahrscheinlichkeitsrechnung*. Ergebnisse der Math., English edition, Chelsea, New York, 1950.

Krickeberg, K. (1965): *Probability Theory*. Addison-Wesley, Reading, Mass.

Neveu J. (1964): *Bases mathématiques du calcul des probabilités*. Masson, Paris. English edition, Holden-Day, New York, 1965.

Ross, S. (1970): *Applied Probability Models with Optimization Applications*. Holden-Day, San Francisco.

Savage, L.J. (1954): *The Foundations of Statistics*. Wiley, New York.

Von Neumann, J. (1932): *Grundlagen der Quantenmechanik*. Springer-Verlag, Berlin. English edition, Princeton University Press, Princeton.

Whittle, P. (1964): Some general results in sequential analysis. *Biometrika*, **51**, 123–141.

Whittle, P. (1969): Refinements of Kolmogorov's inequality. *Teoriya Veroyatnostei*, **14**, 315–317.

Whittle, P. (1970): *Probability*. Penguin, Harmondsworth, U.K. (Original edition of this work.)

Whittle, P. (1971): *Optimisation Under Constraints*. Wiley, New York.

Whittle, P. (1982, 1983): *Optimisation over Time*, Vols. 1 and 2. Wiley, New York.

Whittle, P. (1986): *Systems in Stochastic Equilibrium*. Wiley, Chichester and New York.

Wold, H.O.A. and Whittle, P. (1957): A model explaining the Pareto distribution of wealth. *Econometrica*, **25**, 591–595.

Index

Springer Texts in Statistics *(continued from p. ii)*

Keyfitz	Applied Mathematical Demography Second Edition
Kiefer	Introduction to Statistical Inference
Kokoska and Nevison	Statistical Tables and Formulae
Lindman	Analysis of Variance in Experimental Design
Madansky	Prescriptions for Working Statisticians
McPherson	Statistics in Scientific Investigation: Its Basis, Application, and Interpretation
Nguyen and Rogers	Fundamentals of Mathematical Statistics: Volume I: Probability for Statistics
Nguyen and Rogers	Fundamentals of Mathematical Statistics: Volume II: Statistical Inference
Noether	Introduction to Statistics: The Nonparametric Way
Peters	Counting for Something: Statistical Principles and Personalities
Pfeiffer	Probability for Applications
Santner and Duffy	The Statistical Analysis of Discrete Data
Saville and Wood	Statistical Methods: The Geometric Approach
Sen and Srivastava	Regression Analysis: Theory, Methods, and Applications
Whittle	Probability via Expectation, Third Edition
Zacks	Introduction to Reliability Analysis: Probability Models and Statistical Methods